AF 356539

New Trends in Lyotropic
Liquid Crystals

New Trends in Lyotropic Liquid Crystals

Editors

Antônio Martins Figueiredo Neto
Ingo Dierking

MDPI • Basel • Beijing • Wuhan • Barcelona • Belgrade • Manchester • Tokyo • Cluj • Tianjin

Editors
Antônio Martins Figueiredo Neto
University of São Paulo
Brazil

Ingo Dierking
University of Manchester
UK

Editorial Office
MDPI
St. Alban-Anlage 66
4052 Basel, Switzerland

This is a reprint of articles from the Special Issue published online in the open access journal *Crystals* (ISSN 2073-4352) (available at: https://www.mdpi.com/journal/crystals/special_issues/Lyotropic_Liquid-Crystals).

For citation purposes, cite each article independently as indicated on the article page online and as indicated below:

LastName, A.A.; LastName, B.B.; LastName, C.C. Article Title. *Journal Name* **Year**, *Article Number*, Page Range.

ISBN 978-3-03943-342-1 (Hbk)
ISBN 978-3-03943-343-8 (PDF)

Cover image courtesy of Antônio Martins Figueiredo Neto and Ingo Dierking.

Contents

About the Editors

Antônio Martins Figueiredo Neto is a professor of Physics at the University of São Paulo, São Paulo, Brazil, and is Head of the National Institute of Science and Technology on Complex Fluids, Brazil. He is a member of the organizing committees of more than 20 international conferences and workshops in the field of liquid crystals, magnetic colloids and fluids of biological interest. He has written more than 220 papers in international journals, 1 book published by the Oxford University Press, 5 book chapters, and has given more than 350 presentations at international conferences. He has supervised 22 defended PhD students and 18 Master dissertations. He is a member of the Brazilian Academy of Science and The Academy of Science of the State of São Paulo.

Ingo Dierking is a senior lecturer/associate professor at the Department of Physics and Astronomy of The University of Manchester, UK. His research is focused on soft matter physics, especially liquid crystals and liquid crystal-based composites, including both thermotropic and lyotropic materials. He is the author and editor of several books and book chapters, and has published more than 140 papers on the topic of liquid crystals in high impact peer-reviewed journals. Dierking is the 2009 recipient of the Hilsum medal of the British Liquid Crystal Society (BLCS) and the 2016 recipient of the Samsung Mid-Career Award for Research Excellence awarded by the International Liquid Crystal Society (ILCS). He is the former Chair of the BLCS and the Secretary of the ILCS, and is the current President of the International Liquid Crystal Society.

Preface to "New Trends in Lyotropic Liquid Crystals"

Lyotropic liquid crystals have long led a shadowy existence in LC research, dominated by its much bigger thermotropic LC brother, who is accountable for the multi-billion dollar industry of flat panel screens and information display devices. Nevertheless, it appears that the equilibrium has shifted slightly, as the research community finds increasing interest in lyotropic systems, such as soft matter nanomaterials, biological materials and systems or active liquid crystals. We have thus decided that it is a good idea to collect some overview articles as well as original research papers on these soft matter systems that have gained much increased interest in recent years, in a Special Issue of the journal Crystals. This Special Issue is now also available in printed book form. The first paper of this collection provides an admittedly subjective and personal view from the editors on the current state of lyotropic liquid crystals. The authors of the more detailed and in-depth reviews were chosen such that a large variety of topics would be discussed, authored by the leading figures in their specialty subjects.

We start with an overview about the novel trends in lyotropic liquid crystals, touching on a number of different aspects, which will be presented in a more elaborate form later on. These are, for example, some new observations by the Giesselmann group on amphiphilic molecules that form phases equivalent to the thermotropic ferroelectric liquid crystal SmC* phase. Cellulose nanocrystals, as nanomaterials and equally as biological materials, have gained much interest in recent years, especially with respect to their lyotropic behavior, which is discussed in detail in the papers of the Lagerwall and Godinho groups. Furthermore, chromonic liquid crystals, formed by board-like dye molecules, have been shown to offer new insights, as reported by the Lavrentovich group. Similar board-like systems, albeit on a much larger, macromolecular scale, are represented by the lyotropic nematic liquid crystal phase of graphene oxide, discussed by the Dierking group. This is more generalized in the review by the Davis group about the effects of size and shape dispersity on the phase diagram of lyotropics. This is a topic which may also be related to the observation of lyotropic biaxial nematic systems, as discussed experimentally by the group of Neto, and theoretically explored by Salinas et al. At last, a theoretical approach by Izzo investigates the ordering of rods near surfaces.

We hope that the present collection of papers provides a good basis for the further exploration of modern aspects of lyotropic liquid crystals, both from an experimental and a theoretical point of view. Last, but not least, we would like to thank all of our colleagues who have given up their time and contributed to this Special Issue, as well as the editorial staff of "Crystals" for their professional support.

Antônio Martins Figueiredo Neto, Ingo Dierking
Editors

Review

Novel Trends in Lyotropic Liquid Crystals

Ingo Dierking [1],* and Antônio Martins Figueiredo Neto [2],*

[1]   Department of Physics and Astronomy, University of Manchester, Oxford Road, Manchester M139PL, UK
[2]   Complex Fluids Group, Institute of Physics, University of São Paulo, Rua do Matão, 1371 Butantã,
    São Paulo-SP–Brazil CEP 05508-090, Brazil
*   Correspondence: ingo.dierking@manchester.ac.uk (I.D.); afigueiredo@if.usp.br (A.M.F.N.);
    Tel.: +44-161-275-4067 (I.D.); +55-11-30916830 (A.M.F.N.)

Received: 25 June 2020; Accepted: 10 July 2020; Published: 12 July 2020

Abstract: We introduce and shortly summarize a variety of more recent aspects of lyotropic liquid crystals (LLCs), which have drawn the attention of the liquid crystal and soft matter community and have recently led to an increasing number of groups studying this fascinating class of materials, alongside their normal activities in thermotopic LCs. The diversity of topics ranges from amphiphilic to inorganic liquid crystals, clays and biological liquid crystals, such as viruses, cellulose or DNA, to strongly anisotropic materials such as nanotubes, nanowires or graphene oxide dispersed in isotropic solvents. We conclude our admittedly somewhat subjective overview with materials exhibiting some fascinating properties, such as chromonics, ferroelectric lyotropics and active liquid crystals and living lyotropics, before we point out some possible and emerging applications of a class of materials that has long been standing in the shadow of the well-known applications of thermotropic liquid crystals, namely displays and electro-optic devices.

Keywords: liquid crystal; lyotropic; chromonic; amphiphilic; colloidal; application

1. Introduction

Lyotropic liquid crystals (LLCs) [1,2] are known from before the time of the discovery of thermotropics by Reinitzer in 1888 [3], which is generally (and rightly) taken as the birth date of liquid crystal research. In the work before this time, for example by Virchow [4], Mettenheimer [5], Planer [6], Loebisch [7] or Rayman [8], the liquid crystalline properties were described, but without the explicit realization that this constituted a novel state of matter. The latter was the significant contribution made by Reinitzer [3] in 1888 and Lehmann [9], who coined the term *"liquid crystal"* in 1889 when studying thermotropic phases. Nevertheless, lyotropic liquid crystal research has been present ever since, albeit on a lower quantitative output than that of thermotropic systems, also because of their less obvious potential in applications, being overshadowed by displays and electro-optic devices. But, as thermotropic liquid crystal research surged in the 1970's to the 2000's, so did that of lyotropic liquid crystals (Figure 1), due to the realization of their importance for biological systems and in colloid science.

Over the last two decades, more and more LC researchers have widened the scope of their work to also include lyotropic phases, and to explore systems of both thermotropic and lyotropic behavior. This paper will try and summarize some of the fascinating recent developments, as lyotropics find their way into an increasing number of liquid crystal laboratories. This will by no means be an exhaustive treatment, but will hopefully provide an overview of the current new trends in lyotropic liquid crystals.

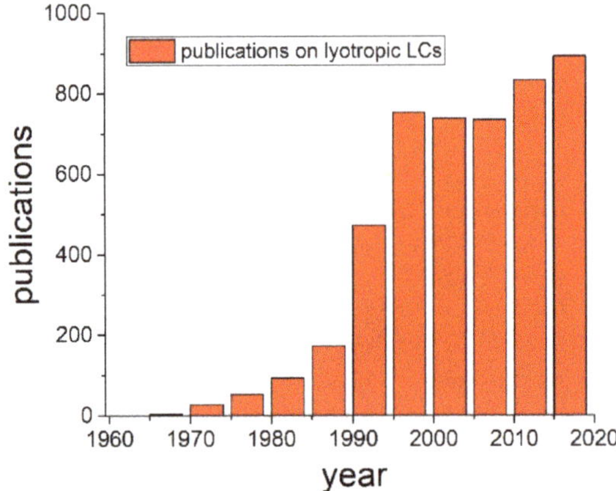

Figure 1. Number of publications in the field of lyotropic liquid crystals (LLCs) shown as bars over a time period of 5 years each.

2. Lyotropic Liquid Crystals

2.1. Classic Lyotropics from Amphiphiles and Polymers

Amphiphilic molecules have the striking property of presenting two antagonistic characteristics within the same molecule, i.e., hydrophobicity and hydrophilicity. In contact with polar and/or nonpolar liquids, under proper temperature and relative concentration conditions, they form lyotropic liquid crystalline phases [1]. Nanoscale molecular segregation (self-assembly) gives rise to different molecular aggregates, from micelles to bicontinuous structures. There are different types of amphiphilic molecules, such as anionic, cationic, zwitterionic and non-ionic amphiphiles, detergents and anelydes [10]. Moreover, other more complex molecules belong to this category, the gemini surfactants [11], the rigid spiro-tensides, phospholipids [12], the facial amphiphiles [13] and the bolaamphiphiles [14]. The lyotropic polymorphism encountered in mixtures with amphiphilic molecules is very rich. Figure 2 shows a sketch of a typical lyotropic liquid crystal phase diagram of a mixture composed by an amphiphile and water. The Krafft line separates the crystalline phase region from the liquid region. The critical micelle concentration line separates the single amphiphilic molecule region from the molecular self-assembled region. By increasing the amphiphile relative concentration, the mixture can present the micellar, hexagonal, lamellar and inverted phase structures. A well-known biological example for lyotropic lamellar structures is the lipid bilayers of cell membranes, as shown in Figure 3a. Other examples are Myelin figures.

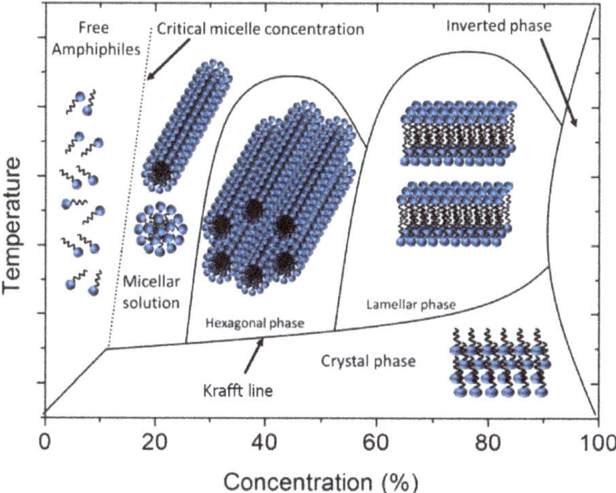

Figure 2. Sketch of a phase diagram of a mixture of an amphiphile and water. In the horizontal axis, the amphiphile concentration is represented. Cubic phases may occur at different areas in the phase diagram.

(a)

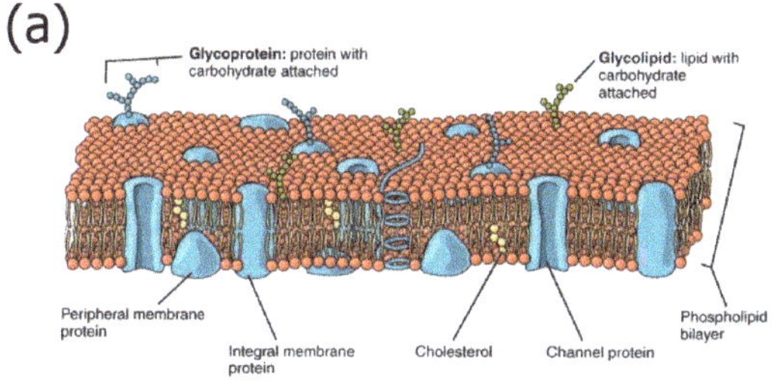

(b)

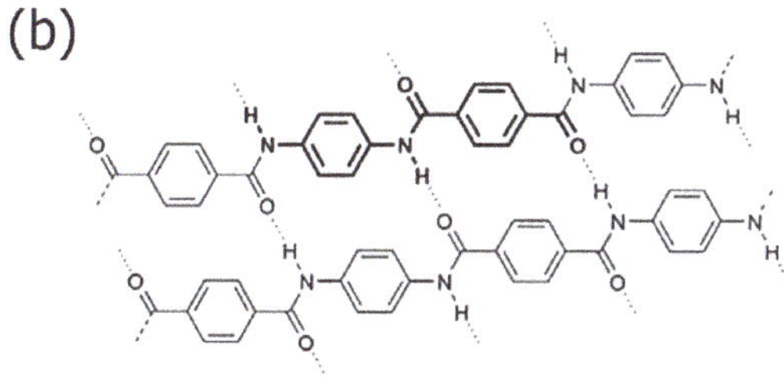

Figure 3. (**a**) Schematic representation of the lyotropic liquid crystalline structure of a cell membrane (reproduced from Wikimedia Commons from OpenStax Anatomy and Physiology). (**b**) Chemical structure of Kevlar®.

Despite the fact that these liquid crystalline phases were extensively studied for many decades, interesting questions still remain to be answered nowadays. Two of these mesophases deserve to be particularly highlighted, the biaxial nematic, N_B [15,16], and the chiral [17,18] mesophases.

Controversies have appeared in the literature over the years about the existence of the N_B phase and its chemical stability. Theoretical [16,19] and experimental [16,20,21] papers were published about the stability of the N_B phase. In this special issue, an extensive discussion about the experimental conditions to stabilize the N_B phase is presented.

Chiral lyotropics is, still, a challenge to be fully understood by physicists and chemists. An intriguing question is how the information about chirality passes from one element of the structure (micelle or lamellae) to the next, since there are water molecules between them. Recently, evidences of an equivalent to the thermotropic (chiral smectic C), SmC*, in lyotropics were reported [18]. These findings further broaden the boundaries of the physical chemistry of lyotropic liquid crystals.

Another class of lyotropic liquid crystals involve polymers in an isotropic, liquid solvent. One classic example found in nature is spider silk, which consists of protein fibers formed from a micellar solution when pushed through a valve at the spiders back under loss of water. This forms oriented crystalline regions of beta-sheets cross-linked via amorphous regions. Another industrially produced high modulus fiber is Kevlar® (Figure 3b), which is produced from the lyotropic liquid crystalline state of the aramide polymer in highly concentrated sulfuric acid. It is used in a wide range of applications, from bullet proof vests to climbing gear.

2.2. Inorganic Liquid Crystals

The most classic example of an inorganic or mineral liquid crystal is vanadium pentoxide, V_2O_5. Its needle-like nanocrystals form a nematic lyotropic liquid crystal, which was first investigated by Diesselhorst and Freundlich [22,23] in the beginning of the 20th century. They reported the occurrence of birefringence of anisotropic crystallites of vanadium pentoxide when subjected to flow or to an electric field. It was concluded that both mechanisms had the same origin, application of a force to orient the elongated, needle-like crystallites that then exhibited a macroscopic birefringence. After the removal of the force, the system relaxed to yield an isotropic appearance. It has often been reported that a newly prepared V_2O_5 sol is initially isotropic, while it takes time to observe the development of nematic tactoids [24], as seen in Figure 4.

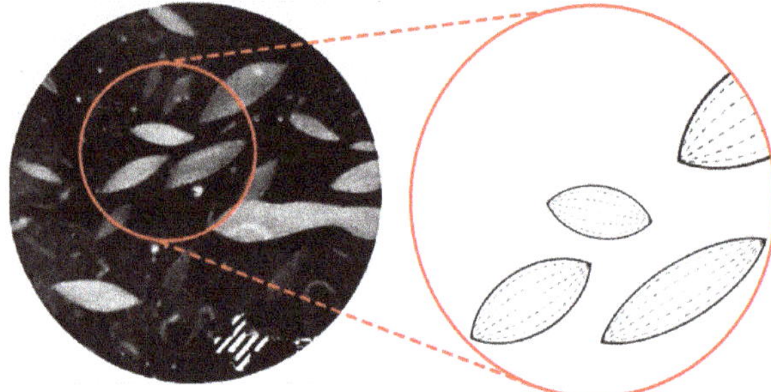

Figure 4. Formation of nematic tactoids in a preparation of vanadium pentoxide, V_2O_5 (reproduced with permission from Reference [24]), and corresponding needle crystallite orientation within some tactoids.

The phase formation largely depends on the preparation conditions and history of V_2O_5. This is related to the aging of freshly prepared sol and depends on concentration, temperature and possible electrolyte addition. The formation of the nematic phase is enhanced for large V_2O_5 concentrations and higher temperatures. The length of the colloidal vanadium pentoxide particle increases at a constant width of approximately 10 nm from the nanometer range to a few micrometers. This is accompanied by a sol-gel transition [25]. Electric field application in the nematic phase indicates that the system has a negative dielectric anisotropy, $\Delta\varepsilon < 0$. This implies that in contrast to standard calamitic nematics with positive anisotropy, the director would switch from homeotropic to planar under electric field application. The phase sequence as a function of increasing concentration of V_2O_5 is isotropic–biphasic–uniaxial nematic for the fluid suspensions. For further concentration increase, a gel transition is observed and the uniaxial nematic gel is transformed into a biaxial nematic gel [26].

Similarly, aluminum oxyhydroxide, AlOOH, forms nematic tactoids, which on merging exhibit a typical nematic Schlieren-texture [27]. $M_2Mo_6X_6$, with the metal from the alkalimetal group M = Li,

Na, K and X = Se, Te, from the chalcogens group, also represents a general group of inorganic LCs with nematic phases. Crystallites exhibit lengths of approximately several micrometers, and Schlieren-textures, as well as sometimes thread-like textures [28], can be observed when these are dispersed in an isotropic solvent, for example methylformamide. The formation of smectic phases has also been observed. This was shown for FeOOH and for tungstic acid H_2WO_4 ($WO_3 \bullet H_2O$) by the observation of steps in the textures of droplets, indicating smectic layering [29]. A more detailed overview about the structures of anisotropic crystallites, their preparation and the conditions employed to form liquid crystalline phases was published in a review by Sonin [30]. It appears that besides phase diagrams, textures and structures, there has been little work on such systems with respect to modern experimental techniques or applications by self-assembly for nanotechnology. Maybe it can be fruitful to revisit these systems from the different perspectives available today.

2.3. Clays

These natural soil materials of micro- and nano-meter dimensions present high shape anisotropy and contain hydrous aluminum phyllosilicates. Typically, the particles are plate-like, with the plate thickness in the nanoscale. In 1995, Mourchid and co-workers [31] reported an interesting study of aqueous suspensions of clay particles. However, they did not clearly identify a liquid crystalline phase. In 2009, Paineau and co-workers [32] published a paper about a highly diluted (5% volume fraction) aqueous suspension of disk-shaped natural beidellite clay (a phyllosilicate), where a first-order isotropic to nematic phase transition was identified. The optical birefringence of these suspensions ($\sim 10^{-4}$) is smaller than that from usual micellar lyotropics. The nematic phase may be aligned by electric and magnetic fields and also by shear. The stabilization of a lyotropic structure is assured in an aqueous medium by the existence of electrical charges on the surface of the particles and the hydration of the particles.

More types of clays in aqueous suspensions showed liquid crystalline behavior: bentonite, an aluminum phyllosilicate clay consisting mostly of montmorillonite [33] (Figure 5a), laponite, a synthetic layered magnesium silicate [34] (Figure 5b), and imogolite, an aluminum silicate [35,36]. The imogolite particle has a hollowed cylindrical shape with diameter of about 2 nm and length of the order of hundreds of nm. An interesting, and until now not fully understood behavior of imogolite suspensions, is the presence of a regular streaked texture, observed in the polarizing optical microscope, that resembles textures from cholesteric ordering. The imogolite particles do not have any chiral component and the chiral arrangement (if demonstrated by other experiments) should be due to a particular packing of the cylinders [36]. This type of texture may also originate by a nematic ordering in gels [37]. Aqueous suspensions of nontronite clay also showed a nematic to isotropic phase transition at low (below 10^{-3} M/L) ionic strengths [38].

From the theoretical point of view, Onsager's approach [39] qualitatively explains the tendency of the plate-like clay particles to align in an aqueous suspension. However, the liquid crystalline behavior is not observed in concentrated suspensions, where phase segregation occurs. Polydispersity is also an issue that must be addressed when a suspension of clay particles in a solvent with liquid crystalline behavior is aimed at. Usually, a size separation procedure is needed before the preparation.

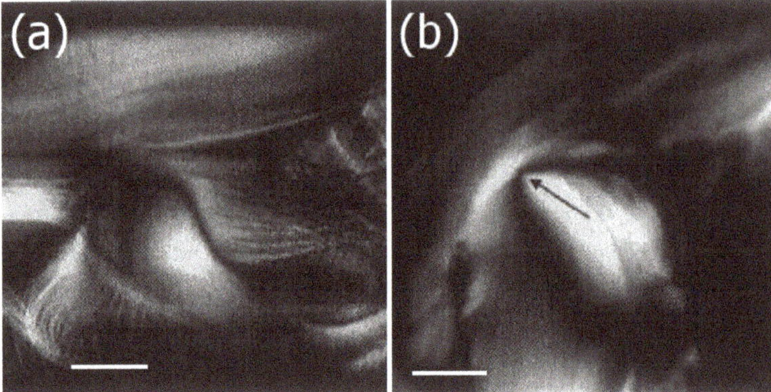

Figure 5. Nematic texture of different clays, (**a**) bentonite and (**b**) laponite (reproduced with permission from Reference [34]). Nematic textures for these materials are generally much less pronounced and typical than those for calamitic thermotropic liquid crystals. White bars correspond to 0.2 mm in (**a**) and 100 µm in (**b**).

2.4. Tobacco Mosaic Virus (TMV) and Other Viruses

The tobacco mosaic virus can be seen as the prototype of a rigid rod system. It is very straight, with a constant length of 18 nm and an often close to monodisperse length distribution round 300 nm. The aspect ratio is thus about 15, and the system is ideally suited to test the Onsager theory [40], see Figure 6. The TMV is a right-handed single-stranded RNA virus which infests the leaves of tobacco, but also other plants, which is clearly visible through a pronounced and characteristic discoloration. Discovered toward the end of the 19th century by Mayer [41], it was first thought to be bacterial, but was later independently shown by Iwanowski [42] and Beijerinck [43] to be of different origin, for which the latter coined the term "virus". It was not until the 1930s that electron microscopic evidence was produced [44], and in 1936, Bawden et al. [45] had already reported the lyotropic liquid crystalline behavior of the tobacco mosaic virus.

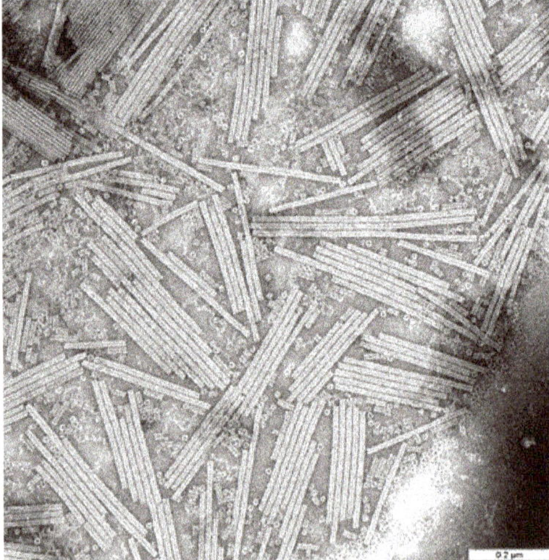

Figure 6. Electron microscopy image of tobacco mosaic viruses (TMV), showing an aspect ratio of approximately 15. The scale bar indicates 0.2 µm (reproduced from Wikimedia Commons, with no author name supplied).

From the magnetic field aligned nematic phase, Oldenbourg et al. [46] produced small-angle diffraction patterns which allowed the determination of the order parameter S. The X-ray scattering for samples of increasing TMV concentration showed transitions from the isotropic phase at low concentrations, passing through a typical isotropic/nematic two-phase region to a pure nematic phase at high concentrations. The order parameter S within the nematic phase changed from about S = 0.7 at the transition to close to perfect order of the long axis of the TMV S = 1 at rather high concentrations. This is indeed in accordance with predictions by the Onsager model. A very detailed study of the liquid crystalline behavior and physical properties of the TMV lyotropic nematic phase was carried out by Fraden et al. [47]. They measured the sample birefringence not only as a function of concentration and temperature, but also for varying ionic strength and different polydispersity. From these investigations, it was concluded that the stability of the nematic phase of tobacco mosaic virus suspensions is predominantly determined by electrostatic repulsion. Attractive van derWaals forces between the TMV rods supposedly play a much less important role. This provides an indication that the transition from isotropic to nematic is practically based on excluded volume effects. This in turn explains why the predictions of Onsager theory work very well for the TMV liquid crystal, because the theory is based on repulsive steric interactions, ignoring attractive forces between the colloidal particles. Graf and Löwen [48] later predicted the detailed phase diagram of the tobacco mosaic virus from theory and the use of computer simulations. They also described a further transition into smectic phases and colloidal crystals. Different virus suspensions, for example rod-like or semiflexible filamentous bacteriophage fd, have been reported to also exhibit chiral nematic or cholesteric order [49], as well as smectic layering [50], respectively. An overview can be found in the reviews of References [51,52].

We shall see below that novel trends in TMV lyotropic phase research have applicational potential in the production of silica nanostructures through templating. Another more fundamental aspect can be the experimental study of the phase behavior of mixtures, for example of rods and spheres [53], but also other systems like rods and plates, and even rod-rod systems with very different aspect ratios. This would be especially of interest in combination with computer simulations. Other novel aspects may be found in biological and chemical sensing or directed drug delivery.

2.5. Lyotropic Phases from DNA

The DNA macromolecule is a charged anionic polyelectrolyte formed by a right-handed double helix. Small fragments of DNA have a cylindrical shape of about 2 nm of diameter and variable lengths (typically ~50 nm). These fragments can be dispersed in water and present lyotropic liquid crystalline phases [54]. Increasing the DNA concentration in aqueous solutions (depending on the ionic strength and DNA persistence length), the phase sequence experimentally observed is: isotropic, blue phase, cholesteric, columnar hexagonal and crystalline (Figure 7). Disclinations and dislocations were observed in textures of aqueous DNA solutions, identifying the cholesteric phase [55]. Besides the texture inspection, measurements of the circular dichroism were performed to identify this mesophase.

More difficult to be identified is the blue phase, since it exists in a narrow range of temperature and DNA concentration, being optically isotropic. The electron microscopy of freeze-fracture replicas was used to identify the macromolecular arrangement in a double-twist ordering within small cylindrical domains. The optically anisotropic columnar phase was identified by different experimental techniques, mainly X-ray diffraction. The transition from the cholesteric to the columnar phase was shown to be of first-order or continuous. The same DNA solution may show both types of phase transition and, until now, the conditions defining one or the other type of transition are not known. Solutions with long DNA fragments (~70% w/v—comparable to that of in vivo systems) showed a cholesteric phase, with concentration-dependent pitch, and another two-dimensional (2D) phase that resembles the smectic thermotropic phase [56].

Figure 7. Transition region of the cholesteric (left) and the high-density region phases (right) in solutions of rod-like DNA molecules (reproduced with permission from Molecular Expressions at the Florida State University Research Foundation. The image can be found at the website https://micro.magnet.fsu.edu/dna/pages/transition3.html). White bar corresponds to about 300 μm.

Not only long DNA fragments were shown to present liquid crystalline phases. Short fragments (about 8 base pairs) in aqueous solutions with RNA presented columnar and cholesteric phases [57]. The local structure is stabilized due to base stacking forces that promote the end-to-end aggregation of duplexes. An interesting behavior was observed in drying droplets of DNA (persistence length of ~50 nm, 48 k pb) aqueous solutions, where "coffee rings" are formed [58]. The DNA macromolecules accumulate in the droplet edge, forming a lyotropic liquid crystal with concentric-chain orientations.

The atomic arrangement and charge distribution present in DNA fragments open many possibilities of liquid crystalline structures with these building blocks. Salamonczyk and co-workers [59] reported an interesting result about the presence of the smectic-A phase in an aqueous suspension of double-stranded DNA fragments. To achieve this, they increased the DNA flexibility by introducing a spacer in the middle of each duplex. Storm and co-workers discussed the formation of a columnar liquid crystalline structure of self-assembled DNA bottlebrushes [60]. The building block of this structure is made of DNA as the backbone molecule and C_4K_{12} protein polymers as the side chains.

Recently, Brach and co-workers reported a study where important differences in the DNA spatial structure were observed between free DNA and DNA organized in a lyotropic liquid crystalline arrangement [61]. The relations between the liquid crystalline structure and the functionality of living processes involving DNA still challenges researchers and opens a fascinating field of investigation. This last aspect inspires researchers to explore the relations between the liquid crystalline structure and the functionality of living processes involving DNA.

2.6. Lyotropic Cholesteric Cellulose Derivatives and Cellulose Nanocrystals

Cellulose is composed of β-D-glucopyranose units covalently linked with (1–4) glycosidic bonds. Cellulose nanocrystals (CNCs) are obtained from natural cellulose fibers. They are hydrophilic but can be surface functionalized to change their properties in the presence of different solvents [62]. CNCs are stiff, lath-like nanoparticles, with a typical diameter as small as ~6 nm, depending on the preparation method [63], and a length of about 100 nm.

Aqueous suspensions of cellulose nanocrystal particles, chemically prepared to avoid electrostatic stabilization and favoring the steric interaction [64], gave rise to a cholesteric mesophase (see Figure 6), with the typical fingerprint texture [65]. The cholesteric liquid crystalline phase occurs at a volume concentration of nanoparticles of about 10% [66]. One interesting application of the CNCs solution showing the cholesteric phase is that the mixture can be dried, maintaining the chiral structure, to make films that acquire photonic band gap properties [67].

Cholesteric properties of suspensions of cellulose nanocrystals can be modified by decorating the nanoparticles with polymers [68]. The surface chemistry of the nanoparticles and interacting forces modifies the phase diagrams and the pitch of the suspensions. Long-pitch chiral mesophases were obtained with a decrease in the surface charge of the particles, decreasing the particle–particle interaction [69]. This mesophase is highly viscous and is located in the vicinity of a biphasic region.

Cellulose-based lyotropic mixtures may also stabilize mesophases [70]. Solutions of cellulose tricarbanilate in methyl acrylate and methyl methacrylate were shown to stabilize nematic and cholesteric mesophases at specific relative component concentrations and temperatures [71,72]. Lyotropic mesophases were also obtained in cellulose derivatives (with hydroxypropylcellulose (Figure 8a–d) and ethyl-cellulose) in inorganic solvents [73]. Cellulose acetate phthalate/hydroxypropyl cellulose blends in *N,N*-dimethylacetamide showed lyotropic polyphormism under proper temperature and relative concentration conditions [74].

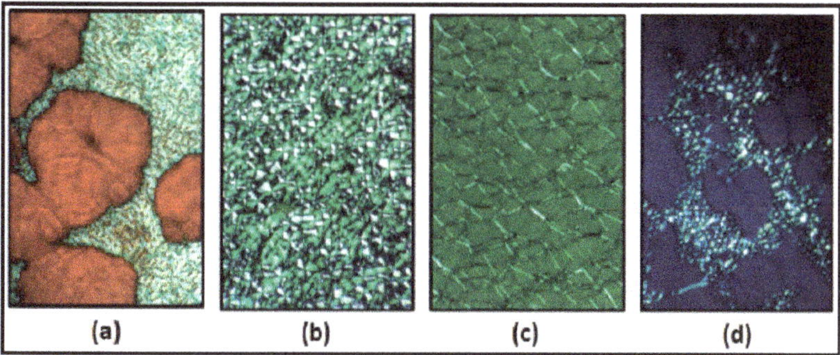

Figure 8. Lyotropic textures from (Hydroxypropyl) cellulose (HPC)/water in a polarized microscope: (**a**) ~45% HPC, planar and focal conic textures, (**b**) 55% HPC, focal conic texture, (**c**) 55% HPC, oily streak texture, (**d**) ~65% HPC, planar and focal conic textures (reproduced with permission from Reference [65]).

2.7. Nanotubes, Nanorods and Nanowires

Most of the systems relating to liquid crystalline behavior and nanotubes, nanorods or nanowires are composites, where the nanomaterial is dispersed in a thermotropic liquid crystal. This is often the nematic phase [75–78], occasionally also a smectic phase, often already with an additional functionality available, such as ferroelectric liquid crystals [79]. Such nanomaterials have been dispersed in lyotropic liquid crystals to a much lesser extent [80–83], often in the hexagonal phase for compatibility reasons. Thermotropic liquid crystals are used with carbon nanotubes to directionally orient the nanotubes or nanorods to exploit their extraordinary properties in a predetermined way as an addition to properties provided by the liquid crystal itself. On the other hand, lyotropic liquid crystals may be used as templates for materials in nanotechnology, often washing the liquid crystal out after the templating process. For example, nanowires and nanorods have been produced by synthesis in the lyotropic liquid crystalline state of TiO_2 [84] and ZnO [85].

In addition to the dispersions of nanotubes, nanorods or nanowires in thermotropic or lyotropic liquid crystal phases, these materials can in fact also form lyotropic liquid crystals by themselves through dispersion in an isotropic solvent. The behavior is often very similar to that observed for needle-like inorganic liquid crystals, or also the tobacco mosaic virus, and largely follows the description by Onsager's theory. At low concentrations of nanomaterials, an isotropic dispersion is observed, that changes to a biphasic region for increasing concentration, eventually forming a nematic lyotropic phase. For nanotubes, this was first theoretically predicted by Somoza et al. [86].

Experimental evidence followed soon for functionalized multi-wall carbon nanotubes (MWNT) in water [87,88], showing a nematic phase above 4 vol% MWNTs. Instead of covalent functionalization, systems of nanotube-adsorbed DNA were also used, providing the electrostatic repulsion favorable for LC formation [89,90]. Electrostatic repulsion for better dispersion was also used by Davis et al. [91] and Rai et al. [92] when choosing strong acids as isotropic solvents, which led to a protonation of the tube walls, instead of nanotube functionalization or decoration.

ZnO is a wide bandgap semi-conductor, which in nanowire form can exhibit liquid crystalline behavior as a lyotropic nematic [93,94]. Similarly, TiO_2 nanowires can assemble into liquid crystal phases [95]. Semiconductor rods of cadmium selenide, CdSe, can be produced with excellent monodispersity and a ratio of length to width of generally 40 to 6 nm, respectively. These are thus ideal candidates to exhibit not only nematic (Figure 9), but also smectic/lamellar ordering of lyotropic liquid crystals, as demonstrated in References [96,97]. The general potential of using nanomaterials in liquid crystals, either to tune the LC properties, to add functionality or to transfer liquid crystal order onto nanomaterials during synthesis or self-assembly in nanotechnology, is enormous. It can be expected that a whole new range of fundamental insights as well as technical applications are still to come in the future.

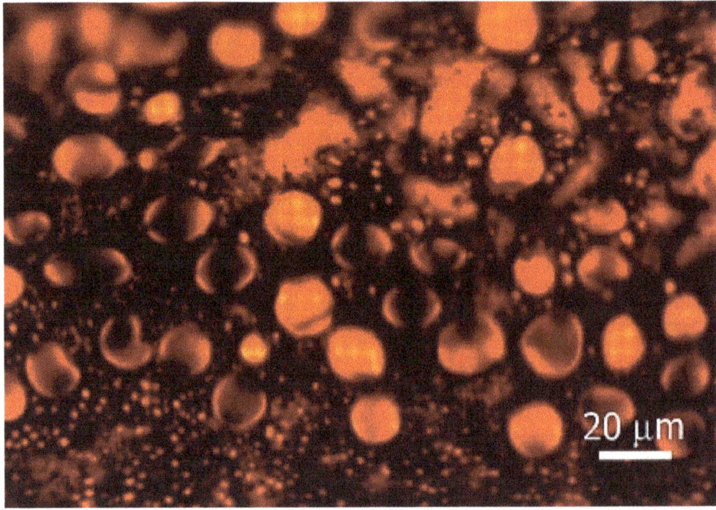

Figure 9. Nematic liquid crystal droplets forming on evaporation of the solvent from a cadmium selenide (CdSe) nano-rod solution (reproduced with permission from Reference [96]).

2.8. Graphene Oxide and Other 2D Materials

Similar to the rod-like and disc-like shapes of calamitic and discotic thermotropic liquid crystals, the analogue to lyotropic nanotubes would be graphene. Graphene itself has been shown to exhibit a nematic liquid crystal phase only in the protonated environment of strong chlorosulphonic acid [98]. A material more similar to the COOH-functionalized nanotubes is graphene oxide (GO), which readily shows liquid crystalline behavior over large concentration ranges [99,100] in a multitude of isotropic solvents, including water. In Figure 10, a well-aligned sample is shown, which exhibits optical properties like a standard calamitic nematic liquid crystal, but with the director in the direction of the sheet normal. The general phase behavior is as discussed for other lyotopic systems: an isotropic phase is followed by a two-phase region, ending in a fully developed nematic phase for increasing graphene oxide concentration. The details of the phase diagram are, on the other hand, dependent on the average size of the GO flakes [101,102], their polydispersity, the polarity of the solvent [101,103] and confinement conditions [101]. It has been shown that graphene oxide liquid crystals respond to applied electric fields [104,105] and that an electro-optic response can be achieved, which is based on a large Kerr effect [106,107], i.e., an induced birefringence proportional to the square of the applied electric field. The Kerr response times are at present longer than those of the well-discussed thermotropic Blue Phases, due to the higher viscosity of GO-LCs. So far, graphene oxide liquid crystals are by far the best studied of the lyotropic liquid crystals made from 2D materials, and their properties have been summarized in a number of review articles [108–110], although it should be mentioned that here, there are also still many open questions. One of these, which has, for example, been addressed via texture and dielectric studies, is the observation of mixtures of thermotropic nematics with GO [111]. There is some evidence [111,112] that on heating such a dispersion into the isotropic phase of the thermotropic LC, this acts as an isotropic solvent to facilitate the formation of a lyotropic GO liquid crystal phase. This is a very interesting topic, as it implies a thermotropic nematic to lyotropic nematic phase transition, which is not accompanied by any transition enthalpy. This should also be attractive for theoretical interpretation.

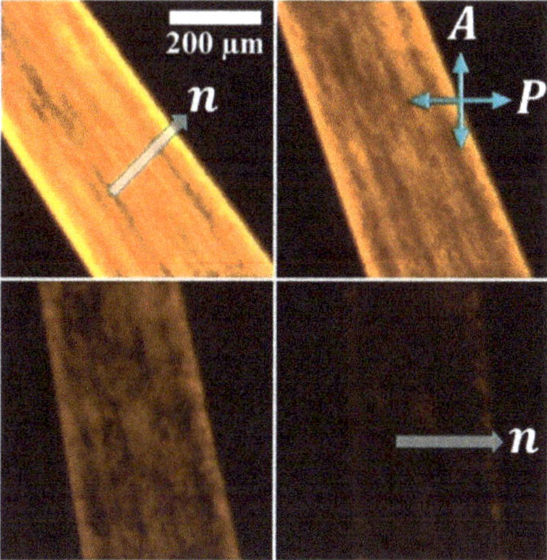

Figure 10. A lyotropic nematic graphene oxide phase can be oriented by a narrow, untreated glass channel. The quality of orientation is demonstrated by rotation of the sample between crossed polarizers. The director is indicated by *n* (reproduced with permission from Reference [37]).

Lastly, it should also be mentioned that other 2D materials can exhibit liquid crystalline phases in isotropic solvents. One of these is reduced graphene oxide (rGO) [113,114], which regains a certain amount of the conductive behavior observed for graphene, which is absent from graphene oxide. Reports have also been published for molybdenum disulphide, MnO_2 [115] and Mxenes [116]. A number of other possible candidates have yet to be investigated further [117]. The topic of graphene oxide liquid crystals, their properties, and dispersions with other lyotropic liquid crystal classes, will certainly be an exciting one over the next years to come. Also, with respect to possible applications, for example as fibers [118,119], in tuneable photonics [120], nanofiltration [121] or reflective displays [122].

2.9. Chromonics

Chromonics are rigid aromatic molecules, with hydrophilic ionic and hydrogen-bond groups located in the peripheries of the molecule [123,124]. π-stacking interactions between these flat molecules favor packing. In the presence of a polar solvent (e.g., water), they stack face to face in columns of different aggregation numbers. The (anisometric) columns consist of the building blocks of lyotropic liquid crystals (Figure 11a–c). Dyes, drugs and even nucleic acids are examples of this type of molecule [125]. Different mesophases were identified in mixtures with this type of molecule: uniaxial nematic [126], hexagonal [127] and lamellar-type structure (proposed for the diethyl ammonium flufenate) [128]. The aggregation of chromonic molecules is isodesmic, where the energy between molecules in a stack is independent of the number of molecules [129]. The addition of salts (e.g., NaCl) to the lyotropic nematic phase of cromolyn aqueous solutions shifted the nematic to isotropic phase boundaries upwards [130].

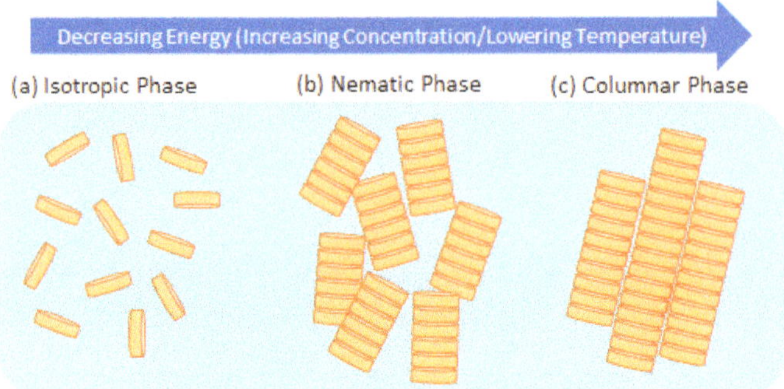

Figure 11. Typical structures of chromonic liquid crystal mesophase (reproduced with permission from https://core.ac.uk/download/pdf/54848113.pdf). (**a**) Isotropic, (**b**) nematic and (**c**) columnar phase.

From the theoretical point of view, besides the classical Onsager's approach, a model considering a competition of charge-like, long-range repulsion and anisotropic short-range attraction among molecules was recently proposed [131]. The nematic and hexagonal phases were obtained in the framework of this model, even at a small volume fraction of molecules, and correlations between the elastic response and stack growth were obtained. The elasticity of chromonic nematics was theoretically investigated (Monte Carlo simulations and Onsager-like model), inspecting the behavior of the Frank elastic constants, K_{11}, K_{22} and K_{33}, as a function of temperature and molecules' volume fraction [132]. The dependence of the elastic constants with temperature and molecules' volume fraction agree with the experiments. The elastic characteristics of chromonics is evidenced in experiments where the liquid crystal is confined in small droplets (typically from 1 to 100 μm) [133]. A very small twist elastic modulus in the nematic phase seems to be responsible for the appearance of a chiral-twisted bipolar configuration in confined conditions. Chromonic nematics are interesting systems to study topological defect cores in disclinations of +1/2 and −1/2 strengths, that extend to micrometric dimensions [134].

Recently, molecules of the chromonic Sunset Yellow (SSY) were added to classical amphiphilic lyotropic mixtures presenting the biaxial nematic phase [135]. It was shown that SSY exhibits a chaotropic character. Moreover, SSY causes an increase of the micellar shape anisotropy and, consequently, an increase of the biaxial nematic phase domain, with respect to the phase domain of the undoped mixture.

2.10. Polar Lyotropic Lamellar Phases

It took about half a century from the discovery of the first ferroelectric crystal, Rochelle salt [136], for the first fluid material, the chiral smectic C* phase, to be discovered [137]. It is already remarkable in itself that a fluid material can exhibit a spontaneous polarization, although, due to the SmC* helix, in its bulk state, it may rather be called helielectric, as true ferroelectricity requires this polarization to be switchable between two stable states. This was demonstrated shortly afterwards with the surface stabilized ferroelectric state [138] and has initiated one of the most active topics in liquid crystal research in the last century. So far, this is all related to thermotropic materials. On the other hand, it was long known that thermotropic and lyotropic phases do show a certain amount of analogy. Both exhibit orientationally ordered nematic phases. The lamellar Lα phase is the lyotropic analogue of the thermotropic SmA phase. So, why was there no tilted lamellar phase, a lyotropic equivalent of the rather common SmC phase? This was observed for the first time by Schaheutle and Finkelmann [139] and was demonstrated clearly by X-ray diffraction. A second report followed quite a number of years later [140]. These appear to have been the only confirmed cases of a tilted fluid lamellar phase. It was not until 2013 when a chiral amphiphilic molecule was shown to exhibit a lyotropic analogue of the ferroelectric SmC* phase [18] when added to an isotropic solvent, water or formamide, over a certain range of concentrations.

The fluid tilted lamellar phase was verified via X-ray diffraction and typical textures could be observed, such as a smectic Schlieren-texture (Figure 12a), broken fan-shaped textures (Figure 12b), and in the surface stabilized geometry, a domain texture, which could mutually be brought to extinction when rotated by twice the tilt angle between crossed polarizers (Figure 12c,d). Chirality could be demonstrated by a typical striped texture due to the helical superstructure, just as observed for the thermotropic SmC* phase. The pitch increased strongly when approaching the orthogonal phase (the SmA* analogue) on heating at a fixed concentration, which is also generally observed for the thermotropic case. But, the most striking evidence can be found in the ferroelectric electro-optic switching of the surface stabilized state. However, a direct measurement of the spontaneous polarization is not possible due to the overall ionic conductivity of the samples. Lastly, it should be mentioned that the electroclinic effect could also be verified for the orthogonal chiral L_α* phase [141], which very much resembles that of the SmA* phase of thermotropics. Other amphiphilic molecules with a similar behavior have been synthesized since [142], and a detailed account of this topic can be found in the review of Reference [143]. The aspect of a conceptual transfer relating to physical properties and structures between thermotropic and lyotropic liquid crystals is certainly a very novel approach and opens up a whole field of future research in lyotropic liquid crystals.

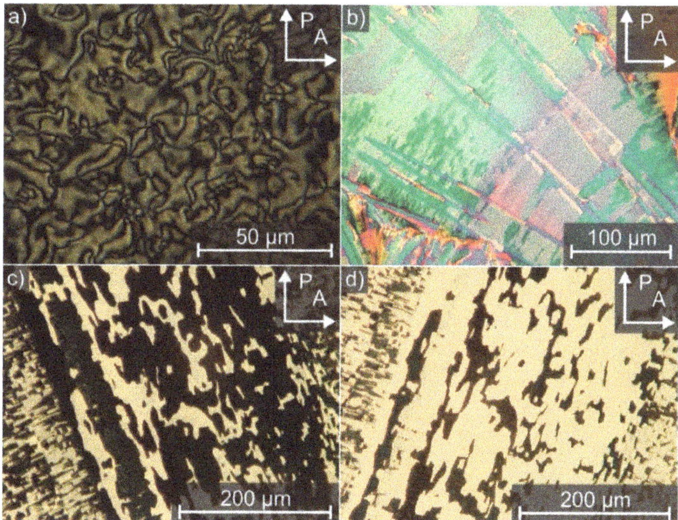

Figure 12. Different textures of the lyotropic SmC* phase. (**a**) SmC* Schlieren-texture, (**b**) broken fan-shaped chevron texture with typical zig-zag lines, and (**c**) and (**d**) surface stabilized SmC* domain texture, rotated by twice the tilt angle respectively, leading to mutually dark and bright domains (reproduced with permission from Reference [18]).

2.11. Active and Living Lyotropic Nematics

An emerging topic in liquid crystal research and especially lyotropic liquid crystals is active matter or living liquid crystals. Active matter [144,145] in general are systems that are composed of a large number of constituents, each consuming or transforming energy for the reason of propulsion. These are thus intrinsic non-equilibrium systems. There is a wide variety of such systems found in soft and biological matter, for example a school of fish, bacteria or microtubules. They all have one property in common, they are self-organizing and exhibit collective, self-propelled motion. Liquid crystal-based active matter has recently attracted much increasing interest [146], due to fascinating phenomena observed that are absent in passive liquid crystals [147].

Active liquid crystal systems that are often studied include bacterial suspensions [148,149], microtubule-motor protein systems [150,151] and actin-motor protein systems [152–154].

A quite different system has been proposed, called "living liquid crystals" [155–158]. These are of particular interest to lyotropic liquid crystals, as they represent swimming, live bacteria in a lyotropic nematic phase (Figure 13). The latter have been shown to be supporting bacteria life [159], which is not the case for thermotropic nematic liquid crystals. Lavrentovich and co-workers [155] demonstrated experimentally that bacteria can sense director field deformations. They showed that for pure splay and pure bend deformations, the motion of the bacteria was equally probable in either direction of the director field. For regions with splay-bend deformations, like in the vicinity of topological defects, the motion was directed towards the positive s = 1/2 defect and avoiding the negative s = −1/2 one. By the use of predetermined director patterns, they directed the motion of bacteria and exerted a directing influence on the otherwise chaotic bacterial motion. Active and living liquid crystals exhibit a plethora of fascinating phenomena, spatio-temporal patterns and prospects for biological and biomedical applications. It is thus very likely that this field of lyotropic liquid crystal research will thrive in the future, not only in experiment but also in theory, simulations and applications.

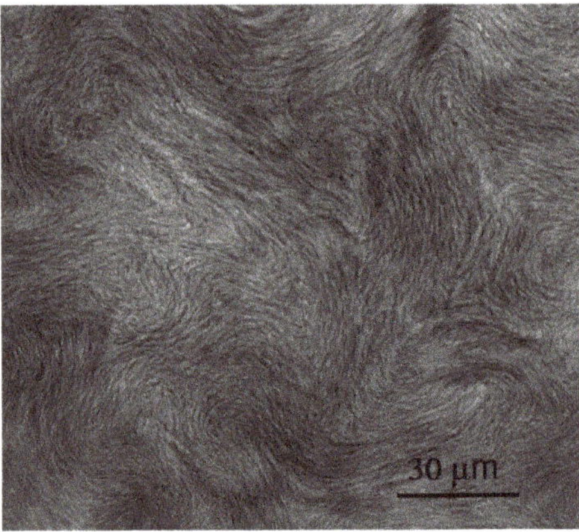

Figure 13. Texture of a living liquid crystal with disclination pairs. The bacteria are aligned along the local nematic director, seen by the lines in the textures (reproduced with permission from Reference [155]).

2.12. Applications

To the general public, liquid crystals are often primarily known through their electro-optic applications in displays, light shutters, or optical light modulators. Lyotropics are generally quite unheard of, despite of the fact that we use them on a daily basis. Obviously, a short paragraph on the applications of lyotropic liquid crystals cannot be all encompassing. Besides the various applications in the food and cosmetics industries, as well as detergents [160–164], we will here give an indicative overview of some other, possibly more modern applications.

One of these topics is drug delivery, which is somewhat related to some aspects encountered in the food and cosmetics industry in terms of the targeted and controlled release of an active ingredient. For this, often lyotropic phases can be exploited to trigger the release, for example through a change of the pH [165], or even by light irradiation [166], where the release is started by molecular switches. For further information, we refer to some relevant reviews [167,168] and citations therein.

Since the seminal work of Abbott and co-workers [169], liquid crystal sensors generally employ a molecularly triggered texture transition of a thermotropic liquid crystal from a dark to bright state (or vice versa) that indicates the absorption of a number of liquid, gas, or biological molecules within the liquid crystal [170]. But, there have also been some reports of lyotropic LC being used for sensing, for example chromonics for biological sensing applications [171], or lyotropic phases of DNA for enzymes [172] and other for antigens [173] and pathogens [174]. In general, particularly the fields of biologically relevant systems like biotechnology, biosensors, drug delivery and biomimetics [175] are of highlighted interest for lyotropic liquid crystals.

Further, the lyotropic phases have found their way into the templated synthesis and self-assembly for functional nanoparticles and nanotechnology. The use of nematic tobacco mosaic viruses has been suggested for the design of silica mesostructures. Exploiting the TMV as a template in the synthesis of inorganic frameworks with ordered porosity was demonstrated by Fowler et al. [176]. They described a method where ordered viruses were first silicated and then thermally removed via biodegradation. This produced silica structures with ordered nanochannels of 20 nm diameter. The method of using a lyotropic structure for the synthesis of nanoparticles [177] and nanostructured materials [178] was quickly picked up and used for a variety of different systems [179,180] (Figure 14a–c). For further information, one may refer to a review article by Hegmann et al. [181] and references therein.

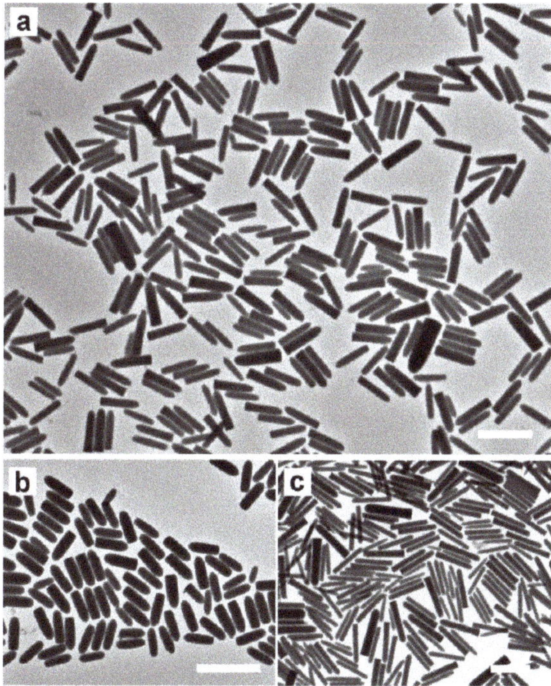

Figure 14. Transmission electron microscopy images of silica rods synthesized by templating a lyotropic phase. The scale bars are 2 μm. Different aspect ratios, L/D can be obtained, for example (**a**) L/D ≈ 5, (**b**) L/D ≈ 3, (**c**) L/D ≈ 8, by running the synthesis for different times. The anisotropy is ascribed to the anisotropic supply of reactants, leading to rod-like growth (reproduced with permission from Reference [182]).

Lastly, one may want to mention some applications of dried cellulose nanocrystals, which firstly transfer liquid crystalline order onto dispersed functional nanomaterials and are subsequently dried to produce thin solid films. These can then be employed as chromonic films [183] due to the residual selective reflection of the cholesteric structure, or as plasmonic films with dispersed gold [184,185] or silver [186] nanorods or nanowires, respectively. From the discussed selected examples, it becomes clear that lyotropic liquid crystals have significant potential for future applications, likely to be most intensified in the areas of biotechnology and biomedicine.

3. Conclusions

We hope that we could provide the reader with a short, yet hopefully interesting overview of some of the recent developments and novel trends in lyotropic liquid crystal research. There are certainly still a lot of unanswered fundamental questions in this field, materials developments to be explored and applications to be found, optimized and introduced to the market. This review of new trends in the field has hopefully sparked the interest of readers to explore some of the topics discussed in much more detail in the papers of this special issue, and maybe this issue will help in further bringing together the two fundamental fields of liquid crystal research, thermotropic and lyotropic systems, in the quest to develop an overarching general understanding of this fascinating aspect of soft matter.

Author Contributions: This review was written in equal parts by I.D. and A.M.F.N. All authors have read and agreed to the published version of the manuscript.

Funding: This research received no external funding.

Conflicts of Interest: The authors declare no conflict of interest.

References

1. Neto, A.M.F.; Salinas, S.R.A. *The Physics of Lyotropic Liquid Crystals: Phase Transitions and Structural Properties*; Oxford University Press: New York, NY, USA, 2005.
2. Petrov, A.G. *The Lyotropic State of Matter: Molecular Physics and Living Matter Physics*; Taylor & Francis: London, UK, 1999.
3. Reinitzer, F. Beiträge zur Kenntniss des Cholesterins. *Monatsh. Chem.* **1888**, *9*, 421–441. [CrossRef]
4. Virchow, R. Myelinformen. *Arch. Pathol. Anatom. Physiol. Klin. Med.* **1854**, *6*, 572. [CrossRef]
5. Mettenheimer, C. Mittheilung in Betreff mikroskopischer Beobachtungen mit polarisirtem Licht. *Correspondenzblatt Vereins Gemeinschaftliche Arb. Förd. Wiss. Heilkd.* **1857**, *24*, 331–332.
6. Planer, J. Notiz über das Cholestearin. *Ann Chem.* **1861**, *118*, 25–27. [CrossRef]
7. Loebisch, W. Zur Kenntniss des Cholesterins. *Ber. Deutsch. Chem. Ges.* **1872**, *5*, 510–514. [CrossRef]
8. Rayman, M.B. Contribution à l'histoire de la cholestérine. *Bull. Soc. Chim. Paris* **1887**, *47*, 898–901.
9. Lehmann, O. Über fliessende Krystalle. *Z. Phys. Chem.* **1889**, *4*, 462. [CrossRef]
10. Markovitsi, D.; Mathis, A.; Simon, J.; Wittman, J.C.; Le Moigne, J. Annelides V: A New Type of Lyotropic Mesomorphic PHASE. *Mol. Cryst. Liq. Cryst.* **1980**, *64*, 121–125. [CrossRef]
11. Menger, F.M.; Littau, C.A. Gemini-surfactants: Synthesis and properties. *J. Am. Chem. Soc.* **1991**, *113*, 1451–1452. [CrossRef]
12. Menger, F.M.; Ding, J. Spiro-Tenside und -Phospholipide: Synthese und Eigenschaften. *Angew. Chem.* **1996**, *108*, 2266–2268. [CrossRef]
13. Schröter, J.A.; Tschierske, C.; Wittenberg, M.; Wendorff, J.H. Formation of Columnar and Lamellar Lyotropic Mesophases by Facial Amphiphiles with Protic and Lipophilic Solvents. *J. Am. Chem. Soc.* **1998**, *120*, 10669–10675. [CrossRef]
14. Fuhrhop, J.H.; Fritsch, D. Bolaamphiphiles form ultrathin, porous and unsymmetric monolayer lipid membranes. *Accounts Chem. Res.* **1986**, *19*, 130–137. [CrossRef]
15. Yu, L.J.; Saupe, A. Observation of a Biaxial Nematic Phase in Potassium Laurate-1-Decanol-Water Mixtures. *Phys. Rev. Lett.* **1980**, *45*, 1000–1003. [CrossRef]
16. Luckhurst, G.R.; Sluckin, T.J. (Eds.) *Biaxial Nematic Liquid Crystals: Theory, Simulation, and Experiment*; John Wiley & Sons: Chichester, UK, 2015.

17. Yu, L.J.; Saupe, A. Liquid crystalline phases of the sodium decylsulfate/decanol/water system. Nematic-nematic and cholesteric-cholesteric phase transitions. *J. Am. Chem. Soc.* **1980**, *102*, 4879. [CrossRef]

18. Bruckner, J.R.; Porada, J.H.; Dietrich, C.F.; Dierking, I.; Giesselmann, F. A Lyotropic Chiral Smectic C Liquid Crystal with Polar Electrooptic Switching. *Angew. Chem. Int. Ed.* **2013**, *52*, 8934–8937. [CrossRef]

19. Mukherjee, P.K.; Sen, K. On a new topology in the phase diagram of biaxial nematic liquid crystals. *J. Chem. Phys.* **2009**, *130*, 141101. [CrossRef]

20. Quist, P.-O. First order transitions to a lyotropic biaxial nematic. *Liq. Cryst.* **1995**, *18*, 623–629. [CrossRef]

21. Akpinar, E.; Otluoglu, K.; Turkmen, M.; Canioz, C.; Reis, D.; Neto, A.M.F. Effect of the presence of strong and weak electrolytes on the existence of uniaxial and biaxial nematic phases in lyotropic mixtures. *Liq. Cryst.* **2016**, *43*, 1693–1708. [CrossRef]

22. Diesselhorst, H.; Freundlich, H. On the double refraction of vanadine pentoxydsol. *Phys. Z.* **1915**, *16*, 419–425.

23. Freundlich, H. Die Doppelbrechung des Vanadinpentoxydsols. *Ber. Bunsenges. Phys. Chem.* **1916**, *22*, 27–33.

24. Zocher, H. Über freiwillige Strukturbildung in Solen. (Eine neue Art anisotrop flüssiger Medien.). *Zeitschrift für Anorganische und Allgemeine Chemie* **1925**, *147*, 91–110. [CrossRef]

25. Davidson, P.; Garreau, A.; Livage, J. Nematic colloidal suspensions of V_2O_5 in water—Or Zocher phases revisited. *Liq. Cryst.* **1994**, *16*, 905–910. [CrossRef]

26. Pelletier, O.; Sotta, P.; Davidson, P. Deuterium Nuclear Magnetic Resonance Study of the Nematic Phase of Vanadium Pentoxide Aqueous Suspensions. *J. Phys. Chem. B* **1999**, *103*, 5427–5433. [CrossRef]

27. Zocher, H.; Török, C. Neuere Beiträge zur Kenntnis der Taktosole. *Colloid Polym. Sci.* **1960**, *173*, 1–7.

28. Davidson, P.; Gabriel, J.-C.P.; Levelut, A.M.; Batail, P. A new nematic suspension based on all-inorganic polymer rods. *Europhys. Lett.* **1993**, *21*, 317–322. [CrossRef]

29. Zocher, H.; Török, C. Crystals of higher order and their relation to other superphases. *Acta Crystallogr.* **1967**, *22*, 751–755. [CrossRef]

30. Sonin, A.S. Inorganic lyotropic liquid crystals. *J. Mater. Chem.* **1998**, *8*, 2557–2574. [CrossRef]

31. Mourchid, A.; Delville, A.; Lambard, J.; Lécolier, E.; Levitz, P. Phase diagram of colloidal dispersions of anisotropic charged particles: Equilibrium properties, structure, and rheology of laponite suspensions. *Langmuir* **1995**, *11*, 1942–1950. [CrossRef]

32. Paineau, E.; Antonova, K.; Baravian, C.; Bihannic, I.; Davidson, P.; Dozov, I.; Impéror-Clerc, M.; Levitz, P.; Madsen, A.; Meneau, F.; et al. Liquid-Crystalline Nematic Phase in Aqueous Suspensions of a Disk-Shaped Natural Beidellite Clay. *J. Phys. Chem. B* **2009**, *113*, 15858–15869. [CrossRef]

33. Langmuir, I. The Role of Attractive and Repulsive Forces in the Formation of Tactoids, Thixotropic Gels, Protein Crystals and Coacervates. *J. Chem. Phys.* **1938**, *6*, 873. [CrossRef]

34. Gabriel, J.-C.P.; Sanchez, C.; Davidson, P. Observation of Nematic Liquid-Crystal Textures in Aqueous Gels of Smectite Clays. *J. Phys. Chem.* **1996**, *100*, 11139–11143. [CrossRef]

35. Kajiwara, K.; Donkai, N.; Hiragi, Y.; Inagaki, H. Lyotropic mesophase of imogolite, 1. Effect of polydispersity on phase diagram. *Makromol. Chem.* **1986**, *187*, 2883–2893. [CrossRef]

36. Kajiwara, K.; Donkai, N.; Fujiyoshi, Y.; Inagaki, H. Lyotropic mesophase of imogolite, 2. Microscopic observation of imogolite mesophase. *Makromol. Chem.* **1986**, *187*, 2895–2907. [CrossRef]

37. Dierking, I.; Al-Zangana, S. Lyotropic Liquid Crystal Phases from Anisotropic Nanomaterials. *Nanomaterials* **2017**, *7*, 305. [CrossRef]

38. Michot, L.J.; Bihannic, I.; Maddi, S.; Baravian, C.; Levitz, P.; Davidson, P. Sol/Gel and Isotropic/Nematic Transitions in Aqueous Suspensions of Natural Nontronite Clay. Influence of Particle Anisotropy. 1. Features of the I/N Transition. *Langmuir* **2008**, *24*, 3127–3139. [CrossRef]

39. Onsager, L. The effects of shape on the interaction of colloidal particles. *Ann. N. Y. Acad. Sci.* **1949**, *51*, 627–659. [CrossRef]

40. Barry, E.; Beller, D.; Dogic, Z. A model liquid crystalline system based on rodlike viruses with variable chirality and persistence length. *Soft Matter* **2009**, *5*, 2563–2570. [CrossRef]

41. Mayer, A. Über die Mosaikkrankheit des Tabaks. *Die Landwirtsch. Versuchsstationen.* **1886**, *32*, 451–467.

42. Iwanowski, D. Über die Mosaikkrankheit der Tabakspflanze. *Bulletin Scientifique Publié Par l'Académie Impériale des Sciences de Saint-Pétersbourg/Nouvelle Serie III* **1892**, *35*, 67–70.

43. Beijerinck, M.W. Über ein Contagium vivum fluidum als Ursache der Fleckenkrankheit der Tabaksblätter. In *Verhandelingen der Koninklijke Akademie van Wetenschappen Te Amsterdam*; J. Müller: Brake, Germany, 1898; Volume 65, pp. 1–22.

44. Kausche, G.A.; Pfankuch, E.; Ruska, H. The visualisation of herbal viruses in surface microscopes. *Naturwissenschaften* **1939**, *27*, 292–299. [CrossRef]

45. Bawden, F.C.; Pirie, N.W.; Bernal, J.D.; Fankuchen, I. Liquid Crystalline Substances from Virus-infected Plants. *Nature* **1936**, *138*, 1051–1052. [CrossRef]

46. Oldenbourg, R.; Wen, X.; Meyer, R.B.; Caspar, D.L.D. Orientational Distribution Function in Nematic Tobacco-Mosaic-Virus Liquid Crystals Measured by X-Ray Diffraction. *Phys. Rev. Lett.* **1988**, *61*, 1851–1854. [CrossRef]

47. Fraden, S.; Maret, G.; Caspar, D.L.D. Angular correlations and the isotropic-nematic phase transition in suspensions of tobacco mosaic virus. *Phys. Rev. E* **1993**, *48*, 2816–2837. [CrossRef] [PubMed]

48. Graf, H.; Löwen, H. Phase diagram of tobacco mosaic virus solutions. *Phys. Rev. E* **1999**, *59*, 1932–1942. [CrossRef]

49. Dogic, Z.; Fraden, S. Cholesteric Phase in Virus Suspensions. *Langmuir* **2000**, *16*, 7820–7824. [CrossRef]

50. Dogic, Z.; Fraden, S. Smectic Phase in a Colloidal Suspension of Semiflexible Virus Particles. *Phys. Rev. Lett.* **1997**, *78*, 2417–2420. [CrossRef]

51. Dogic, Z.; Fraden, S. Ordered phases of filamentous viruses. *Curr. Opin. Colloid Interface Sci.* **2006**, *11*, 47–55. [CrossRef]

52. Dogic, Z. Filamentous Phages as a Model System in Soft Matter Physics. *Front. Microbiol.* **2016**, *7*, 349. [CrossRef]

53. Adams, M.; Dogic, Z.; Keller, S.L.; Fraden, S. Entropically driven microphase transitions in mixtures of colloidal rods and spheres. *Nature* **1998**, *393*, 349–352. [CrossRef]

54. Leforestier, A.; Livolant, F. The Bacteriophage Genome Undergoes a Succession of Intracapsid Phase Transitions upon DNA Ejection. *J. Mol. Boil.* **2010**, *396*, 384–395. [CrossRef] [PubMed]

55. Leforestier, A.; Livolant, F. Supramolecular ordering of DNA in the cholesteric liquid crystalline phase: An ultrastructural study. *Biophys. J.* **1993**, *65*, 56–72. [CrossRef]

56. Strzelecka, T.E.; Davidson, M.W.; Rill, R.L. Multiple liquid crystal phases of DNA at high concentrations. *Nature* **1988**, *331*, 457–460. [CrossRef] [PubMed]

57. Zanchetta, G.; Nakata, M.; Buscaglia, M.; Clark, N.A.; Bellini, T. Liquid crystal ordering of DNA and RNA oligomers with partially overlapping sequences. *J. Phys. Condens. Matter* **2008**, *20*, 494214. [CrossRef]

58. Smalyukh, I.I.; Zribi, O.V.; Butler, J.C.; Lavrentovich, O.D.; Wong, G.C.L. Structure and Dynamics of Liquid Crystalline Pattern Formation in Drying Droplets of DNA. *Phys. Rev. Lett.* **2006**, *96*, 177801. [CrossRef]

59. Salamonczyk, M.; Zhang, J.; Portale, G.; Zhu, C.; Kentzinger, E.; Gleeson, J.T.; Jakli, A.; De Michele, C.; Dhont, J.K.G.; Sprunt, S.; et al. Smectic phase in suspensions of gapped DNA duplexes. *Nat. Commun.* **2016**, *7*, 13358. [CrossRef]

60. Storm, I.M.; Kornreich, M.; Hernandez-Garcia, A.; Voets, I.K.; Beck, R.; Stuart, M.A.C.; Leermakers, F.A.M.; De Vries, R. Liquid Crystals of Self-Assembled DNA Bottlebrushes. *J. Phys. Chem. B* **2015**, *119*, 4084–4092. [CrossRef]

61. Brach, K.; Hatakeyama, A.; Nogues, C.; Olesiak-Banska, J.; Buckle, M.; Matczyszyn, K. Photochemical analysis of structural transitions in DNA liquid crystals reveals differences in spatial structure of DNA molecules organized in liquid crystalline form. *Sci. Rep.* **2018**, *8*, 4528. [CrossRef]

62. George, J.; Sabapathi, S. Cellulose nanocrystals: Synthesis, functional properties, and applications. *Nanotechnol. Sci. Appl.* **2015**, *8*, 45–54. [CrossRef]

63. Pääkkö, M.; Ankerfors, M.; Kosonen, H.; Nykänen, A.; Ahola, S.; Österberg, M.; Ruokolainen, J.; Laine, J.; Larsson, P.T.; Ikkala, O.; et al. Enzymatic Hydrolysis Combined with Mechanical Shearing and High-Pressure Homogenization for Nanoscale Cellulose Fibrils and Strong Gels. *Biomacromolecules* **2007**, *8*, 1934–1941. [CrossRef]

64. Kloser, E.; Gray, D.G. Surface Grafting of Cellulose Nanocrystals with Poly(ethylene oxide) in Aqueous Media. *Langmuir* **2010**, *26*, 13450–13456. [CrossRef] [PubMed]

65. Gray, D.G.; Mu, X. Chiral Nematic Structure of Cellulose Nanocrystal Suspensions and Films; Polarized Light and Atomic Force Microscopy. *Materials* **2015**, *8*, 7873–7888. [CrossRef] [PubMed]

66. Gray, D.G. Recent Advances in Chiral Nematic Structure and Iridescent Color of Cellulose Nanocrystal Films. *Nanomaterials* **2016**, *6*, 213. [CrossRef] [PubMed]

67. Lagerwall, J.P.F.; Schütz, C.; Salajkova, M.; Noh, J.; Park, J.H.; Scalia, G.; Bergström, L. Cellulose nanocrystal-based materials: From liquid crystal self-assembly and glass formation to multifunctional thin films. *NPG Asia Mater.* **2014**, *6*, e80. [CrossRef]

68. Azzam, F.; Heux, L.; Jean, B. Adjustment of the Chiral Nematic Phase Properties of Cellulose Nanocrystals by Polymer Grafting. *Langmuir* **2016**, *32*, 4305–4312. [CrossRef] [PubMed]

69. Abitbol, T.; Kam, D.; Levi-Kalisman, Y.; Gray, D.G.; Shoseyov, O. Surface Charge Influence on the Phase Separation and Viscosity of Cellulose Nanocrystals. *Langmuir* **2018**, *34*, 3925–3933. [CrossRef]

70. Spontak, R.J.; El-Nokaly, M.A.; Bartolo, R.G.; Burns, J.L. *Polymer Solutions, Blends and Interfaces*; Noda, I., Rubingh, D.N., Eds.; Elsevier Science Publishers B.V.: Amsterdam, The Netherlands, 1992.

71. Cowie, J.; Arrighi, V.; Cameron, J.; McEwan, I.; McEwen, I.J. Lyotropic liquid crystalline cellulose derivatives in blends and molecular composites. *Polymer* **2001**, *42*, 9657–9663. [CrossRef]

72. Cowie, J.; Arrighi, V.; Cameron, J.; Robson, D. Lyotropic liquid crystalline cellulose derivatives in blends and molecular composites. *Macromol. Symp.* **2000**, *152*, 107–116. [CrossRef]

73. Kamide, K.; Okajima, K.; Matsui, T.; Kajita, S. Formation of Lyotropic Liquid Crystals of Cellulose Derivatives Dissolved in Inorganic Acids. *Polym. J.* **1986**, *18*, 273–276. [CrossRef]

74. Onofrei, M.-D.; Dobos, A.M.; Stoica, I.; Olaru, N.; Olaru, L.; Ioan, S. Lyotropic Liquid Crystal Phases in Cellulose Acetate Phthalate/Hydroxypropyl Cellulose Blends. *J. Polym. Environ.* **2013**, *22*, 99–111. [CrossRef]

75. Dierking, I.; Scalia, G.; Morales, P.; LeClere, D. Aligning and Reorienting Carbon Nanotubes with Nematic Liquid Crystals. *Adv. Mater.* **2004**, *16*, 865–869. [CrossRef]

76. Dierking, I.; Scalia, G.; Morales, P. Liquid crystal–carbon nanotube dispersions. *J. Appl. Phys.* **2005**, *97*, 44309. [CrossRef]

77. Lynch, M.D.; Patrick, D.L. Organizing Carbon Nanotubes with Liquid Crystals. *Nano Lett.* **2002**, *2*, 1197–1201. [CrossRef]

78. Lagerwall, J.P.F.; Scalia, G. Carbon nanotubes in liquid crystals. *J. Mater. Chem.* **2008**, *18*, 2890. [CrossRef]

79. Yakemseva, M.; Dierking, I.; Kapernaum, N.; Usol'Tseva, N.V.; Giesselmann, F. Dispersions of multi-wall carbon nanotubes in ferroelectric liquid crystals. *Eur. Phys. J. E* **2014**, *37*, 7. [CrossRef] [PubMed]

80. Lagerwall, J.P.F.; Scalia, G.; Haluska, M.; Dettlaff-Weglikowska, U.; Giesselmann, F.; Roth, S. Simultaneous alignment and dispersion of carbon nanotubes with lyotropic liquid crystals. *Phys. Status Solidi B* **2006**, *243*, 3046–3049. [CrossRef]

81. Lagerwall, J.P.F.; Scalia, G.; Haluska, M.; Dettlaff-Weglikowska, U.; Roth, S.; Giesselmann, F. Nanotube Alignment Using Lyotropic Liquid Crystals. *Adv. Mater.* **2007**, *19*, 359–364. [CrossRef]

82. Jiang, W.; Yu, B.; Liu, W.; Hao, J. Carbon Nanotubes Incorporated within Lyotropic Hexagonal Liquid Crystal Formed in Room-Temperature Ionic Liquids. *Langmuir* **2007**, *23*, 8549–8553. [CrossRef]

83. Scalia, G.; Von Bühler, C.; Hägele, C.; Roth, S.; Giesselmann, F.; Lagerwall, J.P.F. Spontaneous macroscopic carbon nanotube alignment via colloidal suspension in hexagonal columnar lyotropic liquid crystals. *Soft Matter* **2008**, *4*, 570–576. [CrossRef]

84. Liu, L.H.; Bai, Y.; Wang, F.M.; Liu, N. Fabrication and Characterizes of TiO_2 Nanomaterials Templated by Lyotropic Liquid Crystal. *Adv. Mater. Res.* **2011**, *399*, 532–537. [CrossRef]

85. Saliba, S.; Davidson, P.; Impéror-Clerc, M.; Mingotaud, C.; Kahn, M.L.; Marty, J.-D. Facile direct synthesis of ZnO nanoparticles within lyotropic liquid crystals: Towards organized hybrid materials. *J. Mater. Chem.* **2011**, *21*, 18191. [CrossRef]

86. Somoza, A.M.; Sagui, C.; Roland, C. Liquid-crystal phases of capped carbon nanotubes. *Phys. Rev. B* **2001**, *63*, 81403. [CrossRef]

87. Song, W.; Kinloch, I.A.; Windle, A.H. Nematic Liquid Crystallinity of Multiwall Carbon Nanotubes. *Science* **2003**, *302*, 1363. [CrossRef] [PubMed]

88. Song, W.; Windle, A.H. Isotropic−Nematic Phase Transition of Dispersions of Multiwall Carbon Nanotubes. *Macromolecules* **2005**, *38*, 6181–6188. [CrossRef]

89. Badaire, S.; Zakri, C.; Maugey, M.; Derré, A.; Barisci, J.N.; Wallace, G.G.; Poulin, P. Liquid Crystals of DNA-Stabilized Carbon Nanotubes. *Adv. Mater.* **2005**, *17*, 1673–1676. [CrossRef]

90. Puech, N.; Blanc, C.; Grelet, E.; Zamora-Ledezma, C.; Maugey, M.; Zakri, C.; Anglaret, E.; Poulin, P. Highly Ordered Carbon Nanotube Nematic Liquid Crystals. *J. Phys. Chem. C* **2011**, *115*, 3272–3278. [CrossRef]

91. Davis, V.A.; Ericson, L.M.; Parra-Vasquez, A.N.G.; Fan, H.; Wang, Y.; Prieto, V.; Longoria, J.A.; Ramesh, S.; Saini, R.K.; Kittrell, C.; et al. Phase behaviour and rheology of SWNTs in superacids. *Macromolecules* **2004**, *37*, 154–160. [CrossRef]

92. Rai, P.K.; Pinnick, R.A.; Parra-Vasquez, A.N.G.; Davis, V.A.; Schmidt, H.K.; Hauge, R.H.; Smalley, R.E.; Pasquali, M. Isotropic–Nematic Phase Transition of Single-Walled Carbon Nanotubes in Strong Acids. *J. Am. Chem. Soc.* **2006**, *128*, 591–595. [CrossRef]

93. Zhang, S.; Majewski, P.W.; Keskar, G.; Pfefferle, L.D.; Osuji, C.O. Lyotropic Self-Assembly of High-Aspect-Ratio Semiconductor Nanowires of Single-Crystal ZnO. *Langmuir* **2011**, *27*, 11616–11621. [CrossRef]

94. Zhang, S.; Pelligra, C.I.; Keskar, G.; Majewski, P.W.; Ren, F.; Pfefferle, L.D.; Osuji, C.O. Liquid Crystalline Order and Magnetocrystalline Anisotropy in Magnetically Doped Semiconducting ZnO Nanowires. *ACS Nano* **2011**, *5*, 8357–8364. [CrossRef]

95. Ren, Z.; Chen, C.; Hu, R.; Mai, K.; Qian, G.; Wang, Z. Two-Step Self-Assembly and Lyotropic Liquid Crystal Behavior of TiO2 Nanorods. *J. Nanomater.* **2012**, *2012*, 180989. [CrossRef]

96. Li, L.-S.; Walda, J.; Manna, L.; Alivisatos, A.P.; Alivisatos, A.P. Semiconductor Nanorod Liquid Crystals. *Nano Lett.* **2002**, *2*, 557–560. [CrossRef]

97. Li, L.-S.; Alivisatos, A.P. Semiconductor Nanorod Liquid Crystals and Their Assembly on a Substrate. *Adv. Mater.* **2003**, *15*, 408–411. [CrossRef]

98. Behabtu, N.; Lomeda, J.R.; Green, M.J.; Higginbotham, A.L.; Sinitskii, A.; Kosynkin, D.V.; Tsentalovich, D.; Parra-Vasquez, A.N.G.; Schmidt, J.; Kesselman, E.; et al. Spontaneous high-concentration dispersions and liquid crystals of graphene. *Nat. Nanotechnol.* **2010**, *5*, 406–411. [CrossRef] [PubMed]

99. Xu, Z.; Gao, C. Graphene chiral liquid crystals and macroscopic assembled fibres. *Nat. Commun.* **2011**, *2*, 571. [CrossRef]

100. Kim, J.E.; Han, T.H.; Lee, S.H.; Kim, J.Y.; Ahn, C.W.; Yun, J.M.; Kim, S.O.; Lee, J. Graphene Oxide Liquid Crystals. *Angew. Chem. Int. Ed.* **2011**, *50*, 3043–3047. [CrossRef]

101. Al-Zangana, S.; Iliut, M.; Turner, M.; Vijayaraghavan, A.; Dierking, I. Confinement effects on lyotropic nematic liquid crystal phases of graphene oxide dispersions. *2D Mater.* **2017**, *4*, 041004. [CrossRef]

102. Dan, B.; Behabtu, N.; Martinez, A.; Evans, J.S.; Kosynkin, D.V.; Tour, J.M.; Pasquali, M.; Smalyukh, I.I. Liquid crystals of aqueous, giant graphene oxide flakes. *Soft Matter* **2011**, *7*, 11154. [CrossRef]

103. Jalili, R.; Aboutalebi, S.H.; Esrafilzadeh, D.; Shepherd, R.L.; Chen, J.; Aminorroaya-Yamini, S.; Konstantinov, K.; Minett, A.I.; Razal, J.M.; Wallace, G.G. Scalable One-Step Wet-Spinning of Graphene Fibers and Yarns from Liquid Crystalline Dispersions of Graphene Oxide: Towards Multifunctional Textiles. *Adv. Funct. Mater.* **2013**, *23*, 5345–5354. [CrossRef]

104. Kim, J.Y.; Kim, S.O. Liquid crystals: Electric fields line up graphene oxide. *Nat. Mater.* **2014**, *13*, 325–326. [CrossRef]

105. Hong, S.-H.; Shen, T.-Z.; Song, J.-K. Electro-optical Characteristics of Aqueous Graphene Oxide Dispersion Depending on Ion Concentration. *J. Phys. Chem. C* **2014**, *118*, 26304–26312. [CrossRef]

106. Shen, T.-Z.; Hong, S.-H.; Song, J.-K. Electro-optical switching of graphene oxide liquid crystals with an extremely large Kerr coefficient. *Nat. Mater.* **2014**, *13*, 394–399. [CrossRef] [PubMed]

107. Ahmad, R.T.M.; Hong, S.-H.; Shen, T.-Z.; Song, J.-K. Optimization of particle size for high birefringence and fast switching time in electro-optical switching of graphene oxide dispersions. *Opt. Express* **2015**, *23*, 4435–4440. [CrossRef] [PubMed]

108. Narayan, R.; Kim, J.E.; Kim, J.Y.; Lee, K.-E.; Kim, S.O. Graphene Oxide Liquid Crystals: Discovery, Evolution and Applications. *Adv. Mater.* **2016**, *28*, 3045–3068. [CrossRef]

109. Sasikala, S.P.; Lim, J.; Kim, I.H.; Jung, H.J.; Yun, T.; Han, T.H.; Kim, S.O. Graphene oxide liquid crystals: A frontier 2D soft material for graphene-based functional materials. *Chem. Soc. Rev.* **2018**, *47*, 6013–6045. [CrossRef] [PubMed]

110. Draude, A.; Dierking, I. Lyotropic Liquid Crystals from Colloidal Suspensions of Graphene Oxide. *Crystals* **2019**, *9*, 455. [CrossRef]

111. Al-Zangana, S.; Iliut, M.; Turner, M.; Vijayaraghavan, A.; Dierking, I. Properties of a Thermotropic Nematic Liquid Crystal Doped with Graphene Oxide. *Adv. Opt. Mater.* **2016**, *4*, 1541–1548. [CrossRef]

112. Al-Zangana, S.; Iliut, M.; Boran, G.; Turner, M.; Vijayaraghavan, A.; Dierking, I. Dielectric spectroscopy of isotropic liquids and liquid crystal phases with dispersed graphene oxide. *Sci. Rep.* **2016**, *6*, 31885. [CrossRef]

113. Zamora-Ledezma, C.; Puech, N.; Zakri, C.; Grelet, E.; Moulton, S.E.; Wallace, G.G.; Gambhir, S.; Blanc, C.; Anglaret, E.; Poulin, P. Liquid Crystallinity and Dimensions of Surfactant-Stabilized Sheets of Reduced Graphene Oxide. *J. Phys. Chem. Lett.* **2012**, *3*, 2425–2430. [CrossRef]

114. Kim, M.J.; Park, J.H.; Yamamoto, J.; Kim, Y.S.; Scalia, G. Electro-optic switching with liquid crystal graphene. *Phys. Status Solidi Rapid Res. Lett.* **2016**, *10*, 397–403. [CrossRef]

115. Jalili, R.; Aminorroaya-Yamini, S.; Benedetti, T.R.B.; Aboutalebi, S.H.; Chao, Y.; Wallace, G.G.; Officer, D.L. Processable 2D materials beyond graphene: MoS_2 liquid crystals and fibres. *Nanoscale* **2016**, *8*, 16862–16867. [CrossRef] [PubMed]

116. Xia, Y.; Mathis, T.S.; Zhao, M.-Q.; Anasori, B.; Dang, A.; Zhou, Z.; Cho, H.; Gogotsi, Y.; Yang, S. Thickness-independent capacitance of vertically aligned liquid-crystalline MXenes. *Nature* **2018**, *557*, 409–412. [CrossRef] [PubMed]

117. Nicolosi, V.; Chhowalla, M.; Kanatzidis, M.G.; Strano, M.S.; Coleman, J.N. Liquid Exfoliation of Layered Materials. *Science* **2013**, *340*, 1226419. [CrossRef]

118. Liu, Y.; Xu, Z.; Gao, W.; Cheng, Z.; Gao, C. Graphene and Other 2D Colloids: Liquid Crystals and Macroscopic Fibers. *Adv. Mater.* **2017**, *29*, 1606794. [CrossRef]

119. Xin, G.; Yao, T.; Sun, H.; Scott, S.M.; Shao, D.; Wang, G.; Lian, J. Highly thermally conductive and mechanically strong graphene fibers. *Science* **2015**, *349*, 1083–1087. [CrossRef]

120. Li, P.; Wong, M.; Zhang, X.; Yao, H.; Ishige, R.; Takahara, A.; Miyamoto, M.; Nishimura, R.; Sue, H. Tunable Lyotropic Photonic Liquid Crystal Based on Graphene Oxide. *ACS Photon.* **2014**, *1*, 79–86. [CrossRef]

121. Akbari, A.; Sheath, P.; Martin, S.T.; Shinde, D.B.; Shaibani, M.; Banerjee, P.C.; Tkacz, R.; Bhattacharyya, D.; Majumder, M. Large-area graphene-based nanofiltration membranes by shear alignment of discotic nematic liquid crystals of graphene oxide. *Nat. Commun.* **2016**, *7*, 10891. [CrossRef]

122. He, L.; Ye, J.; Shuai, M.; Zhu, Z.; Zhou, X.; Wang, Y.; Li, Y.; Su, Z.; Zhang, H.; Chen, Y.; et al. Graphene oxide liquid crystals for reflective displays without polarizing optics. *Nanoscale* **2015**, *7*, 1616–1622. [CrossRef]

123. Lydon, J.E. Chromonics. In *Handbook of Liquid Crystals, Vol 2B*; Demus, D., Goodby, J., Gray, G.W., Speiss, H.-W., Vi, V., II, Eds.; Wiley-VCH: Weinheim, Germany, 1998; pp. 981–1007.

124. Lydon, J.E. Chromonic liquid crystal phases. *Curr. Opin. Colloid Interface Sci.* **1998**, *3*, 458–466. [CrossRef]

125. Tam-Chang, S.-W.; Huang, L. Chromonic liquid crystals: Properties and applications as functional materials. *Chem. Commun.* **2008**, *17*, 1957. [CrossRef]

126. Nastishin, Y.A.; Liu, H.; Shiyanovskii, S.V.; Lavrentovich, O.; Kostko, A.; Anisimov, M.A. Pretransitional fluctuations in the isotropic phase of a lyotropic chromonic liquid crystal. *Phys. Rev. E* **2004**, *70*, 051706. [CrossRef] [PubMed]

127. Hartshorne, N.H.; Woodard, G.D. Mesomorphism in the System Disodium Chromoglycate-Water. *Mol. Cryst. Liq. Cryst.* **1973**, *23*, 343–368. [CrossRef]

128. Kustanovich, I.; Poupko, R.; Zimmermann, H.; Luz, Z.; Labes, M.M. Lyomesophases of the diethylammonium flufenamate-water system studied by deuterium NMR spectroscopy. *J. Am. Chem. Soc.* **1985**, *107*, 3494–3501. [CrossRef]

129. Horowitz, V.R.; Janowitz, L.A.; Modic, A.L.; Heiney, P.A.; Collings, P.J. Aggregation behavior and chromonic liquid crystal properties of an anionic monoazo dye. *Phys. Rev. E* **2005**, *72*, 041710. [CrossRef]

130. Kostko, A.; Cipriano, B.H.; Pinchuk, O.A.; Ziserman, L.; Anisimov, M.A.; Danino, D.; Raghavan, S.R. Salt Effects on the Phase Behavior, Structure, and Rheology of Chromonic Liquid Crystals. *J. Phys. Chem. B* **2005**, *109*, 19126–19133. [CrossRef] [PubMed]

131. Sidky, H.; Whitmer, J.K. The Emergent Nematic Phase in Ionic Chromonic Liquid Crystals. *J. Phys. Chem. B* **2017**, *121*, 6691–6698. [CrossRef]

132. Romani, E.; Ferrarini, A.; De Michele, C. Elastic Constants of Chromonic Liquid Crystals. *Macromolecules* **2018**, *51*, 5409–5419. [CrossRef]

133. Jeong, J.; Davidson, Z.S.; Collings, P.J.; Lubensky, T.C.; Yodh, A.G. Chiral symmetry breaking and surface faceting in chromonic liquid crystal droplets with giant elastic anisotropy. *Proc. Natl. Acad. Sci. USA* **2014**, *111*, 1742–1747. [CrossRef] [PubMed]

134. Zhou, S.; Shiyanovskii, S.V.; Park, H.-S.; Lavrentovich, O.D. Fine structure of the topological defect cores studied for disclinations in lyotropic chromonic liquid crystals. *Nat. Commun.* **2017**, *8*, 14974. [CrossRef]

135. Akpinar, E.; Topcu, G.; Reis, D.; Neto, A.M.F. Effect of the presence of the anionic azo dye Sunset Yellow in lyotropic mixtures with uniaxial and biaxial nematic phases. Submitted.

136. Valášek, J. Piezo-Electric and Allied Phenomena in Rochelle Salt. *Phys. Rev.* **1921**, *17*, 475–481. [CrossRef]

137. Meyer, R.; Liebert, L.; Strzelecki, L.; Keller, P. Ferroelectric liquid crystals. *J. Phys. Lett.* **1975**, *36*, 69–71. [CrossRef]

138. Clark, N.A.; Lagerwall, S.T. Submicrosecond bistable electro-optic switching in liquid crystals. *Appl. Phys. Lett.* **1980**, *36*, 899–901. [CrossRef]

139. Schafheutle, M.A.; Finkelmann, H. Shapes of Micelles and Molecular Geometry Synthesis and Studies on the Phase Behaviour, Surface Tension and Rheology of Rigid Rod-Like Surfactants in Aqueous Solutions. *Liq. Cryst.* **1988**, *3*, 1369–1386. [CrossRef]

140. Ujiie, S.; Yano, Y. Thermotropic and lyotropic behavior of novel amphiphilic liquid crystals having hydrophilic poly(ethyleneimine) units. *Chem. Commun.* **2000**, *1*, 79–80. [CrossRef]

141. Harjung, M.D.; Giesselmann, F. Electroclinic effect in the chiral lamellar α phase of a lyotropic liquid crystal. *Phys. Rev. E* **2018**, *97*, 032705. [CrossRef] [PubMed]

142. Harjung, M.D.; Schubert, C.P.J.; Knecht, F.; Porada, J.H.; Lemieux, R.P.; Giesselmann, F. New amphiphilic materials showing the lyotropic analogue to the thermotropic smectic C* liquid crystal phase. *J. Mater. Chem. C* **2017**, *5*, 7452–7457. [CrossRef]

143. Bruckner, J.R.; Giesselmann, F. The Lyotropic Analog of the Polar SmC* Phase. *Crystals* **2019**, *9*, 568. [CrossRef]

144. Menzel, A.M. Tuned, driven, and active soft matter. *Phys. Rep.* **2015**, *554*, 1–45. [CrossRef]

145. Golestanian, R.; Ramaswamy, S. Active Matter. *Eur. Phys. J. E* **2013**, *36*, 54002. [CrossRef] [PubMed]

146. Bukusoglu, E.; Pantoja, M.B.; Mushenheim, P.C.; Wang, X.; Abbott, N.L. Design of Responsive and Active (Soft) Materials Using Liquid Crystals. *Annu. Rev. Chem. Biomol. Eng.* **2016**, *7*, 163–196. [CrossRef]

147. Doostmohammadi, A.; Ignés-Mullol, J.; Yeomans, J.M.; Sagués, F. Active nematics. *Nat. Commun.* **2018**, *9*, 3246. [CrossRef] [PubMed]

148. Dombrowski, C.; Cisneros, L.; Chatkaew, S.; E Goldstein, R.; Kessler, J.O. Self-Concentration and Large-Scale Coherence in Bacterial Dynamics. *Phys. Rev. Lett.* **2004**, *93*, 098103. [CrossRef] [PubMed]

149. Li, H.; Shi, X.-Q.; Huang, M.; Chen, X.; Xiao, M.; Liu, C.; Chaté, H.; Zhang, H.P. Data-driven quantitative modeling of bacterial active nematics. *Proc. Natl. Acad. Sci. USA* **2018**, *116*, 777–785. [CrossRef] [PubMed]

150. Sanchez, T.; Chen, D.T.N.; DeCamp, S.J.; Heymann, M.; Dogic, Z. Spontaneous motion in hierarchically assembled active matter. *Nature* **2012**, *491*, 431–434. [CrossRef] [PubMed]

151. Henkin, G.; DeCamp, S.J.; Chen, D.T.N.; Sanchez, T.; Dogic, Z. Tunable dynamics of microtubule-based active isotropic gels. *Philos. Trans. R. Soc. A Math. Phys. Eng. Sci.* **2014**, *372*, 20140142. [CrossRef] [PubMed]

152. Schaller, V.; Weber, C.; Semmrich, C.; Frey, E.; Bausch, A.R. Polar patterns of driven filaments. *Nature* **2010**, *467*, 73–77. [CrossRef]

153. Zhang, R.; Kumar, N.; Ross, J.L.; Gardel, M.L.; De Pablo, J.J. Interplay of structure, elasticity, and dynamics in actin-based nematic materials. *Proc. Natl. Acad. Sci.* **2017**, *115*, E124–E133. [CrossRef]

154. Kumar, N.; Zhang, R.; De Pablo, J.J.; Gardel, M.L. Tunable structure and dynamics of active liquid crystals. *Sci. Adv.* **2018**, *4*, eaat7779. [CrossRef]

155. Zhou, S.; Sokolov, A.; Lavrentovich, O.D.; Aranson, I.S. Living Liquid Crystals. *Biophys. J.* **2014**, *106*, 420a. [CrossRef]

156. Peng, C.; Turiv, T.; Guo, Y.; Wei, Q.-H.; Lavrentovich, O.D. Command of active matter by topological defects and patterns. *Science* **2016**, *354*, 882–885. [CrossRef] [PubMed]

157. Genkin, M.M.; Sokolov, A.; Lavrentovich, O.D.; Aranson, I.S. Topological Defects in a Living Nematic Ensnare Swimming Bacteria. *Phys. Rev. X* **2017**, *7*, 011029. [CrossRef]

158. Genkin, M.M.; Sokolov, A.; Aranson, I.S. Spontaneous topological charging of tactoids in a living nematic. *New J. Phys.* **2018**, *20*, 043027. [CrossRef]

159. Woolverton, C.J.; Gustely, E.; Li, L.; Lavrentovich, O.D. Liquid crystal effects on bacterial viability. *Liq. Cryst.* **2005**, *32*, 417–423. [CrossRef]

160. Mezzenga, R.; Seddon, J.M.; Drummond, C.J.; Boyd, B.J.; Schröder-Turk, G.E.; Sagalowicz, L. Nature-Inspired Design and Application of Lipidic Lyotropic Liquid Crystals. *Adv. Mater.* **2019**, *31*, e1900818. [CrossRef]

161. Sadeghpour, A. Lyotropic Liquid Crystalline Phases for the Formulation of Future Functional Foods. *J. Nutr. Health Food Eng.* **2016**, *5*, 553–557. [CrossRef]

162. Garti, N.; Libster, D.; Aserin, A. Lipid polymorphism in lyotropic liquid crystals for triggered release of bioactives. *Food Funct.* **2012**, *3*, 700–713. [CrossRef]

163. Larsson, K. Lyotropic liquid crystals and their dispersions relevant in foods. *Curr. Opin. Colloid Interface Sci.* **2009**, *14*, 16–20. [CrossRef]

164. Engels, T.; Von Rybinski, W. Liquid crystalline surfactant phases in chemical applications. *J. Mater. Chem.* **1998**, *8*, 1313–1320. [CrossRef]

165. Negrini, R.; Mezzenga, R. pH-Responsive Lyotropic Liquid Crystals for Controlled Drug Delivery. *Langmuir* **2011**, *27*, 5296–5303. [CrossRef]

166. Aleandri, S.; Speziale, C.; Mezzenga, R.; Landau, E.M. Design of Light-Triggered Lyotropic Liquid Crystal Mesophases and Their Application as Molecular Switches in "On Demand" Release. *Langmuir* **2015**, *31*, 6981–6987. [CrossRef]

167. Kim, N.-H.; Jahn, A.; Cho, S.-J.; Kim, J.S.; Ki, M.-H.; Kim, D.-D. Lyotropic liquid crystal systems in drug delivery: A review. *J. Pharm. Investig.* **2014**, *45*, 1–11. [CrossRef]

168. Guo, C.; Wang, J.; Cao, F.; Lee, R.J.; Zhai, G. Lyotropic liquid crystal systems in drug delivery. *Drug Discov. Today* **2010**, *15*, 1032–1040. [CrossRef]

169. Gupta, V.K. Optical Amplification of Ligand-Receptor Binding Using Liquid Crystals. *Science* **1998**, *279*, 2077–2080. [CrossRef]

170. Luan, C.; Luan, H.; Luo, D. Luan Application and Technique of Liquid Crystal-Based Biosensors. *Micromachines* **2020**, *11*, 176. [CrossRef] [PubMed]

171. Shiyanovskii, S.V.; Lavrentovich, O.D.; Schneider, T.; Ishikawa, T.; Smalyukh, I.I.; Woolverton, C.J.; Niehaus, G.D.; Doane, K.J. Lyotropic Chromonic Liquid Crystals for Biological Sensing Applications. *Mol. Cryst. Liq. Cryst.* **2005**, *434*, 259–270. [CrossRef]

172. Tan, H.; Yang, S.; Shen, G.; Yu, R.; Wu, Z. Signal-Enhanced Liquid-Crystal DNA Biosensors Based on Enzymatic Metal Deposition. *Angew. Chem. Int. Ed.* **2010**, *49*, 8608–8611. [CrossRef]

173. Popov, P.; Honaker, L.; Kooijman, E.E.; Mann, E.K.; Jákli, A. A liquid crystal biosensor for specific detection of antigens. *Sens. Bio-Sens. Res.* **2016**, *8*, 31–35. [CrossRef]

174. Oton, E.; Oton, J.M.; Caño-García, M.; Escolano, J.M.; Quintana, X.; Geday, M.A. Rapid detection of pathogens using lyotropic liquid crystals. *Opt. Express* **2019**, *27*, 10098–10107. [CrossRef]

175. De Souza, J.F.; Pontes, K.D.S.; Alves, T.F.; Amaral, V.; Rebelo, M.A.; Hausen, M.; Chaud, M.V. Spotlight on Biomimetic Systems Based on Lyotropic Liquid Crystal. *Molecules* **2017**, *22*, 419. [CrossRef]

176. Fowler, C.E.; Shenton, W.; Stubbs, G.; Mann, S. Tobacco Mosaic Virus Liquid Crystals as Templates for the Interior Design of Silica Mesophases and Nanoparticles. *Adv. Mater.* **2001**, *13*, 1266–1269. [CrossRef]

177. Dellinger, T.M.; Braun, P.V. Lyotropic Liquid Crystals as Nanoreactors for Nanoparticle Synthesis. *Chem. Mater.* **2004**, *16*, 2201–2207. [CrossRef]

178. Wang, C.; Chen, D.; Jiao, X. Lyotropic liquid crystal directed synthesis of nanostructured materials. *Sci. Technol. Adv. Mater.* **2009**, *10*, 23001. [CrossRef] [PubMed]

179. Umadevi, S.; Umamaheswari, R.; Ganesh, V. Lyotropic liquid crystal-assisted synthesis of micro- and nanoparticles of silver. *Liq. Cryst.* **2017**, *44*, 1–12. [CrossRef]

180. Salili, S.M.; Worden, M.; Nemati, A.; Miller, D.W.; Hegmann, T. Synthesis of Distinct Iron Oxide Nanomaterial Shapes Using Lyotropic Liquid Crystal Solvents. *Nanomaterials* **2017**, *7*, 211. [CrossRef] [PubMed]

181. Hegmann, T.; Qi, H.; Marx, V.M. Nanoparticles in Liquid Crystals: Synthesis, Self-Assembly, Defect Formation and Potential Applications. *J. Inorg. Organomet. Polym. Mater.* **2007**, *17*, 483–508. [CrossRef]

182. Kuijk, A.; Van Blaaderen, A.; Imhof, A. Synthesis of Monodisperse, Rodlike Silica Colloids with Tunable Aspect Ratio. *J. Am. Chem. Soc.* **2011**, *133*, 2346–2349. [CrossRef] [PubMed]

183. Park, J.H.; Noh, J.; Schütz, C.; Salazar-Alvarez, G.; Scalia, G.; Bergström, L.; Lagerwall, J.P.F. Macroscopic Control of Helix Orientation in Films Dried from Cholesteric Liquid-Crystalline Cellulose Nanocrystal Suspensions. *ChemPhysChem* **2014**, *15*, 1477–1484. [CrossRef]

184. Querejeta-Fernández, A.; Chauve, G.; Méthot, M.; Bouchard, J.; Kumacheva, E. Chiral Plasmonic Films Formed by Gold Nanorods and Cellulose Nanocrystals. *J. Am. Chem. Soc.* **2014**, *136*, 4788–4793. [CrossRef]

185. Liu, Q.; Campbell, M.G.; Evans, J.S.; Smalyukh, I.I. Orientationally Ordered Colloidal Co-Dispersions of Gold Nanorods and Cellulose Nanocrystals. *Adv. Mater.* **2014**, *26*, 7178–7184. [CrossRef]
186. Chu, G.; Wang, X.; Chen, T.; Gao, J.; Gai, F.; Wang, Y.; Xu, Y. Optically Tunable Chiral Plasmonic Guest–Host Cellulose Films Weaved with Long-range Ordered Silver Nanowires. *ACS Appl. Mater. Interfaces* **2015**, *7*, 11863–11870. [CrossRef] [PubMed]

 *crystals*

Review

The Lyotropic Analog of the Polar SmC* Phase

Johanna R. Bruckner * and Frank Giesselmann *

Institute of Physical Chemistry, University of Stuttgart, Pfaffenwaldring 55, 70569 Stuttgart, Germany
* Correspondence: johanna.bruckner@ipc.uni-stuttgart.de (J.R.B.); frank.giesselmann@ipc.uni-stuttgart.de (F.G.);
Tel.: +49-711-685-64136 (J.R.B.); Tel.: +49-711-685-64460 (F.G.)

Received: 10 October 2019; Accepted: 25 October 2019; Published: 29 October 2019

Abstract: Only six years ago, the first clear-cut example of a ferroelectric, lyotropic liquid crystal was discovered. Since then, ongoing investigations in this new research field provided numerous instances of the missing pieces to complete the formerly blank picture of the lyotropic smectic C* (SmC*) phase. In this review we wanted to combine these new results and put them into a wider historical and scientific context. We start by giving an introduction about characteristic features of the well-known thermotropic SmC* phase and why it is so difficult to find a lyotropic equivalent of this fascinating phase. After discussing early examples of achiral lyotropic and swollen SmC phases, we recap the discovery of the first lyotropic SmC* phase. The molecular features necessary for its formation and its properties are analyzed. We place special emphasis on discussing the long-range orientational order of the tilt direction and the corresponding chirality effects. By comparing these exceptional features with thermotropic and swollen SmC* phases, we aim to improve not only the understanding of the lyotropic SmC* phase, but also of the relationship between thermotropic and lyotropic systems in general.

Keywords: lyotropic liquid crystals; SmC* phase; chirality; ferroelectricity; hydrogen bonds; hydration forces

---

1. Introduction

In the present article, the discovery of the lyotropic analogue of the thermotropic polar smectic C* phase (lyo-SmC*), its properties and the prerequisites for its formation are reviewed. We cover the very first examples of fluid and tilted lyotropic lamellar phases and the latest developments in the newly evolved research field. In order to illuminate the special status of the lyo-SmC* phase among the lyotropic liquid crystal phases, we begin this review with a short recapitulation of the unique features of the thermotropic polar and ferroelectric SmC* phase, and discuss the obstacles to overcome on the way to finding lyotropic analogues of the former.

1.1. Ferroelectricity in Liquid Crystals

The term ferroelectricity is used in analogy to ferromagnetism and describes the property of certain dielectric materials to have a spontaneous electric polarization, the direction of which can be changed—in most cases reversed—by the action of an external electric field [1]. After the first discovery of ferroelectricity in Rochelle salt by the physicist Joseph Valasek in 1920 [2], it was long believed that ferroelectricity could only be found in solid crystals of low symmetry; namely, those crystals which belong to the ten polar crystallographic point groups (crystal classes) C_1, C_s, C_n and C_{nv} with $n = 2, 3, 4$ and 6. In the 1970s, however, the Harvard physicist Robert B. Meyer realized that a thermotropic smectic C liquid crystal which is composed of chiral molecules has polar C_2 symmetry, and might, thus, be the first example of a fluid medium with a spontaneous electric polarization [3,4]. Meyer's discovery initiated one of the most active fields in soft matter research ranging from ferro- and antiferroelectric liquid crystals to the plethora of polar liquid crystal structures formed by bent-core mesogens [5–7].

The structures of the thermotropic fluid smectic liquid crystals, smectic A (SmA) and smectic C (SmC), can be considered as 1D-periodic stacks of 2D-fluid layers formed by elongated molecules, the long axes of which are orientationally ordered along the liquid crystal director **n** (Figure 1). While **n** is parallel to the smectic layer normal **k** (the stacking direction) in SmA (Figure 1a), the director **n** is tilted in SmC by the tilt angle θ with respect to **k** (Figure 1b). The tilt of the director breaks the full rotational symmetry of the uniaxial SmA phase around **k** and makes the SmC phase biaxial. In SmC, the principal symmetry axis is the C_2-axis normal to the tilt plane (the plane spanned by **k** and **n**; see Figure 2). In addition, the tilt plane itself is a mirror plane. The non-chiral SmC structure, thus, belongs to the non-polar point group C_{2h}. Chirality, however, excludes the presence of any mirror planes, such that the symmetry of the chiral SmC phase (SmC*) is reduced to the polar point group C_2 [5].

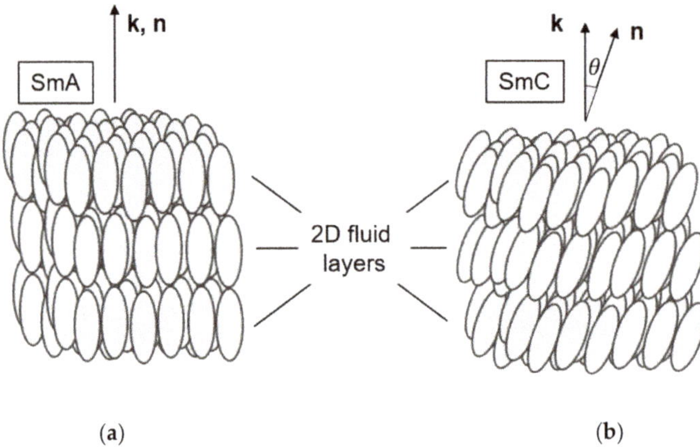

Figure 1. Structures of the fluid smectic phases (**a**) smectic A (SmA) and (**b**) smectic C (SmC). Both phases are 1D stacks of 2D fluid layers formed by orientationally ordered mesogenic molecules. In SmA, the director **n** (the mean direction of the long molecular axes) is parallel to the smectic layer normal **k** (the stacking direction of the layers). In SmC, **n** is tilted with respect to **k** by the director tilt angle θ.

The absence of a mirror plane normal to the C_2 axis makes this axis a polar axis, in the direction of which the presence of vectorial properties, such as a spontaneous electric polarization **P$_s$**, is allowed by symmetry. Since the polar C_2 axis is normal to both the director **n** and the layer normal **k**, the direction of **P$_s$** is expressed by [8]:

$$\mathbf{P_s} \propto \mathbf{k} \times \mathbf{n}, \tag{1}$$

which implies that the magnitude of spontaneous polarization $P_s = |\mathbf{P_s}|$ increases with the director tilt θ as $P_s \propto \sin\theta \approx \theta$ in the first approximation. In conclusion, both the direction and the magnitude of **P$_s$** depend on the direction and magnitude of tilt. This is known as the polarization-tilt coupling in chiral smectic liquid crystals.

The point group symmetries $D_{\infty h}$ of non-chiral SmA and D_∞ of chiral SmA* are too high to allow ferroelectricity and a spontaneous polarization in any direction. In SmA*, however, an electric polarization induced by an electric field **E** along the smectic layers is linearly coupled to an induced director tilt $\delta\theta(E)$ via the polarization-tilt coupling. This so-called electroclinic effect increases in amplitude towards the transition temperature T_c from the SmA* into the SmC* phase in a Curie–Weiss-like anomaly [9–11].

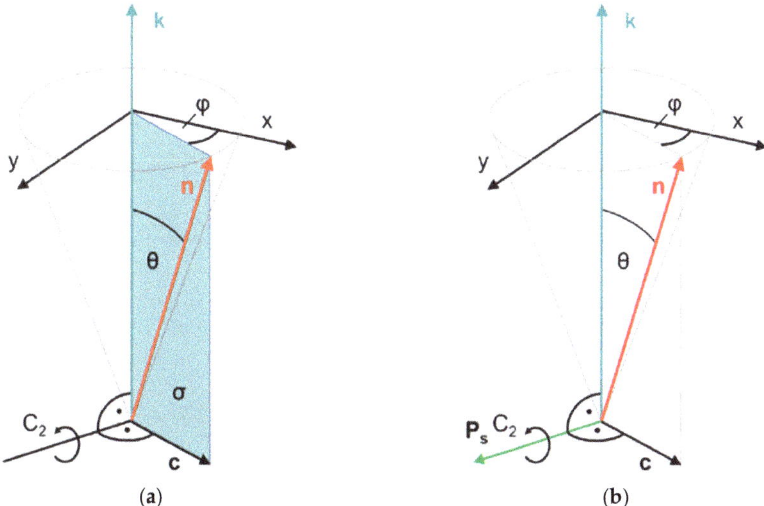

(a)   (b)

Figure 2. Symmetries of (**a**) smectic C (SmC) and (**b**) chiral smectic C (SmC*) liquid crystals. The director **n** is tilted with respect to the smectic layer normal **k** by the tilt angle θ. The tilt direction is indicated by the **c** director and specified by the azimuth angle φ. The directions of **k** and **n** define the tilt plane. Normal to the tilt plane, we find a twofold symmetry axis C_2, since the directions of $+\mathbf{n}$ and $+\mathbf{k}$ are physically equivalent to the directions of $-\mathbf{n}$ and $-\mathbf{k}$, respectively. In addition, the tilt plane is a mirror plane in the case of non-chiral SmC. Both symmetry elements combine to the non-polar point group C_{2h}. In the presence of chiral molecules, however, mirror symmetry is expelled and the C_2 axis remains the only symmetry element. Thus, the chiral SmC* phase has polar C_2 symmetry, and a spontaneous polarization vector $\mathbf{P_s}$ along the C_2 axis is allowed by symmetry.

In addition to the spontaneous electric polarization of each smectic layer, chiral smectic C forms a helical superstructure in such a way that the tilt direction—and thus the direction of $\mathbf{P_s}$ as well—slightly twists from layer to layer along the layer normal **k** (Figure 3). The pitch p of the SmC* helix is typically in the order of several microns and is, therefore, several orders of magnitude larger than the thickness of the smectic layers d which are typically in the range of a few nanometers. As a result of this helical superstructure, the spontaneous polarization is macroscopically cancelled out over a full pitch length.

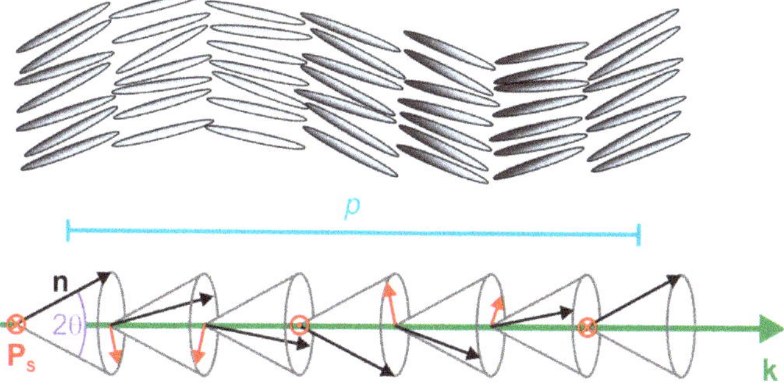

Figure 3. Helical superstructure of a chiral smectic C (SmC*) phase depicted with hard rods (top) and vectors (bottom). The director **n** (black) and the spontaneous polarization $\mathbf{P_s}$ (red) are twisted along the direction of the layer normal **k** (green) from layer to layer with a helical pitch p (blue). The latter is typically in the micron range, and thus, several orders of magnitude larger than the layer spacing d.

In 1980, Noel A. Clark and Sven T. Lagerwall discovered that the SmC* helix formation can be easily suppressed if the phase is confined between two planarly aligning glass plates, the distance between which is less or similar to the SmC* pitch p (Figure 4a) [12]. In this so-called surface stabilized ferroelectric liquid crystal (SSFLC) configuration, the SmC* phase behaves very much like a classic ferroelectric material. The SSFLC film forms two kinds of domains with opposite tilt directions parallel to the glass surfaces (Figure 4). As a result of the polarization-tilt coupling, these two kinds of tilt domains also have opposite directions of $\mathbf{P_s}$ and are, therefore, the liquid crystal equivalents to the ferroelectric domains in solid ferrolelectrics. An electric field $\mathbf{E}$ is applied across the cell switches' tilt directions such that $\mathbf{P_s}$ is parallel to $\mathbf{E}$ (Figure 4b). Observed between crossed polarizers, the field-induced reversal of the SmC* tilt direction gives rise to a very fast and bistable electro-optic effect which attracted tremendous interest for future display applications and initiated a boost of research in the field of ferroelectrics (FLCs), and later, the antiferroelectric (AFLC) liquid crystal field as well [8]. Today, hundreds of thermotropic FLC and AFLC materials are known.

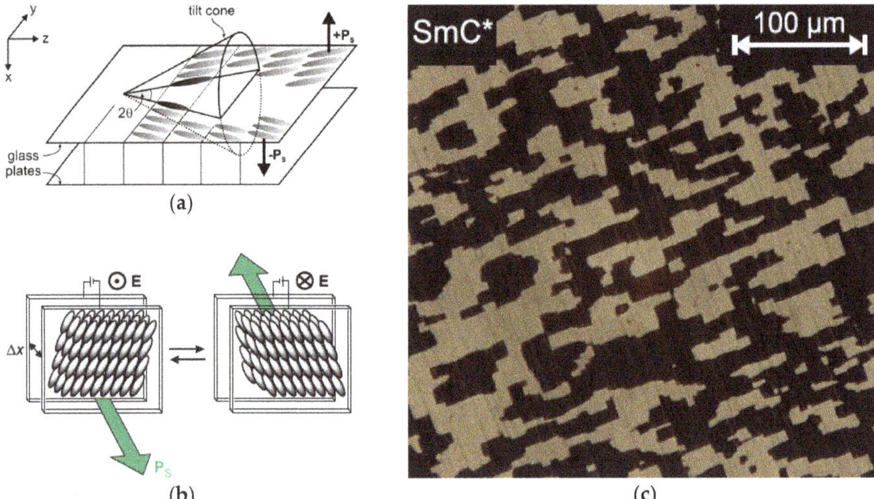

Figure 4. Surface stabilized ferroelectric liquid crystals (SSFLCs). (**a**) SmC* phase confined between two glass plates (with conductive indium tin oxide (ITO) electrodes) of a liquid crystal cell which aligns the SmC* director parallel to the glass surfaces. All possible tilt directions of the SmC* director are represented by the so-called tilt cone. Only the two tilt directions parallel to the glass surfaces meet the surface anchoring condition, and thus, the helical SmC* structure is suppressed if the gap between the glass plates is sufficiently small. The two possible tilt directions ($\pm$y) are coupled to opposite directions of the spontaneous polarization $\mathbf{P_s}$ (up and down). (**b**) The application of an electric field $\mathbf{E}$ across the cell reorients the $\mathbf{P_s}$ vectors of all domains into the direction of $\mathbf{E}$. The field-induced reversal of $\mathbf{P_s}$ also reverses the tilt direction of the SmC* director, which gives rise to a fast electro-optic effect if seen between crossed polarizers. (**c**) In the polarizing optical microscope, the ferroelectric domain structure of the virgin SSFLC configuration is seen between crossed polarizers as an array of bright and dark domains with opposite tilt directions, and thus opposite $\mathbf{P_s}$ directions. (Reprinted with permission (**a**) from Springer Nature: [13], redrawn after [12]; (**b**,**c**) from A. Bogner: [14].)

Two examples of thermotropic FLC materials are shown in Figure 5. The chiral Schiff base (*S*)-*p*-(*n*-decyloxybenzylidene)-*p*-amino-(2-methylbutyl) cinnamate, code named DOBAMBC, shown in Figure 5a, was actually the first thermotropic SmC* material which was recognized as a ferroelectric liquid crystal [3] and much of the pioneering work was done with this material. DOBAMBC undergoes a second-order transition from the paraelectric SmA* to the ferroelectric SmC* phase below $T_{AC} = 95\,^{\circ}\mathrm{C}$.

Below T_{AC} the director tilt angle θ and the spontaneous polarization $\mathbf{P_s}$ continuously increase from zero and reach far below T_{AC} values of 28° and 6 nC cm^{-2}, respectively (Figure 5a) [15]. Even though many FLCs have second-order SmA*–SmC* transitions like DOBAMBC, materials with first-order ferroelectric transitions are also known. The first example was the chiral biphenyl (*S,S*)-4-(3-methyl-2-chloropentanoyloxy)-4'-heptyloxybiphenyl, code named C7 (Figure 5b) [16]. At the transition temperature T_{AC} = 55°C the paraelectric SmA* phase coexists with the SmC* phase which has 15° tilt and 100 nC cm^{-2} polarization at the transition point. In the SmC* phase below, T_{AC} tilt and polarization reach values of up to 23° and 300 nC cm^{-2}, respectively (Figure 5b). In the (*T,E*)-phase diagram, the first-order SmA*–SmC* transition terminates under the action of a strong electric field E at a critical point [17].

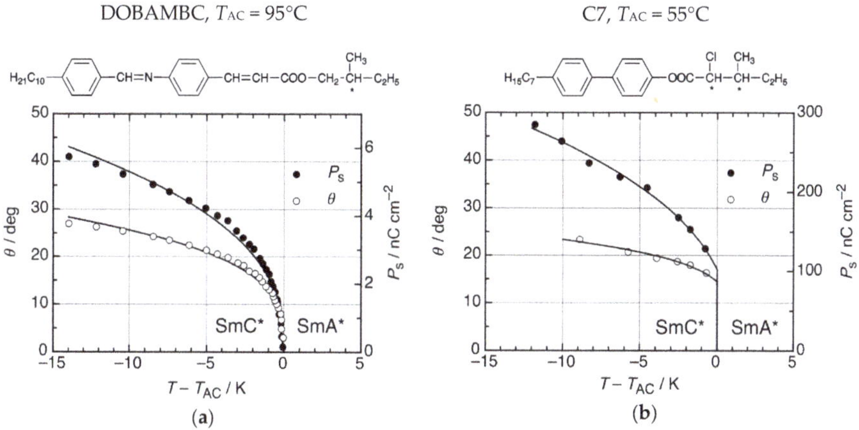

Figure 5. Two examples of ferroelectric SmC* materials. (**a**) The chiral Schiff base DOMBAMBC has a second-order SmA*–SmC* transition at the transition temperature T_{AC} = 95 °C. At increasing temperatures $T < T_{AC}$, the director tilt angle θ and the spontaneous electric polarization P_s continuously decrease to zero values at $T = T_{AC}$. At that critical point, the SmC* phase is identical to SmA*. (Data taken from [16].) (**b**) The chiral biphenyl C7 has a first-order SmA*–SmC* transition. At T_{AC}, the paraelectric SmA* phase coexists with a ferroelectric SmC* phase with non-zero θ and P_s. Values of θ and P_s further increase in the SmC* phase at decreasing temperatures below T_{AC}. (Data taken from [17].)

1.2. The Challenge of a Lyotropic SmC* Phase

Most thermotropic liquid crystal phases find a counterpart with equivalent structure and same symmetry in the world of lyotropic liquid crystals. The lyotropic counterpart of the thermotropic SmA phase, for instance, is the well-known lamellar α-phase (L$_\alpha$) which consists of 2D-fluid surfactant bilayers which are separated from each other by fluid solvent layers (Figure 6a). An important exception, however, is the family of tilted fluid smectics. Even though SmC is among the most common phases found in thermotropics, the lyotropic equivalent to the SmC phase (Figure 6b) is almost unknown. To the best of our knowledge, only two examples of lyo-SmC phases and not a single example of a chiral lyo-SmC* phase were clearly confirmed, e.g., by 2D X-ray diffraction in the literature until 2013.

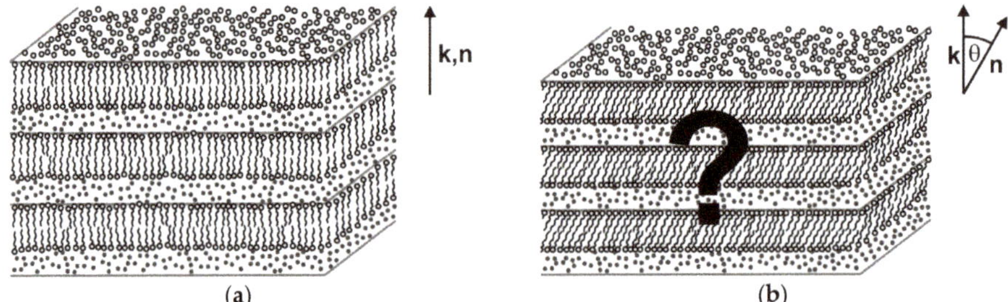

Figure 6. Lyotropic equivalents to the thermotropic fluid smectic phases SmA and SmC. (**a**) The lamellar α-phase (L_α) is a 1D-stack of 2D-fluid surfactant bilayers (lamellae) which are separated from each other by interlamellar solvent layers. Since the director **n** is parallel to the bilayer normal **k**, L_α is the lyotropic equivalent to thermotropic SmA. (**b**) The lyotropic equivalent to the thermotropic SmC phase should have the same fundamental structure as L_α except that the surfactant molecules are tilted by the tilt angle θ into the same direction in all bilayers (synclinic tilt correlation). (Reprinted by permission from Springer Nature: [13].)

We believe that this obvious dissymmetry between thermotropic and lyotropic phases is mainly related to the issue of long-range correlations of tilt directions in a lyotropic medium on different levels:

1. Intra-layer tilt correlation: In order to form a lyotropic C-phase the surfactant bilayers must be in a 2D-fluid state, the hydrophobic tails in each single bilayer must be tilted and the direction of tilt must be more or less the same throughout the bilayer. A uniform tilt direction in each single fluid bilayer might be hardly achieved with the flexible alkyl-tails of classic surfactants. The incorporation of rigid core segments (mesogenic cores) into the tails might, however, promote a collective tilt direction, e.g., by steric interactions between the rigid segments, and might lead to uniformly tilted amphiphiles in the bilayer.

2. Inter-layer tilt correlation: The formation of a lyo-SmC structure further requires that in a 1D stack of bi-layers, each of which has a uniform tilt direction, the tilt directions become correlated between the bi-layers such that they all point into the same direction (synclinic correlation). In thermotropic SmC, where the smectic layers are in direct contact with each other, the synclinic tilt correlation is explained by short-range interactions, such as steric interactions or out-of-layer fluctuations, which align the tilt directions of molecules in adjacent layers [18]. In the lyotropic case, however, where the by-layers are separated from each other by solvent layers, these short-range interaction mechanisms are less relevant. Probably, the inter-layer correlation of tilt directions is the most critical step in the formation of lyo-SmC phases.

3. Helical correlation: In addition to the first two points, the formation of a chiral lyo-SmC* structure requires that also subtle chiral perturbations of the synclinic tilt correlation, namely the helical precession of the tilt direction from bilayer to bilayer, are transmitted through the solvent layers.

In view of these general requirements it seems understandable that the counterparts of the thermotropic SmC and SmC* phases are very rare in lyotropic liquid crystals.

2. Examples of Swollen Thermotropic and Lyotropic SmC Phases

In general, SmC phases containing solvent can be distinguished into one of three categories: the swollen, the hyper-swollen and the true lyotropic SmC phases. In the following, examples for each type and their properties are reviewed, briefly. Regrettably, there are not a lot of examples for hyper-swollen or lyotropic SmC phases, and frequently, they are not investigated in detail. This leaves a big gap of knowledge about the properties of the hyper-swollen, and especially, lyotropic SmC phase and its correlation with the conventional thermotropic SmC phase.

The simplest way to obtain a lyotropic SmC phase, is by mixing an amphiphilic mesogen, which exhibits a SmC phase in the neat state already, with a protic solvent. However, in most cases, the tilted phase will only be stable up to a few weight percent of solvent [19–21]. Instead it is replaced by the L_α phase in which there is no macroscopic tilt angle. These phases are called swollen SmC phases. Typical examples for such swollen SmC phases are shown in Figure 7. Due to the strongly destabilizing impact of the solvent, they are normally not investigated further.

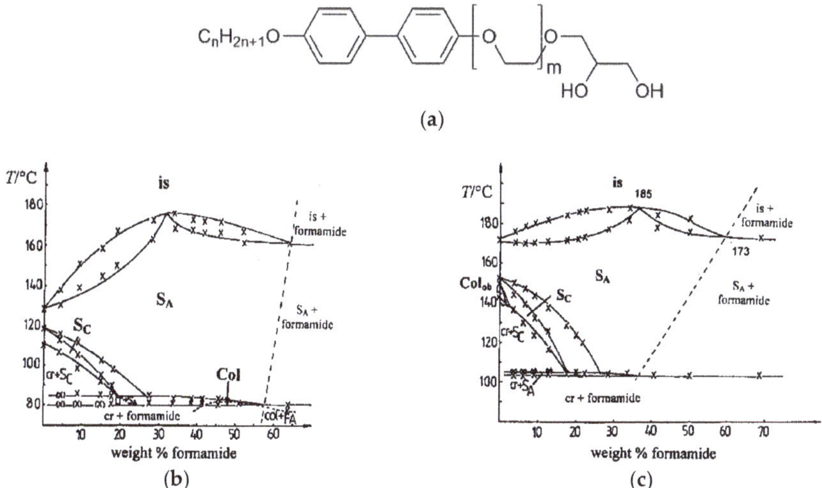

Figure 7. (**a**) Molecular constitution of a calamitic mesogen which incorporates a rigid biphenyl core and a diol moiety, the latter providing solubility in protic solvents. For (**b**) n = 12 and m = 3 and for (**c**) n = 10 and m = 1, the amphiphiles show a swollen SmC phase, denoted with S_C in the phase diagram. In neither case, is the phase stable with mass fractions of formamide larger than approximately 15 wt%. (Phase diagrams reprinted with permission from Spie [19], and [20]. Copyright 1998 American Chemical Society.)

Thermotropic SmC phases, which can take up considerable amounts of solvent, are the so-called hyper-swollen phases. Kanie et al. [22] discovered that combining a phospholipid with an aromatic cyano azobenzene unit produces an SmC phase, which tolerates a water amount of up to 50 wt% before transforming into the orthogonal L_α phase. This observation was, however, solely based on the detection of characteristic textures; i.e., oily streaks and Schlieren textures.

Tan et al. [23] reported in 2015 that another azobenzene mesogen, this time combined with a hydrophilic polyethylene glycol tail, showed a thermotropic and a lyotropic SmC phase in mixtures with water, formamide and ethylene glycol. They investigated the system by both POM and XRD. For ethylene glycol, the maximum solvent uptake before phase separation was found to be about 50 wt%. The phase diagram shows, that the SmC phase is even stabilized by the addition of ethylene glycol, increasing the thermal phase width from 13 to 18 K. While the molecule behaves similar in mixtures with formamide, the SmC phase is reduced to a thermal width of only 2 K in a water saturated mixture. Investigations of the layer spacing at different concentrations of ethylene glycol show the typical behavior known from thermotropic SmA to SmC phase transitions with an increase of the layer spacing in L_α phase, due to the decreasing orientational order at increasing temperature, and a decrease of the layer spacing in the SmC phase which is connected to the increasing tilt angle. Most remarkable, is that the layer spacing at the phase transition temperature is increased from roughly 5.4 nm in the neat state to 11.7 nm in the ethylene glycol saturated state, giving evidence that a considerable solvent layer is formed between the amphiphile bilayers.

In the same year, an interesting report about hyper-swollen SmA and SmC phases was published by Murase et al. [24]. Once again, the mesogen included a rigid aromatic core and showed amphiphilic behavior. However, contrary to the examples discussed before, the mesogen had no hydrophilic, but instead, a fluorophilic tail, as shown in Figure 8a. Thus, mixtures of the amphiphile with three different perflourinated solvents (e.g., Figure 8b) were investigated. Phase diagrams, such as the one in Figure 8c, were recorded by observation of characteristic textures; i.e., the dark homeotropic texture of the SmA or L_α phase with a few defects and the Schlieren texture of the SmC phase. In all cases, the SmC phase got destabilized by the addition of the perfluorinated solvent, but still appeared in a narrow temperature range at solvent mass fractions higher than 50 wt%. They investigated the temperature-dependent layer spacing for different concentrations and found striking differences for the three solvents. Mixtures of the mesogen with elongated solvents, as shown in Figure 8b, exhibit an almost monotonic increase of the layer spacing with increasing solvent concentration. Within the concentration range of the SmC phase, the layer spacing is roughly tripled in dimension. In contrast to this, mixtures of the more spherically shaped solvent perfluorodecalin did not show a substantial increase of the layer spacing with increasing concentration of the solvent. The authors showed by careful processing of their X-ray data, that the elongated solvent molecules were incorporated in between the partial bilayers of the amphiphiles, while in the perfluorodecalin molecules were localized within the amphiphile bilayers. This case does not meet the common understanding of a lamellar phase, while the others do. Furthermore, the authors speculated about possible mechanisms for the long-range correlation of the tilt across the solvent layers, but did not find a conclusive explanation.

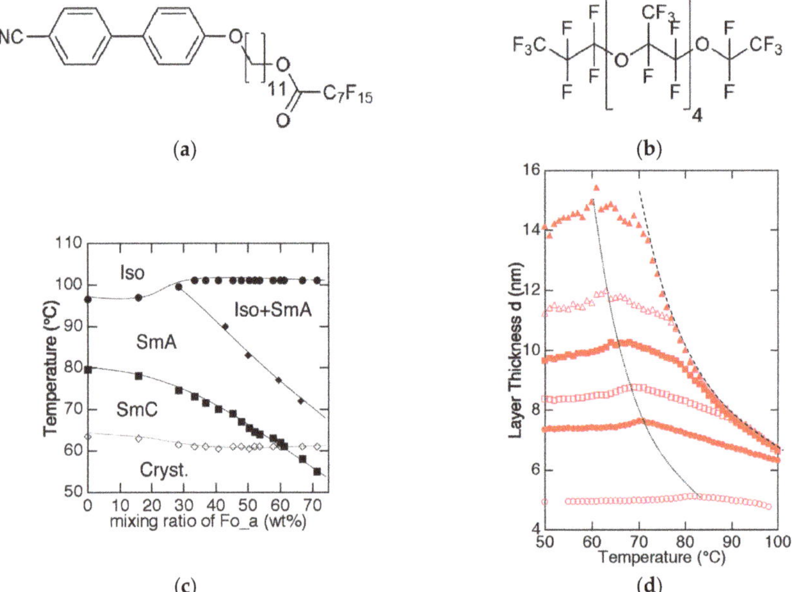

Figure 8. (**a**) Chemical structure of the mesogenic perflourinated cyano biphenyl. (**b**) Examples of one of the perflourinated solvents used, (**c**) the corresponding phase diagram and (**d**) the temperature-dependent layer spacing $d(T)$. The mesogen to solvent ration in (**d**) gradually changes from 1:2 (filled triangles) to 2:1 (filled circles). The layer spacing of the neat mesogen in depicted with open circles. (Reprinted by permission from RCS Publishing: [24].)

Finally, "true" lyotropic SmC-phases, which are formed only in the presence of a solvent, and thus, do not exist in the neat state of the amphiphile, are the least common case. In this case, the presence of the solvent is a necessary condition for the formation of the tilted lamellar mesophase and not just a destabilizing factor of a formerly thermotropic SmC phase, as in the cases of swollen and hyper-swollen smectics. Therefore, these truly lyotropic SmC phases are the most interesting to study and are in the focus of this review.

Until 2013, there were only two proven reports of truly lyotropic SmC phases. The first publication was by Schaheutle and Finkelmann and was published in 1988 [25]. They investigated a series of amphiphiles with rigid hydrophobic moieties and varying lengths of hydrophilic ethylene glycol units (see Figure 9a). Originally interested in the effects of packing constrains on the shape of micelles, they found that the amphiphiles formed lamellar mesophases only. In all three cases, SmC phases were observed in the presence of water, while there were no liquid crystalline phases in the neat states. They verified the structural analogy of this lyotropic phase with the known thermotropic SmC phase by X-ray diffraction measurements. In Figure 9b, a reprint of this measurement is shown. The two-dimensional diffraction pattern shows the typical features of a fluid, lamellar and tilted phase, i.e., diffuse scattering maxima in the wide angle regime, sharp layer peaks in the small angle region and an azimuth angle differing from 90° between the two of them. Even though the tilt angle measurable from this diffraction pattern is only a couple of degrees, it is clearly observable. Comparing amphiphiles with varying hydrophilic-hydrophobic balances, the authors found that the molecule with the shortest polyethylene glycol chain forms the most stable lyotropic SmC phase in a concentration regime between roughly 30 and 70 wt% of surfactant. By increasing the chain length, the concentration range, in which the tilted phase occurs, is diminished and shifted to higher mass fractions of surfactant. In Figure 9c the phase diagram of the investigated amphiphile with an intermediate chain length and water is shown as an example.

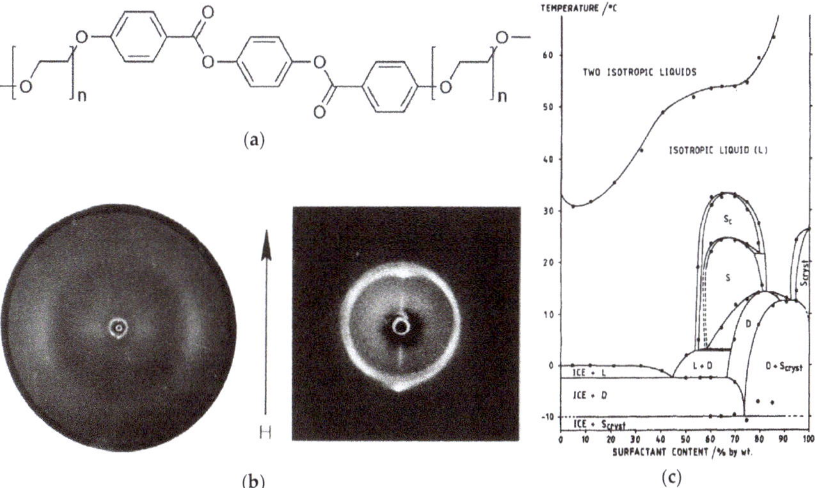

(a)

(b)   (c)

Figure 9. (**a**) A rigid aromatic core is combined with two polyethylene glycol units on each end, varying from n = 5 to n = 7. (**b**) Two-dimensional X-ray diffraction pattern of a magnetic-field aligned sample with 70 wt% percent of the amphiphile with n = 6 at 26 °C. The arrow indicates the direction of the magnetic field (H). The phase diagram corresponding to this system is shown in (**c**). The lyotropic SmC phase is denoted as 'S_C' in this diagram; 'S' and 'D' stand either for a supposedly higher ordered smectic or dystetic (i.e., L_α) phase, respectively. (Reprinted by permission form Taylor and Francis Ltd.: [25].)

In spite of Schafheutle's and Finkelmann's outstanding discovery, it took slightly more than a decade until a further example of a truly lyotropic SmC phase was published. Ujiie and Yano [26] reported the occurrence of a lyotropic SmC phase in mixtures of an ionic amphiphile and water. The amphiphile incorporates a rigid hydrophobic azobenzene unit which is connected to a polyethylene imine chain by a hexamethylene linker. To every imine moiety, a 2-hydroxy ethane group is attached, providing the necessary hydrophilicity. The molecule exhibits a thermotropic SmA phase at elevated temperatures. By adding water, the melting point of the system decreases significantly, but the SmA phase, which turns into a L_α phase, stays stable up to a mass fraction of surfactant as low as 20 wt%. At lower temperatures, in the concentration range between roughly 15 and 75 wt%, the lyotropic SmC phase occurs. The authors provide evidence of the lamellar and fluid nature of this phase by a one-dimensional X-ray diffraction pattern, and for the director tilt by a picture of the typical Schlieren texture.

In addition to those two cases, a further amphiphile was reported to show a true lyotropic SmC phase [27]. However, later investigations revealed that the supposed lyotropic SmC phase was in fact an oblique columnar phase [28].

In view of the very small number of lyotropic SmC phases reported, which only form in the presence of a solvent, one can easily understand why the search for a chiral variant of this rare phase took so long. On the one hand, the missing experience with the design of lyotropic SmC phases, and on the other hand, the more demanding synthesis of chiral components, made it quite difficult to find promising amphiphiles. Furthermore, the knowledge from the often separated research fields of lyotropic and thermotropic liquid crystals had to be combined, to handle the more complicated sample preparation of lyotropic systems and notice the characteristic features of a chiral lyotropic SmC* phase.

3. The Recognition of a First Lyotropic SmC* Phase

In 2013, the discovery of a chiral, lamellar tilted and fluid phase—a true lyotropic SmC* phase—was reported for the first time [29]. The phase is formed by the chiral amphiphile shown in Figure 10a. Without solvent, the amphiphile does not form any stable liquid crystal phase, but after the addition of water or formamide, several lyotropic mesophases appear (Figure 10b).

The two-dimensional X-ray diffraction pattern of a magnetically aligned sample with 64 wt% water which is depicted in Figure 10c, proves without doubt that this is truly a structural equivalent to the thermotropic SmC or SmC* phase. The sharp, Bragg-like peaks of 1st and 2nd order in the small angle region clearly show that the phase is lamellar; the diffuse maxima in the wide angle region attest that it is fluid; and the azimuth angle between the two of them deviates by 90°, confirming that the surfactant molecules within the fluid bilayers are tilted with respect to the layer normal **k**. Evidently, the tilt directions are long-range correlated, at least over the macroscopic length scale of the scattering volume. A distinction between the chiral and achiral SmC phase is not possible from X-ray diffraction alone.

Furthermore, the lyotropic SmC* phase exhibits the same characteristic textures as those known from its thermotropic counterpart. In Figure 11, striking examples of this similarity in terms of textures are given. The Schlieren texture in Figure 11a underlines that the phase is biaxial and the zigzag defects [30] in Figure 11b indicate a substantial layer shrinkage at the L_α to lyo-SmC* phase transition. Most remarkable is that the lyo-SmC* phase can be surface stabilized (Figure 11c,d), just as the thermotropic SmC* phase (cf. Figure 4c). The rotation at about two times the optical tilt angle, exchanges the brightness of the tilt domains between crossed polarizers.

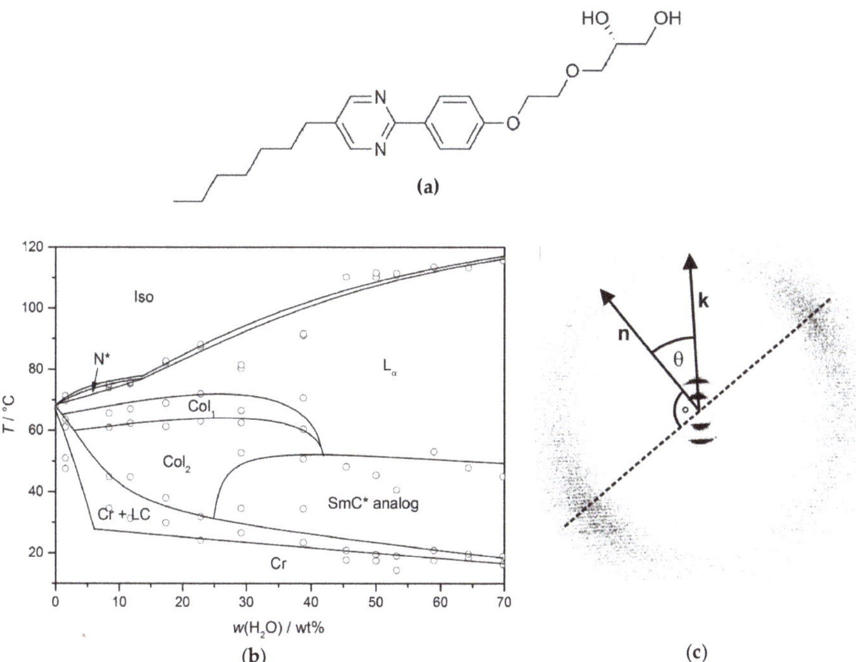

(a)

(b)

(c)

Figure 10. The chiral diol depicted in (**a**) forms true lyotropic SmC* phases in mixtures with water and formamide. (**b**) The phase diagram with water exhibits a cholesteric (N*), two different columnar phases (Col$_1$ and Col$_2$), a L$_\alpha$ phase and the lyotropic SmC* phase. (**c**) A two-dimensional X-ray diffraction pattern of the aligned phase provides clear evidence that the probed lyotropic phase possesses a structure analogous to the thermotropic SmC* phase. (Phase diagram reprinted by permission from John Wiley and Sons: [29]; X-ray pattern reprinted by permission from Springer Nature: [13].)

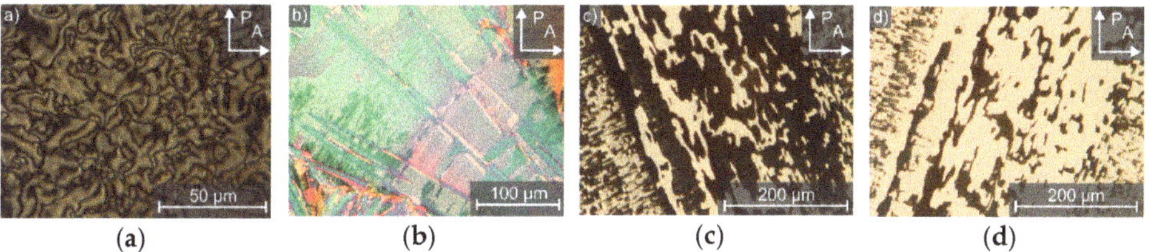

(a) **(b)** **(c)** **(d)**

Figure 11. Typical textures of the lyotropic SmC* phase observed by polarized optical microscopy. In thicker samples (**a**) Schlieren textures and (**b**) broken fans with zigzag defects appear. (**c**) In the surface stabilized state, bright and dark domains in the range of several hundred micrometers with opposite tilt directions can be observed. (**d**) By rotating the sample between crossed polarizers (P and A), the brightness of the tilt domains reverses. (Reprinted by permission from John Wiley and Sons: [29].)

Even more impressive than the long-range correlation of the director tilt, is that even the subtler precession of the tilt direction along the layer normal **k** is transmitted across the interlamellar solvent layers, resulting in a macroscopically observable chirality. This helical director modulation causes a striped pattern, as exemplarily shown in Figure 12a for a sample with 32 wt% of formamide. The distance between two of these so-called pitch lines is equal to the full pitch, the magnitude of which is in the same order as in thermotropic SmC* phases. By plotting the pitch versus the temperature,

(Figure 12b) an exponential increase of the values is observed when approaching the L_α phase. This is again, a typical behavior known from the thermotropic SmC* phase [31–33].

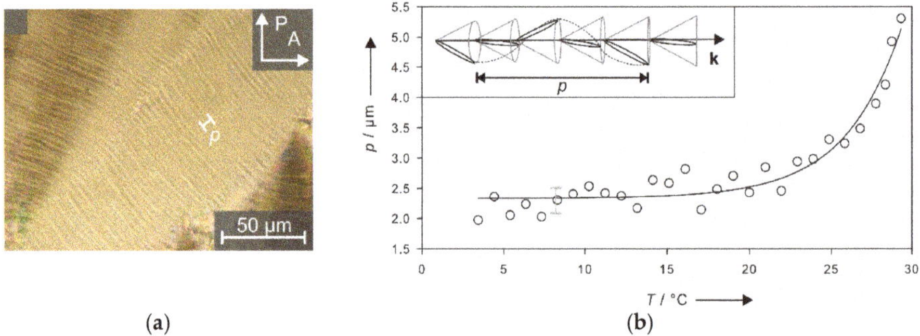

(a) **(b)**

Figure 12. The striped texture shown in (**a**) originates from a helical precession of the tilt direction along the layer normal **k** which is depicted in the inset of (**b**). The distance necessary for a full rotation of the director corresponds to the helical pitch p, which is plotted versus the temperature T in (**b**). (Reprinted by permission from John Wiley and Sons: [29].)

Measurements of the helical twist in dependence of the formamide mass fraction show a trend which is at the first glance counterintuitive, as the twist (the inverse pitch) increases with increasing solvent concentration (Figure 13). Osipov et al. [34] presented a possible explanation, as follows: The twisting is promoted by the chiral centers and counteracted by an elastic force, the effective elastic constant of which is known to be proportional to $\sin^2\theta$. If we add solvent, the volume density of chiral centers decreases linearly, while the tilt angle decreases at least linearly with the solvent concentration (cf. Figure 19). In this case, the restoring force decays more strongly than the twisting power, and thus, an increasing twist is observed. Nevertheless, this explanation has still to be verified by further experiments.

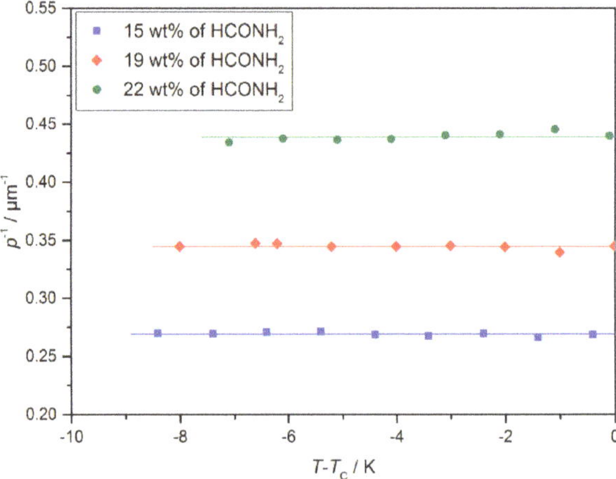

Figure 13. Helical twist p^{-1} of the lyo-SmC* phase for different mass fractions of formamide measured by the Cano method [34]. While a strong increase of the twist was measured with increasing solvent content, no temperature dependence was observed. Most likely, the latter is due to a pinning effect between liquid crystalline phase and the glass surfaces of the measurement set up. (Reprinted by permission from Springer Nature: [13].)

The most prominent chirality effect in thermotropic SmC* is the spontaneous electric polarization of its smectic layers (cf. Figure 3). So far it was not possible to directly measure this spontaneous polarization in a lyotropic SmC* phase, since its current response to an electric field is dominated by its high electric conductivity. The relatively high conductivity of the lyotropic phase originates from residual ions which cannot be fully avoided in the solvent layers containing water (autoprotolysis) or formamide (hydrolysis to NH_4^+ and $HCOO^-$ [35,36]). Instead, the presence of a spontaneous polarization in lyo-SmC* phase was indirectly confirmed by its polar electro-optic switching in the SSFLC state between crossed polarizers. Just as in the thermotropic case, the transmittance of the sample depends on the sign of the electric field (Figure 14), while any dielectric effect is proportional to E^2, and thus, insensitive to the sign of E [29].

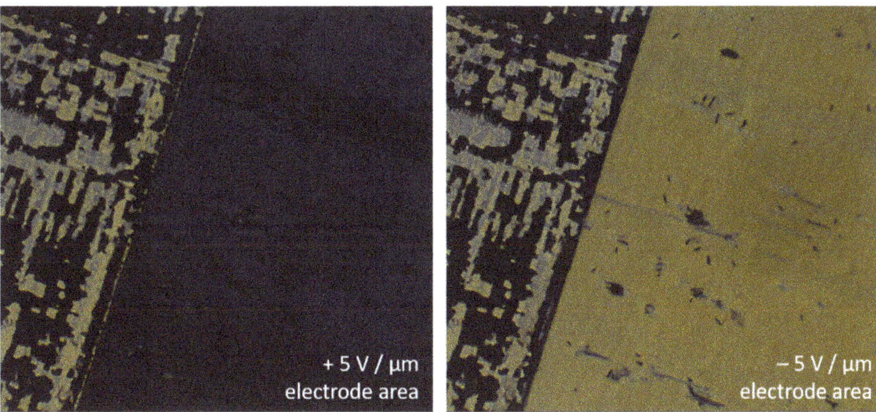

Figure 14. Polar electro-optic effect of a lyotropic SmC* phase in the SSFLC geometry (**4** with 20 wt% formamide at 28 °C in a 1.6 μm thick cell). The directions of the spontaneous polarization and director tilt depend on the sign of the applied electric field. Therefore, the transmittance of the sample depends on the sign of E. Outside the electrode area (upper left corner) the tilt domain texture of the virgin sample can be seen.

4. Prerequisites and Properties of Lyotropic SmC* Phases

4.1. Lyotropic SmC* Systems—A Delicate Balance

Since and even before the discovery of the first example of a lyotropic SmC* phase, there were constant research efforts—especially in and around the soft matter group in Stuttgart—to find a lyotropic analog of the thermotropic SmC* phase. As discussed in Section 1.2, the first and fundamental challenge was to identify the structural features necessary on a molecular basis for the formation of the lyo-SmC phase. For this, the early examples—namely, the lyotropic systems from Schafheutel and Finkelmann [25], Ujiie and Yano [26] and Pietschmann et al. [27]—were used as templates, although the last example turned out not to form a lyotropic SmC phase later on [28]. All of these model systems share certain common features, which are:

- A rigid aromatic core;
- A polar head group which is attached to the core by a slightly hydrophilic linker (e.g., ethylene glycol units);
- Another flexible chain attached to the other side of the core;
- Water as solvent.

Based on those observations, the amphiphiles listed in Table 1 were synthesized by Porada [28,29] and the Lemieux group [37–39]. All of the molecules incorporate a diol head group, a rigid aromatic core and a hydrophobic alkyl chain. Furthermore, the 1,2-diol head group introduces chirality to the molecules. Often, the amphiphiles only differ in the linker length between the diol group and the aromatic core, the number of oxygen atoms or the nature of the aromatic core. These slight variations allow a systematic analysis of the factors necessary for the formation of the lyotropic SmC* phase.

The molecules **1** to **8** all possess a phenyl pyrimidine core which is a promoter for the formation of tilted phases in thermotropic liquid crystals [40,41], a hydrophobic heptyl chain or a diol head group. Variations were only made in the linker between the head group and the aromatic core. Amphiphile **4** is the one presented in Section 3 already. Next to water (cf. Figure 10b) it forms a lyotropic SmC* phase with formamide, too. The phase diagram depicted in Figure 15, reveals that the same mesophases occur as found in mixtures with water. However, the N*, the Col1, the Col2 and the lyo-SmC* phases appear in a much more confined concentration range for the benefit of the L_α phase. While the lyo-SmC* phase is stabilized between roughly 25 and 70 wt% of water, it appears in mixtures with formamide only between 7 and 25 wt% of formamide. This finding underlines the importance of the solvent for the stability of the tilted phase.

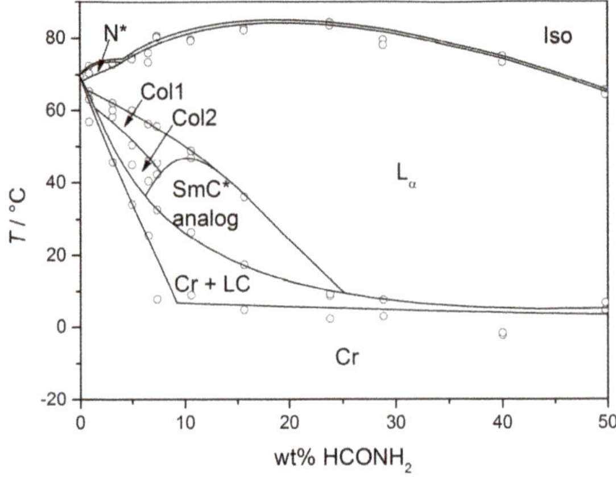

Figure 15. Phase diagram of the system **4**/formamide measured in heating. Next to the lyotropic SmC* phase (SmC* analog), a cholesteric (N*), two columnar (Col1, Col2) and a rather pronounced lamellar L_α phase occur. (Reprinted by with permission from Springer Nature: [13].)

Elongating the linker by one ethylene glycol unit leads to **5**. Instead of a monotoropic N* phase, this amphiphile forms a monotropic SmA* phase in the neat state which is substantially stabilized by the addition of water (Figure 16(a-a)). Moreover, a lyotropic SmC* phase forms, which was identified by the typical broken fan and schlieren textures (Figure 16(a-b–a-d)). No columnar or cholesteric phases occur in the phase diagram shown in Figure 16b. Another difference to the parent amphiphile **4** is that the lyotropic SmC* phase is only stable in mixtures with water, but not with formamide. Furthermore, the phase occurs only between 10 to 20 wt% of water. Overall, the lyotropic SmC* phase is destabilized in comparison to the parent systems with amphiphile **4**.

Table 1. Summary of the amphiphiles which were screened for the lyotropic SmC* phase. Thermo-tropic liquid crystalline (TLC) phases that occur are listed, as are solvents tested during screening which enable the formation of a lyotropic SmC* phase.

No°	Structure		TLC [1]	Lyo-SmC* [1]	Ref.
1	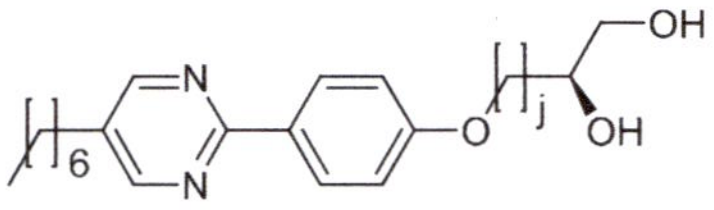	j = 1	-	-	[13]
2		j = 3	-	-	[28]
3		j = 4	-	-	[13]
4	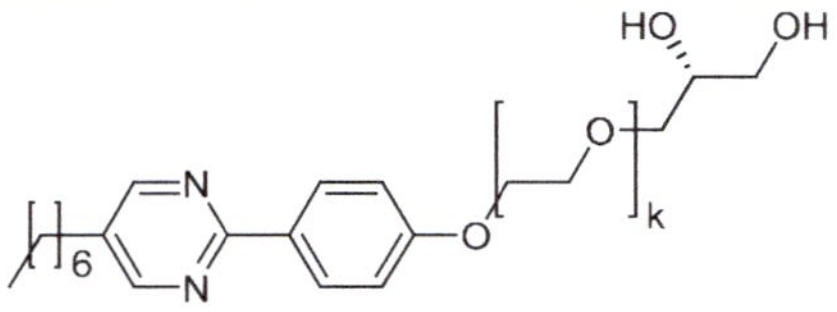	k = 1	(N*)	H_2O, formamide	[13,29]
5		k = 2	(SmA*)	H_2O	[37]
6		k = 3	SmA*	-	[37]
7	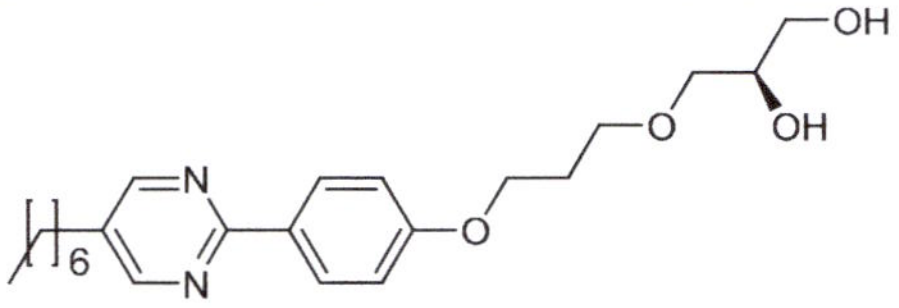		SmA*	-	[13]

Table 1. *Cont.*

No°	Structure	TLC [1]	Lyo-SmC* [1]	Ref.
8	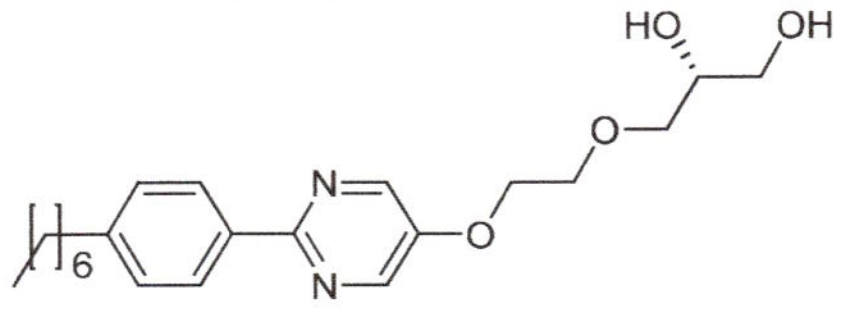	SmA*, SmC*	H$_2$O, formamide	[37]
9	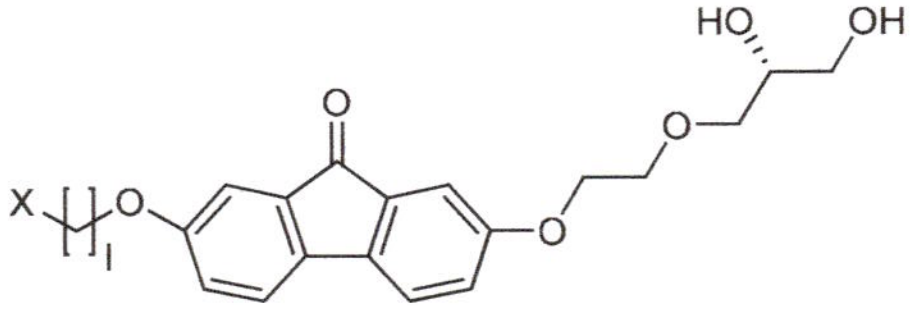 l = 4, X = H	-	-	[37,39]
10	l = 5, X = H	SmA*	-	[37,39]
11	l = 6, X = H	SmA*, SmC*	H$_2$O, formamide	[37]
12	l = 6, X = Cl	SmA*	(H$_2$O, formamide)	[37,39]

[1] Brackets indicate monotropic phases.

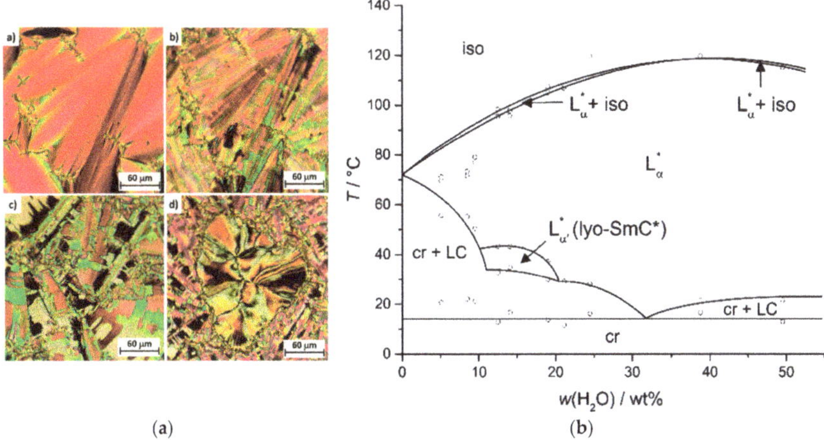

Figure 16. Mixtures of the amphiphile **5** with 19 wt% water show characteristic textures, as is known from thermotropic systems while cooling: (**a-a**) fan texture of the L$_\alpha$* phase, (**a-b**) broken fan texture of the lyo-SmC* phase, (**a-c**) well defined tilt domains of the lyo-SmC* phase at lower temperatures and (**a-d**) Schlieren texture of the lyo-SmC* phase surrounded by the broken fan texture. (**b**) Phase diagram of **5** with water. (Reprinted by permission from the Royal Society of Chemistry: [37].)

If the linker consists of three ethylene glycol units (**6**), no lyotropic SmC* phase is detected at all. Instead an enatiotropic mesophase—SmA* phase—is formed in the neat state for the first time. Comparing the amphiphiles **4, 5** and **6** suggests that the length of the linker is essential for the formation of the lyotropic SmC* phase. The linker in **4** seems to have the optimum length for stabilizing the lyo-SmC* phase.

For the amphiphiles **1, 2** and **3**, the ethylene glycol unit is replaced by a simple alkyl chain which is attached to the aromatic core by a single ether unit. Even though all three amphiphiles form a variety of lyotropic liquid crystal phases, they neither form any thermotropic phases nor a lyotropic SmC* phase. Especially for **3**, this is remarkable, considering that the only difference to **4** is the exchange of an oxygen atom in the linker by a CH$_2$-moiety. The change of the molecular length and the flexibility of the linker due to this alteration are negligible. Thus, the most reasonable explanation is, that the modification in the hydrophilicity—namely the ability of the oxygen atom in the linker to accept hydrogen bonds—has a major impact on the formation of the lyotropic SmC* phase.

The amphiphile **8** is composed of the same alkyl tail, diol head group and linker as the amphiphile **4**. However, the aromatic core is inverted compared to the original molecule. This simple switch in the position of the pyrimidine and the phenyl rings has a major impact on the mesophase behavior of the amphiphile. In the solvent free state, **8** exhibits a SmA* and a SmC* phase. Both mesophases stay stable upon the addition of water or formamide up to roughly 50 wt%. No further lyotropic liquid crystal phases were detected.

The amphiphiles **9–12** all incorporate the same diol head group and linker as the amphiphile **5**. The phenyl pyrimidine core, however, is substituted by a 2,7-fluorenone core which is known to be an even stronger SmC-promoter in thermotropic smectics [42,43]. From **9** to **11**, the hydrophobic alkoxy tail is elongated. With the shortest hydrophobic chain—a butoxy chain (**9**), neither a thermotropic mesophase nor a lyotropic SmC* phase forms. Elongating the chain by one methylene unit (**10**) leads to the appearance of a SmA* phase in the neat state, but only the elongation by two carbon atoms (**11**) allows the formation of lyotropic SmC* phase with water and formamide. The latter amphiphile already exhibits a SmC* and a SmA* phase in the solvent free state. The phase diagram of the system **11**/formamide is presented in Figure 17a and reveals that the thermal stability of the SmC* phase is gradually decreased by the addition of formamide.

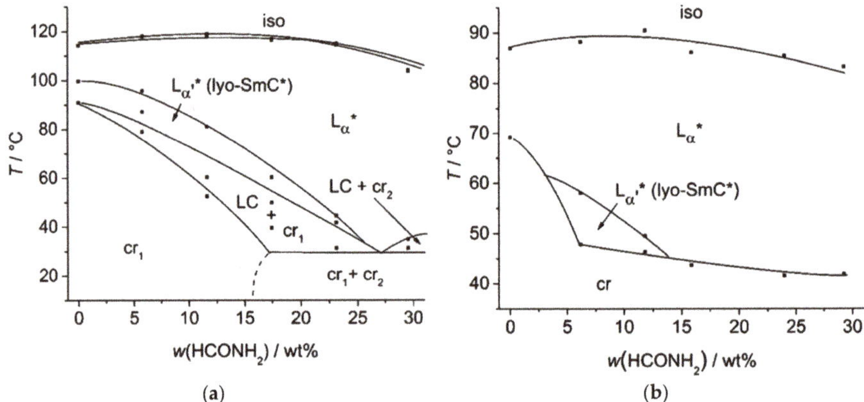

(**a**) (**b**)

Figure 17. (**a**) Already in the neat state, the amphiphile **11** exhibits a SmA* and a SmC* phase. In mixtures with formamide, the orthogonal phase is stabilized while the tilted phase is destabilized, but remains present up to 25 wt% of solvent. (**b**) The amphiphile **12** shows only a monotropic lyo-SmC* phase. Thus, the phase diagram measured while cooling is presented here. During heating, the temperature and concentration range in which the phase occurs is superposed by a two-phase region of crystalline and liquid crystalline phases. (Reprinted by permission from the Royal Society of Chemistry: [37].)

In the amphiphile **12**, one terminal hydrogen atom of the alkyl tail in **11** is replaced by a chlorine atom. The terminal chloro substituent, which favors orthogonal versus tilted phases [44], suppresses the thermotropic SmC* phase and increases the thermal stability of the SmA* phase. The lyotropic SmC* phase is reduced to a monotropic phase. Thus, the phase diagram shown in Figure 17b was measured during cooling.

Concluding this section, it is obvious that every structural subunit of the amphiphiles plays an important role in the formation of the lyotropic SmC(*) phase. Minor changes in the hydrophilic linker, the hydrophobic chain or the aromatic core easily lead to the appearance or disappearance of the tilted lamellar phase.

4.2. Structure and Phase Transitions of Lyotropic SmC* Phases

In Section 3 of this review, a two-dimensional X-ray pattern of the lyotropic SmC* phase (cf. Figure 10c) proved that the structure of the lyotropic SmC* phase is an analog to its thermotropic equivalent. To address the question of how the amount of solvent affects the structural parameters of the phase, temperature and concentration-dependent measurements of the system 4/formamide are shown in Figure 18 [13]. For all mixtures investigated a temperature-dependent $d(T)$ characteristic of thermotropic SmA(*) to SmC(*), phase transitions were found: In the L$_\alpha$ phase the lamellar repeat unit d increases with increasing temperatures due to the decreasing orientational order. The maximum is reached at the transition from the lamellar L$_\alpha$ to the lyotropic SmC* phase. By further heating the lamellar repeat unit decreases again, which can be explained by an increasing value of the tilt angle.

A second observation is, that the addition of solvent shifts the $d(T)$ curves to higher values but does not change their principal shape. To describe these observations in a more quantitative way, we assume that the lamellar repeat unit d is composed of two additive contributions:

- The thickness of the solvent layer d_s, which mainly depends on the mass fraction w of solvent;
- The thickness of the amphiphile bilayer d_{bl}, which changes with the temperature T.

The total repeat unit $d(T,w)$ can, thus, be written as:

$$d(T,w) = d_s(w) + d_{bl}(T), \tag{2}$$

with

$$d_s(w) = mw, \tag{3}$$

where the slope m denotes the increase in solvent layer thickness per wt% of solvent.

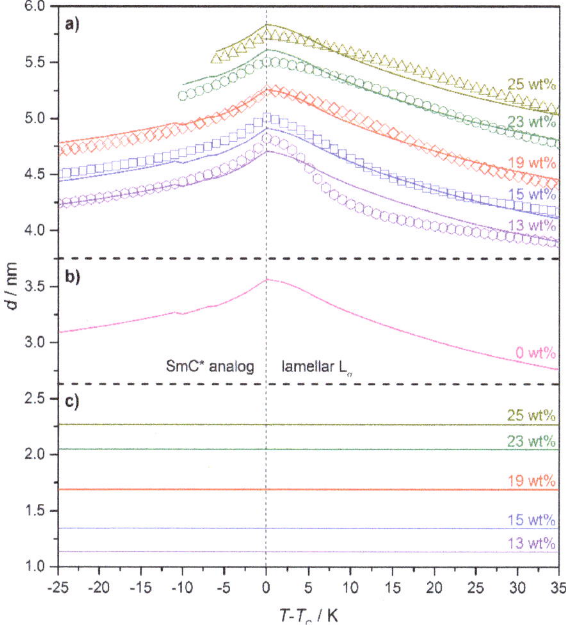

Figure 18. (**a**) Temperature and concentration-dependent measurements of the lamellar repeat unit d of the system **4**/formamide. The weight fraction w of formamide is given in the graph. Measured data points are depicted as symbols, while the lines represent fits according to the Equations (2) and (3). (**b**) Calculated thickness of the amphiphile bilayer $d_{bl}(T)$ of a hypothetic SmC* phase in the neat state. (**c**) Calculated solvent layer thickness $d_s(w)$ for the different weight fractions w of formamide. (Reprinted by permission from Springer Nature: [13].)

The measured data points were fitted to these equations resulting in the continuous lines in Figure 18a. From that, the thickness of the pure bilayer $d_{bl}(T)$ (Figure 18b) and the thickness of the solvent layers $d_s(T)$ for the different mass fractions (Figure 18c) were separated from each other. The reasonable agreement of the measured data and the fits in Figure 18a imply that the solvent molecules form a separate layer between the amphiphile bilayers and do not penetrate them. Furthermore, it allows an estimation of the solvent layer thickness. For example, in a sample of **4** with 25 wt% of formamide, the solvent layer is roughly 2.3 nm thick, while the amphiphile bilayer ranges between 3.3 and 3.6 nm.

The foremost difference between L$_\alpha$ and the lyotropic SmC* phase, is the collective tilt of the elongated molecules relative to the layer normal. The presence of the tilt was already verified by X-ray diffraction (cf. Figure 10c) and polarizing optical microscopy (cf. Figure 11c,d). In Figure 19 the optical tilt angle θ_{opt} of the system **4**/formamide is plotted against the relative temperature for different mass fraction of formamide [45]. For all samples investigated, the tilt angle seems to saturate at lower temperatures. This saturation value decreases with increasing mass fractions of formamide, from about 27° at 13 wt% to 12.5° at 25 wt% of formamide. Moreover, the course of the measured values in dependence of the temperature changes with the concentration. At high solvent concentrations, the tilt discontinuously drops to zero in a first order phase transition to the L$_\alpha$ phase, while the transition is continuously second order at low concentrations [13]. This is a remarkable example of a solvent-induced change in the nature of a phase transition.

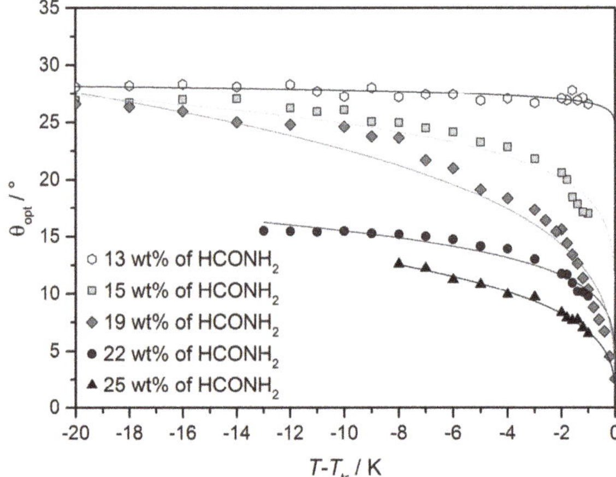

Figure 19. Tilt angle θ_{opt} of the system **4**/formamide measured by rotating a surface stabilized sample between crossed polarizers. The values were plotted relative to the temperature T_{tr} of the phase transition from the high temperature phase to the lyotropic SmC* phase. (Reprinted by permission from John Wiley and Sons: [45].)

A more detailed investigation of the phase transition from the SmC* analog phase to the high temperature phase was done by differential scanning calorimetry (DSC) [13]. The heat flows measured while cooling and heating are displayed in Figure 20; the discussed transition enthalpies are highlighted in yellow. For the lowest solvent concentration measured in the system 4/formamide, the mesophase above the lyotropic SmC* phase is a columnar phase. As already required by symmetry arguments, the heat flow measured confirms a 1st order phase transition. At formamide mass fractions between 12 and 22 wt%, the high temperature phase is the lamellar L_α phase. At 12 wt%, the peak is less pronounced than in the former case, but still suggests a phase transition close to 1st order. By increasing the formamide content further, the peak starts to flatten out more and more, only showing a step in the curves for 18 and 22 wt%. This suggests that the order of phase transition is shifted from 1st to 2nd with an increasing amount of solvent. With 27 wt% of formamide, only a non-tilted phase is formed.

From the measured tilt angles and lamellar repeat units, and the estimated thickness of the amphiphile bilayer and the solvent layer a structural model of the lyotropic SmC* phase was sketched [13]. In Figure 21, a true to scale model of the lyotropic SmC* phase of the system 4/formamide with 19 wt% of formamide 10 K below the phase transition from the L_α phase is shown. As depicted, the amphiphile bilayers consist of partial bilayers with a significant interdigitation of the hydrophobic parts. The formamide is not mixed within the amphiphile bilayers but forms separate, well-defined solvent layers. Thus, the amphiphile molecules from different bilayers are not in direct contact with each other, which raises the question once again, of how the long-range correlation of the director tilt across the solvent layers takes place.

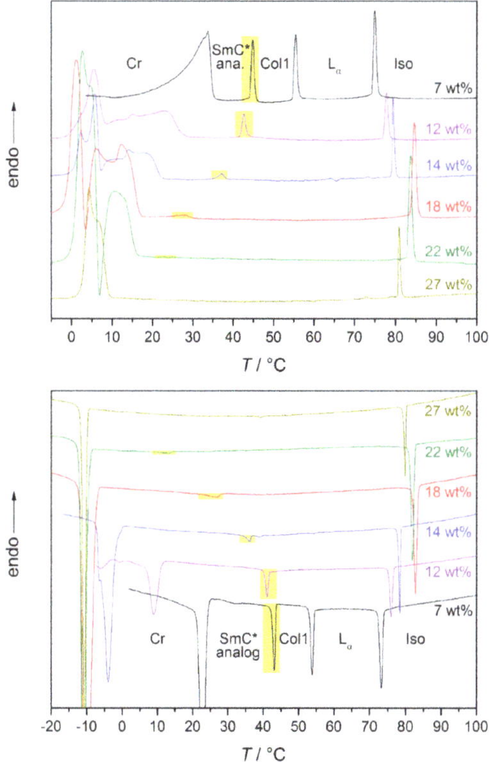

Figure 20. DSC curves obtained while heating (top) and cooling (bottom) the system **4**/formamide. The weight fraction of formamide is indicated in the graphs. The transition from the high temperature phase to the lyotropic SmC* phase is highlighted in yellow. In the sample with the smallest amount of formamide the high temperature phase is a columnar (Col1) phase. For the samples with 12 to 22 wt% of formamide, it is a L_α phase. In case of the sample with the highest formamide concentration, the only mesophase occurring is a L_α phase (cf. Figure 15). (Reprinted by permission from Springer Nature: [13].)

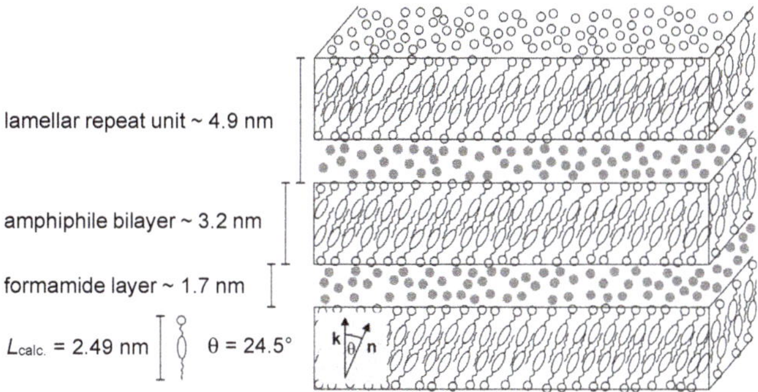

Figure 21. Model of the lyotropic SmC* phase for the system **4**/formamide. The model suggested by Bruckner et al. is based on their measurements shown in Figures 18 and 19 and the length $L_{calc.}$ of the amphiphile calculated by molecular modeling. The sketch corresponds to a sample with 19 wt% of formamide (•) at $T - T_{tr} = -10$ K. (Reprinted by permission from Springer Nature: [13].)

4.3. Origin of the Director Tilt

To learn more about the origin of the director tilt and its long-range correlation, the orientation of the C–N pyrimidine vibration which is directed along the long molecular axis of **4** was selectively probed by polarized micro-Raman spectroscopy [45]. For this a mixture of amphiphile **4**, 15 wt% of formamide was used. The mean direction of the C–N vibration is found to be tilted with respect to **k** and its tilt angle is in good agreement with the director tilt angle optically measured under the same conditions. Thus, the authors concluded that the origin of the tilt in the lyotropic SmC* phase is a tilt of the aromatic 2-phenylpyrimidine core relative to the layer normal.

However, this still does not explain how the tilt direction is transmitted between adjacent bilayers. To clarify the long-range correlation of the director tilt, the authors made use of an observation mentioned in the Dection 4.1 already: the appearance or disappearance of the lyotropic SmC* phase by simply changing the solvent. Thus, different solvents were tested for their ability to stabilize the lyotropic SmC* phase. The solvents and some of their physical data are listed in Table 2.

Table 2. Physical data of the solvents which were tested for the formation of a lyotropic SmC* with the amphiphile **4**. (Adapted from [13].)

Solvent	Dielectric Permittivity ε_r [46]	Dipole Moment μ/D [46]	Number of H-bond Donor Atoms	Number Density [1] $n_\rho/10^{22} \cdot cm^{-3}$
Water	80.1	1.85	2	3.27
Formamide	111.0	3.73	2	1.52
Ethylene glycol	41.4	2.36	2	1.08
NMF	189.0	3.83	1	1.03
DMF	38.3	3.82	0	0.78
PEG 200	22.1 [47]	3.28 [47]	2	0.34
PEG 300	19.2 [47]	3.91 [47]	2	0.23

[1] Calculated according to $n_\rho = \rho \cdot N_A/M$.

As discussed before, the amphiphile **4** forms a lyotropic SmC* phase in mixtures with water and formamide. The corresponding phase diagrams are depicted in Figure 22a,b once again, for comparability. By changing the solvent from water to formamide, the concentration and temperature range in which the tilted phase occurs is reduced significantly. Exchanging the solvent to *N*-methyl formamide (NMF), leads to the disappearance of the lyotropic SmC* phase and the two columnar phases (Figure 22c). Moreover, the stability of the lamellar L_α phase is reduced with increasing mass fraction of NMF. When the solvent is replaced by *N*,*N*-dimethyl formamide (DMF), only a N* phase appears (Figure 22d).

Neither the dielectric permittivities, nor the dipole moments listed in Table 2 may explain this trend. Therefore, pure electrostatic interactions between adjacent amphiphile bilayers cannot explain the long-range correlation of the director tilt in an obvious way. A quantity which follows the trend observed experimentally is the number of hydrogen bond donor atoms. Since all four solvents can accept up to two hydrogen bonds, water and formamide are known to form extended hydrogen bond networks in the liquid state, while NMF only forms chains of hydrogen bonds and DMF no hydrogen bonds with itself. Thus, the authors conclude, that anisotropic hydration interactions which are mediated by a dense three dimensional hydrogen bond network might be responsible for the correlation of the tilt direction in adjacent bilayers [45]. In general, hydration forces have already been known for phospholipids forming lyotropic lamellar phases [48–54], but have never been observed in combination with lamellar, fluid and tilted phases. Several studies show that the orientational ordering of water molecules by the phospholipid head groups can bridge a distance of up to 1 nm [55–60].

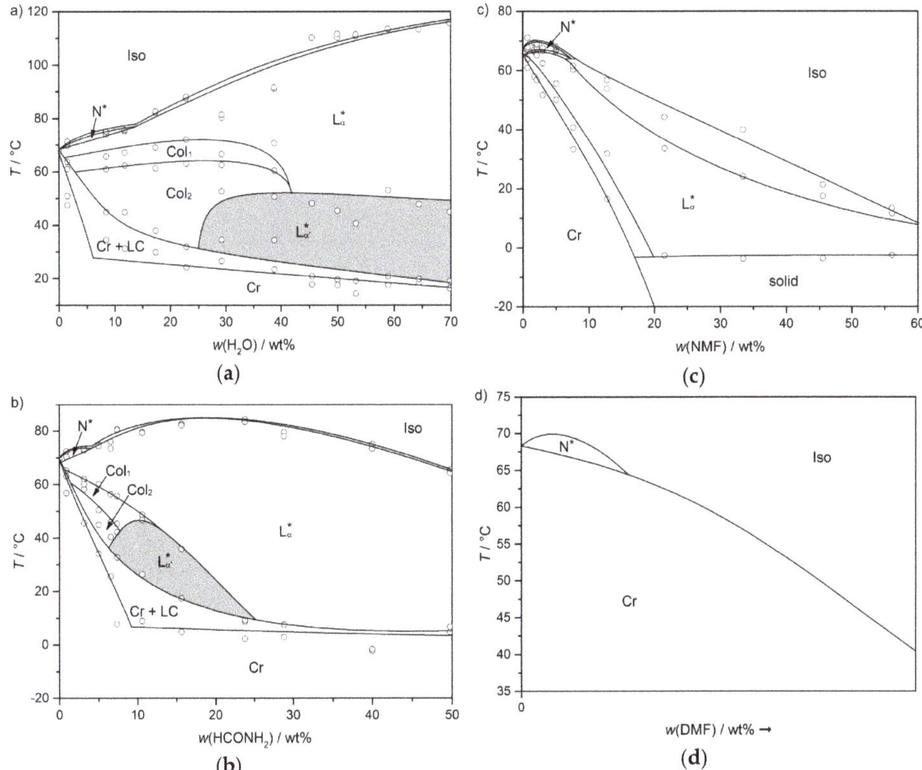

Figure 22. Phase diagrams of the amphiphile **4** with (**a**) water, (**b**) formamide, (**c**) *N*-methylformamide and (**d**) *N,N,*-dimethylformamide. The ability of the solvent to stabilize lyotropic mesophases in general—and the lyotropic SmC* phase in particular—is diminished in the order of presentation. For the system **4**/DMF, only a schematic phase diagram is presented, which was derived by investigation of a contact sample. (Reprinted by permission from John Wiley and Sons: [45].)

Solvents other than water and formamide which hold two hydrogen bond donor atoms are ethylene glycol and its polymers (PEG). Consequently, these solvents were investigated too [13]. In mixtures of amphiphile **4** and ethylene glycol, two columnar mesophases a N* and a L_α phase appear. Yet, a lyotropic SmC* phase is not stabilized. In mixtures with PEG 200 or PEG 300, the only stable mesophase is a cholesteric phase next to a monotropic L_α phase. In conclusion, the number of hydrogen bonds per solvent molecule cannot be the sole determining factor for the stability of the lyotropic SmC* phase. A second important factor is the density of the hydrogen bond network which is directly related to the number densities of the solvent molecules (Table 2) and the average number of hydrogen bonds per molecule.

To underline the importance of the hydrogen bond network further, the tilt angle in mixtures with ordinary formamide and deuterated formamide were compared [45], since deuteration selectively modifies the strength and dynamics of the hydrogen bond network. The result is shown in Figure 23. Even though the investigated mole fractions of formamide were the same, the saturation value of the sample with deuterated solvent was roughly 15% lower than in the equivalent sample with ordinary formamide. This is indeed, a quite substantial isotope effect which clearly confirms that the correlation mechanism is sensitive to the structure and dynamics of the hydrogen bond network in the solvent layers.

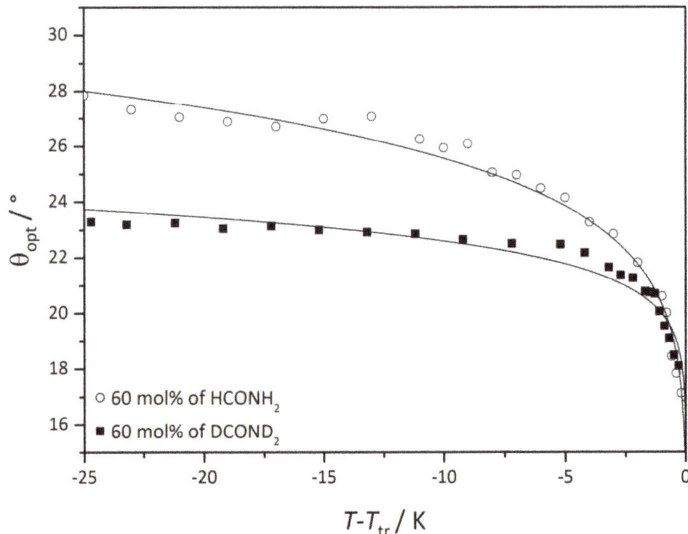

Figure 23. Optically measured tilt angle of the system **4**/formamide (circles) and **4**/deuterated formamide (squares) at the same mole fraction relative to the phase transition temperature T_{tr} from the L_α phase. (Reprinted by permission from John Wiley and Sons: [45].)

An example which seems to contradict the importance of a strong hydrogen bond network for the long-range correlation of the tilt direction across the solvent layers, is the case of the hyper-swollen perfluorinated smectics [24] (cf. Section 2). The perfluorinated oils which are used for swelling do not form any hydrogen bonds at all. Nonetheless, the hyper-swollen SmC phase stays stable up to an oil content of 60 wt%. We suggest that the mechanism for the long-range tilt correlation with perfluorinated solvents is completely different. Due to the size of the fluorine atoms, a free rotation of the perfluorinated carbon chains is hindered [61,62]. Therefore, perfluorinated polymers are rather stiff and possess a significantly higher persistence length than unfluorinated polymers [63–65]. Therefore, we propose, that the two elongated perfluorinated polymeric oils (cf. Figure 8b) might possess enough rigidity to transmit the tilt direction sterically. This of course, has to be confirmed by experiments and/or simulations.

4.4. Electroclinic Effect

As we previously discussed in Section 1.1, the D_∞-symmetry of a chiral SmA* phase excludes polar properties, such as a spontaneous polarization. The polarization-tilt coupling, however—present in all chiral smectics—gives rise to the so-called electroclinic effect [9,10], the properties of which resemble, in some aspects, the (inverse) piezoelectricity in non-centrosymmetric solid crystals. An electric field **E** along the SmA* layers induces an electric polarization **P** in a non-parallel direction to the layer normal **k**. In a chiral medium, the plane spanned by **k** and **P** is no mirror plane. As a result of **E**, the free energy and the distribution of molecular tilt directions are no longer symmetric about the **k**,**P**-plane and the macroscopic director **n** is, therefore, observed to tilt away from **k** in a direction normal to the **k**,**P**-plane. Similar to the field-induced mechanical strain in the inverse piezoelectric effect, the field-induced tilt angle $\delta\theta$ is linear in E and the direction of tilt reverses upon reversal of the field direction [9–11].

According to these symmetry arguments, the electroclinic effect must be present in all SmA* phases even though the effect might be very small. Indeed, a measurable electroclinic effect was only observed in the pretransitional regime of a second-order (or weak first-order) transition from the SmA* into the tilted SmC* phase. In that temperature regime, the electroclinically induced tilt $\delta\theta$ grows hyperbolically towards the critical temperature $T_c \leq T_{AC}$ according to a Curie–Weiss-like behavior [10,11,66].

Since the symmetry arguments leading to the electroclinic effect in SmA* apply in the very same way to the lyotropic case of a chiral lamellar α-phase, there were several attempts to detect an electroclinic response in L_α*. A first indication was the observation by Jakli et al. that the chirality of phospholipids makes fluid lamellar phases piezoelectric [67,68]. The discovery of the lyo-SmC* phase however sheds new light on this issue: in 2017 Harjung et al. observed an electroclinic response in the L_α* phase of the amphiphile 4 with 23 wt% formamide at temperatures close to its transition into the newly discovered lyo-SmC* phase [69]. The samples were poured into a liquid crystal cell and a square-wave electric field was applied to its transparent ITO electrodes (Figure 24a). The field $E(t)$ electroclinically induced an alternating tilt $\delta\theta(t)$, which gave rise to an essentially square-wave modulation $I(t)$ of the light intensity transmitted through the cell between crossed polarizers (Figure 24b). The electroclinic electro-optic response completely vanished in the case of the corresponding non-chiral L_α phase composed of the racemic version of the amphiphile 4. At least beyond a certain threshold field in the order of 1 V/μm, the electroclinic tilt increased linearly with the field amplitude (Figure 25a) and grew hyperbolically if the temperature was lowered towards the transition temperature into the tilted ferroelectric lyo-SmC* phase (Figure 25b).

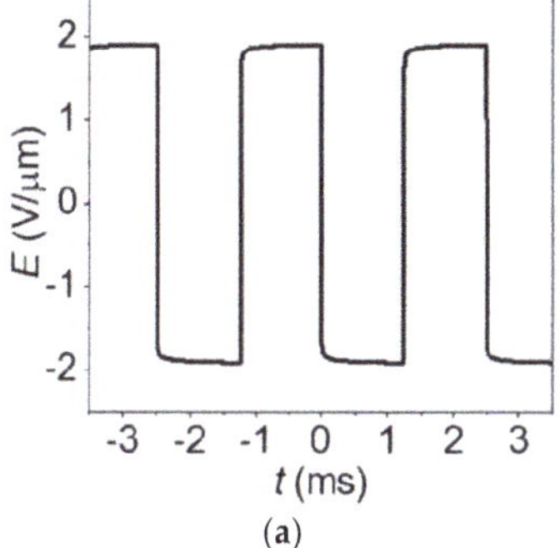
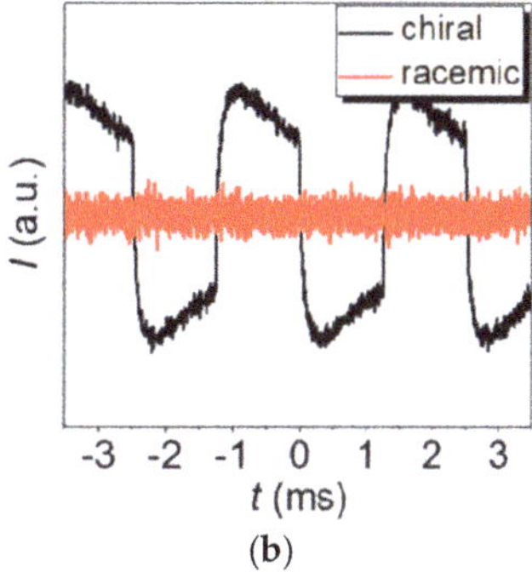

(a) (b)

Figure 24. Electroclinic electro-optic effect of a chiral lamellar α-phase in the vicinity to its transition into a tilted lyo-SmC* phase. (**a**) Square-wave electric field E(t) applied to lyotropic $L_\alpha^{(*)}$ samples in a direction normal to the layer normal **k**. (**b**) Corresponding electro-optic response of the lyotropic samples between crossed polarizers measured by the transmitted light intensity $I(t)$ at a temperature of 0.2 K above the transition into the tilted phase; black line: chiral L_α* phase of 4 with 23 wt% of formamide; red line: nonchiral L_α phase of racemic 4 with 23 wt% of formamide. (Reprinted by permission from [69] Copyright 2018 by the American Physical Society.)

The electroclinic effect in the L_α* phase, thus shows all signatures of the electroclinic effect in thermotropic SmA* phases, namely the effect (i) is chiral in nature, (ii) is essentially linear in the sign and magnitude of the electric field, and (iii) shows Curie–Weiss-like behavior in the pretransitional regime of a tilting transition into lyo-SmC* [69]. Beyond these striking similarities between the electroclinic effects in SmA* and L_α* there are also specific deviations, namely, the slow decay of the electro-optic response in Figure 24b, after switching and the non-linearity at low field strength in Figure 24a. The experiments in [69] indicate that these deviations are related to the comparatively high electric conductivity of lamellar phases which originates from the inevitable presence of ionic impurities in the highly polar and protic solvent layers. Under the action of an electric field, these ionic impurities accumulate at the interfaces between the solid electrodes and the liquid-crystalline electrolyte and form electric double layers which screen the external electric field, and thus reduce the effective field inside the liquid crystal layer. Since the electroclinic effect probes the effective field inside the lyotropic phase, the screening effect of electric double layers considerably complicates the dynamics of the electroclinic effect in L_α^* phases.

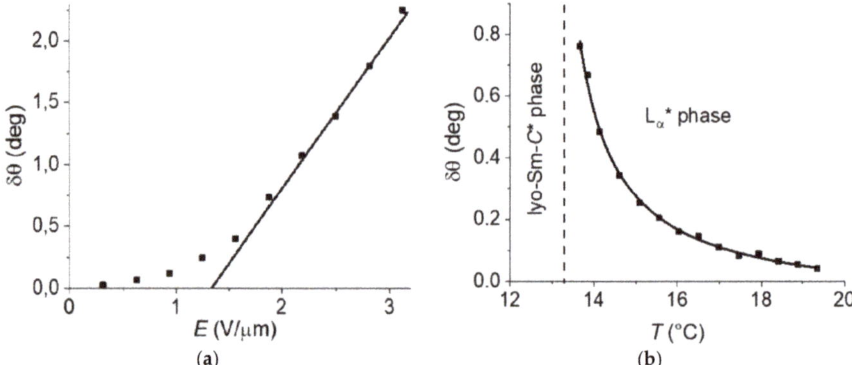

Figure 25. Signatures of the electroclinic effect in the L_α^* phase of the chiral amphiphile 4 with 23 wt.% of formamide. (**a**) Electroclinically induced tilt angle $\delta\theta$ versus amplitude E of a 400 Hz square-wave electric field at a fixed temperature of 0.5 K above the transition temperature into the lyo-SmC* phase. The straight line shows the linear regime of $\delta\theta(E)$. (**b**) $\delta\theta$ versus temperature T at a fixed amplitude $E = 1.9$ V/μm of the 400 Hz square-wave electric field. The solid line shows the Curie–Weiss-like hyperbolic divergence of $\delta\theta(T)$. (Reprinted by permission from [69] Copyright 2018 by the American Physical Society.)

All in all, 40 years after the discovery of the electroclinic effect in thermotropic SmA* it has now become clear that the same effect exists in chiral lyotropic L_α^* phases as well. Since phospholipid cell membranes are in a fluid L_α^* state, this observation might also have certain implications in biophysics.

5. Conclusions and Outlook

After the discovery of ferroelectricity in thermotropic liquid crystals in the 1970s, it has become evidently clear, in recent years, that the lyotropic equivalent to the thermotropic ferroelectric smectic C* phase indeed exists. The new lyo-SmC* phase shares many characteristic properties of its thermotropic counterpart:

1. It has the same fundamental SmC structure; namely, a 1D periodic layer structure of 2D-fluid bilayers with the liquid crystal director macroscopically tilted from the layer normal by typically 10–30 degrees.
2. The lyo-SmC* phase has a helical ground state with a pitch typically in the micron range and textures showing pitch lines familiar from thermotropic SmC* phase.
3. In thin samples the lyo-SmC* phase can be surface stabilized, showing ferroelectric tilt domains, sometimes prevaded by zig-zag defects, and it can be electrically switched between the two stable states just like a thermotropic ferroelectric SmC* sample.
4. At higher temperatures and/or solvent concentrations, the lyo-SmC* phase often transforms into the chiral L_α^* phase, the lyotropic equivalent of thermotropic SmA*. As in thermotropic SmC*–SmA* materials, the transition can be second order and an electroclinic effect is observed in the L_α^* phase.

Beyond these convincing similarities, the new lyotropic SmC* phase is clearly distinguished from its thermotropic counterpart by the presence of water or water-like solvent layers between the bilayers of amphiphilic mesogens. This raises a lot of new questions which all address the specific role of the solvent. In comparison to thermotropic SmC*, the solvent concentration is an additional thermodynamic degree of freedom, and thus, an additional "control variable" in lyo-SmC*. Exploring how the solvent concetration changes properties, such as order parameters, elasticity, viscosity, pitch and polarization, will certainly involve new physics and lead to an improved understanding of interactions in lyotropic systems. In view of the surprising long-range tilt correlation in lyo-SmC*, the most challenging question is how not only the elastic interactions (tilt) but also the much weaker chiral interactions (twist) are transmitted through the interlamellar solvent layers. Evidently these interactions do transform the disordered solvent into partially ordered solvent. The further investigation of this subtle order is a most interesting subject and key to the detailed understanding of lyotropic SmC* phases. Even though there are already strong experimental indications that hydrogen bond networks in the solvent layers play an important role in the interlamellar correlation of tilt directions, more experimental and theoretical work is needed to really understand how elastic and chiral interactions are "communicated" across nanoscopic solvent layers. This might also be helpful in our further understanding of chirality effects in other soft matter and biological systems.

The tailored design of new lyo-SmC* phases remains a non-trivial task, since it relies on a detailed balance between hydrophilic, hydrophobic and tilt-promoting molecular interactions. As far as we know today, the formation of lyo-SmC* phases requires, on the one hand, protic water-like solvents which are able to form extended hydrogen bond networks, and on the other hand, rod-shaped amphiphiles terminated by a hydrogen-bonding hydrophilic head group that is able to connect to the hydrogen bond network of the solvent. The molecular structure of the amphiphile further contains a rigid tilt-promoting core which is linked to the head group by a slightly hydrophilic spacer chain. The further proof and extension of this rather specific design concept is a future challenge for synthetic liquid crystal chemistry.

Last but not least, all attempts to measure the actual values of spontaneous polarization in lyo-SmC* phases by field-reversal, dielectric or pyroelectric techniques have failed so far, due to the high electrolytic conductivity of lyo-SmC* phases originating from the mobile ionic impurities located in the solvent layers. Since P_s is the key parameter of any ferroelectric material, this situation is highly unsatisfactory and it remains another future challenge to directly measure the spontaneous polarization in an electrolytically conductive medium.

In terms of possible applications, the use of the lyo-SmC* phases in displays—as known from its thermotropic counterpart—namely, in SSFLC devices [70], is not very promising, considering the high electrolytic conductivity and the potential leakage of the solvents. At the same time, the solvent opens new perspectives; e.g., in chirality sensing. For instance, the enantiomeric excess of physiologically-important, watersoluble drug molecules might be probed by their effect on the electo-optic switching time. In addition, the discovery of the elctroclinic effect in L_α* has certain implications in the biophysics of cell membranes, since these are in a lamellar α state and contain substantial amounts of chiral inclusions; e.g., cholesterols and membrane proteins.

All in all, the recent discovery of the lyotropic equivalent to the thermotropic ferroelectric SmC* phase has not only bridged a long-standing gap between the worlds of thermotropic and lyotropic liquid crystals, it also opened new research directions to the understanding of solvent-mediated chirality effects in lyotropic and biological systems.

Author Contributions: Both authors contributed equally.

Funding: Parts of our research were funded by *Deutsche Forschungsgemeinschaft*, DFG Gi 243/4, and *Landesgraduiertenförderung* of the state Baden-Württemberg.

Acknowledgments: Mikhail Osipov, Friederike Knecht and Marc Harjung are gratefully acknowledged for valuable discussions and input.

Conflicts of Interest: The authors declare no conflict of interest.

References

1. Lines, M.E.; Glass, A.M. *Principles and Applications of Ferroelectrics and Related Materials*; Oxford University Press: Oxford, UK, 2001; ISBN 978-0-19850778-9.
2. Valasek, J. Piezo-Electric and Allied Phenomena in Rochelle Salt. *Phys. Rev.* **1921**, *17*, 475–481. [CrossRef]
3. Meyer, R.B.; Liebert, L.; Strzelecki, L.; Keller, P. Ferroelectric Liquid Crystals. *J. Phyique Lett.* **1975**, *36*, 69–71. [CrossRef]
4. Meyer, R.B. Ferroelectric Liquid Crystals; a Review. *Mol. Cryst. Liq. Cryst.* **1977**, *40*, 33–48. [CrossRef]
5. Lagerwall, S.T. Ferroelectric Liquid Crystals. In *Handbook of Liquid Crystals*; Goodby, J.W., Collings, P.J., Kato, T., Tschierske, C., Gleeson, H.F., Raynes, P., Eds.; Wiley-VCH: Weinheim, Germany, 2014; Volume 4, ISBN 978-3-527-32773-7.
6. Takezoe, H.; Cepic, M. Antiferroelectric Liquid Crystals. In *Handbook of Liquid Crystals*; Goodby, J.W., Collings, P.J., Kato, T., Tschierske, C., Gleeson, H.F., Raynes, P., Eds.; Wiley-VCH: Weinheim, Germany, 2014; Volume 4, ISBN 978-3-527-32773-7.
7. Takezoe, H.; Eremin, A. *Bent-Shaped Liquid Crystals*; CRC Press: Boca Raton, FL, USA, 2017; ISBN 978-1-4822-4759-6.
8. Lagerwall, S.T. *Ferroelectric and Antiferroelectric Liquid Crystals*; Wiley-VCH: Weinheim, Germany, 1999; p. 129. ISBN 3-527-29831-2.
9. Garoff, S.; Meyer, R.B. Electroclinic Effect at the A—C Phase Change in a Chiral Smectic Liquid Crystal. *Phys. Rev. Lett* **1977**, *38*, 848–851. [CrossRef]
10. Garoff, S.; Meyer, R.B. Electroclinic Effect at the A—C Phase Change in a Chiral Smectic Liquid Crystal. *Phys. Rev. A* **1979**, *19*, 338–347. [CrossRef]
11. Bahr, C.; Heppke, G. Optical and Dielectric Investigations on the Electroclinic Effect Exhibited by a Ferroelectric Liquid Crystal with High Spontaneous Polarization. *Liquid Cryst.* **1987**, *2*, 825–831. [CrossRef]
12. Clark, N.A.; Lagerwall, S.T. Submicrosecond bistable electro-optic switching in liquid crystals. *Appl. Phys. Lett.* **1980**, *36*, 899–901. [CrossRef]
13. Bruckner, J.R. *A First Example of a Lyotropic Smectic C* Analog Phase—Design, Properties and Chirality Effects*; Springer: Cham, Switzerland, 2015; ISBN 978-3-319-27202-3.
14. Bogner, A. Maßgeschneiderte smektische Flüssigkristalle vom ‚de Vries'-Typ: Struktur-Eigenschaftsvariationen in nanosegregierenden Organosiloxanen und Organocarbosilanen. Ph.D. Thesis, University of Stuttgart, Stutgart, Germany, 2015.
15. Dumrongrattana, S.; Huang, C.C. Polarization and Tilt-Angle Measurements Near the Smectic-a-Chiral-Smectic-C Transition of P-(N-Decyloxybenzylidene)-P-Amino-(2-Methyl-Butyl)Cinnamate (DOBAMBC). *Phys. Rev. Lett* **1986**, *56*, 464–467. [CrossRef]
16. Bahr, C.; Heppke, G. Ferroelectric Liquid-Crystals: Properties of Binary Mixtures and Pure Compounds with High Spontaneous Polarization. *Mol. Cryst. Liq. Cryst.* **1987**, *148*, 29–43. [CrossRef]
17. Bahr, C.; Heppke, G. Influence of Electric Field on a First-Order Smectic-A–Ferroelectric-Smectic-C Liquid-Crystal Phase Transition: A Field-Induced Critical Point. *Phys. Rev. A* **1990**, *41*, 4335–4342. [CrossRef]
18. Glaser, M.A.; Clark, N.A. Fluctuations and Clinicity in Tilted Smectic Liquid Crystals. *Phys. Rev. E* **2002**, *66*, 021711–021714. [CrossRef] [PubMed]
19. Tschierske, C.; Schroeter, J.A.; Lindner, N.; Sauer, C.; Diele, S. Formation of columnar mesophases by rodlike molecules. *SPIE* **1998**, *3319*, 8–13. [CrossRef]
20. Lindner, N.; Kölbel, M.; Sauer, C.; Diele, S.; Jokiranta, J.; Tschierske, C. Formation of Columnar and Cubic Mesophases by Calamitic Molecules: Novel Amphotropic Biphenyl Derivatives. *J. Phys. Chem. B* **1998**, *102*, 5261–5273. [CrossRef]
21. Neumann, B.; Sauer, C.; Diele, S.; Tschierske, C. Molecular design of amphotropic materials: Influence of oligooxyethylene groups on the mesogenic properties of calamitic liquid crystals. *J. Mater. Chem.* **1996**, *6*, 1087–1098. [CrossRef]
22. Kanie, K.; Sekiguchi, J.; Zeng, X.; Ungar, G.; Atsushi, M. Phospholipids with a stimuli-responsive thermotropic liquid-crystalline moiety. *Chem. Commun.* **2011**, *47*, 6885–6887. [CrossRef]

23. Tan, X.; Zhang, R.; Guo, C.; Cheng, X.; Gao, H.; Liu, F.; Bruckner, J.R.; Giesselmann, F.; Prehm, M.; Tschierske, C. Amphotropic azobenzene derivatives with oligooxyethylene and glycerol based polar groups. *J. Mater. Chem. C* **2015**, *3*, 11202–11211. [CrossRef]

24. Murase, M.; Takanishi, Y.; Nishiyama, I.; Yoshizawa, A.; Yamamoto, J. Hyper swollen perfluorinated smectic liquid crystal by perfluorinated oils. *RSC Adv.* **2015**, *5*, 215–220. [CrossRef]

25. Schafheutle, M.A.; Finkelmann, H. Shapes of Micelles and Molecular Geometry Synthesis and Studies on the Phase Behaviour, Surface Tension and Rheology of Rigid Rod-Like Surfactants in Aqueous Solutions. *Liq. Cryst.* **1988**, *3*, 1369–1386. [CrossRef]

26. Ujiie, S.; Yano, Y. Thermotropic and lyotropic behavior of novel amphiphilic liquid crystals having hydrophilic poly(ethyleneimine) units. *Chem. Commun.* **2000**, 79–80. [CrossRef]

27. Pietschmann, N.; Lunow, A.; Brezesinski, G.; Tschierske, C.; Kuschel, F.; Zaschke, H. The first liquid crystalline diol compound which exhibits nematic, smectic A^+, and smectic C^+ mesophases in the presence of water. *Colloid Polym. Sci.* **1991**, *269*, 636–639. [CrossRef]

28. Bruckner, J.R.; Krueerke, D.; Porada, J.H.; Jagiella, S.; Blunk, D.; Giesselmann, F. The 2D-correlated structures of a lyotropic liquid crystalline diol with a phenylpyrimidine core. *J. Mater. Chem.* **2012**, *22*, 18198–18203. [CrossRef]

29. Bruckner, J.R.; Porada, J.H.; Dietrich, C.F.; Dierking, I.; Giesselmann, F. A Lyotropic Chiral Smectic C Liquid Crystal with Polar Electrooptic Switching. *Angew. Chem. Int. Ed.* **2013**, *52*, 8934–8937. [CrossRef] [PubMed]

30. Clark, N.A.; Rieker, T.P.; Maclennan, J.E. Director and layer structure of SSFLC cells. *Ferroelectrics* **1988**, *85*, 79–97. [CrossRef]

31. Kuczyński, W. Behavior of the helix in some chiral smectic-C* liquid crystals. *Phys. Rev. E* **2010**, *2*, 021708. [CrossRef]

32. Martinot-Lagarde, P. Flexo and ferro-electricity observation of ferroelectrical monodomains in the chiral smectic liquid crystal. *J. Phys. Colloques* **1976**, *37*, 129–132. [CrossRef]

33. Brunet, M.; Isaert, N. Periodic structures in smectic C*—Pitch and unwinding lines. *Ferroelectrics* **1988**, *84*, 25–52. [CrossRef]

34. Osipov, M.A.; Bruckner, J.R.; Giesselmann, F. On the Theory of Helical Twisting in Ferroelectric Liquid Crystals, 2. In Proceedings of the Joint German British Liquid Crystal Conference, Würzburg, Germany, 3–5 April 2017.

35. Gorb, L.; Asensio, A.; Tuñón, I.; Ruiz-López, M.F. The Mechanism of Formamide Hydrolysis in Water from Ab Initio Calculations and Simulations. *Chem. Eur. J.* **2005**, *11*, 6743. [CrossRef]

36. Notley, J.M.; Spiro, M. The Purification of Formamide, and the Rate of its Reaction with Dissolved Water. *J. Chem. Soc. B* **1966**, 362. [CrossRef]

37. Harjung, M.D.; Schubert, C.P.J.; Knecht, F.; Porada, J.H.; Lemieux, R.P.; Giesselmann, F. New amphiphilic materials showing the lyotropic analogue to the thermotropic smectic C* liquid crystal phase. *J. Mater. Chem. C* **2017**, *5*, 7452–7457. [CrossRef]

38. Knecht, F. Neue Einblicke in die Lyotrope Smektische C* Phase: Untersuchungen zur Phasenstabilität und zum Mechanismus Interlamellarer Direktorkorrelation. Ph.D. Thesis, University of Stuttgart, Stutgart, Germany, 2019.

39. Harjung, M.D. Chirale Lyotrop-Lamellare Flüssigkristalle: Phasenverhalten Neuer Chiraler Amphiphile und Nachweis des Elektroklinen Effektes. Ph.D. Thesis, University of Stuttgart, Stutgart, Germany, 2019.

40. Li, L.; Jones, C.D.; Magolan, J.; Lemieux, R.P. Siloxane-terminated phenylpyrimidine liquid crystal hosts. *J. Mater. Chem.* **2007**, *17*, 2313–2318. [CrossRef]

41. Roberts, J.C.; Kapernaum, N.; Giesselmann, F.; Lemieux, R.P. Design of liquid crystals with "de Vries-like" properties: Organosiloxane mesogen with a 5-phenylpyrimidine core. *J. Am. Chem. Soc.* **2008**, *130*, 13842–13843. [CrossRef] [PubMed]

42. McCubbin, J.A.; Tong, X.; Wang, R.; Zhao, Y.; Snieckus, V.; Lemieux, R.P. Directed metalation route to ferroelectric liquid crystals with a chiral fluorenol core: The effect of restricted rotation on polar order. *J. Am. Chem. Soc.* **2004**, *126*, 1161–1167. [CrossRef] [PubMed]

43. Takatoh, K.; Sunohara, K.; Sakamoto, M. Mesophase Transition of Series Materials Containing Fluorene, Fluorenone and Biphenyl Structures with Chiral End Groups. *Mol. Cryst. Liq. Cryst.* **1988**, *164*, 167–178. [CrossRef]

44. Roberts, J.C.; Kapernaum, N.; Song, Q.; Nonnenmacher, D.; Ayub, K.; Giesselmann, F.; Lemieux, R.P. Design of liquid crystals with "de Vries-like" properties: Frustration between SmA- and SmC-promoting elements. *J. Am. Chem. Soc.* **2010**, *132*, 364–370. [CrossRef]

45. Bruckner, J.R.; Knecht, F.; Giesselmann, F. Origin of the Director Tilt in the Lyotropic Smectic C* Analog Phase: Hydration Interactions and Solvent Variations. *ChemPhysChem* **2016**, *17*, 86–92. [CrossRef]

46. Rumble, J.R. (Ed.) *CRC Handbook of Chemistry and Physics*, 99th ed.; CRC Press: Boca Raton, FL, USA, 2018; ISBN 1138561630.

47. Sengwa, R.J.; Kaur, K.; Chaudhary, R. Dielectric properties of low molecular weight poly(ethylene glycol)s. *Polym. Int.* **2000**, *49*, 599–608. [CrossRef]

48. LeNeveu, D.M.; Rand, R.P.; Parsegian, V.A. Measurement of forces between lecithin bilayers. *Nature* **1976**, *259*, 601–603. [CrossRef]

49. Lis, L.J.; McAlister, D.; Fuller, N.; Rand, R.P.; Parsegian, V.A. Interactions between neutral phospholipid bilayer membranes. *Biophys. J.* **1982**, *37*, 657–666.

50. Leikin, S.; Parsegian, V.A.; Rau, D.C.; Rand, R.P. Hydration Forces. *Ann. Rev. Phys. Chem.* **1993**, *44*, 369–395. [CrossRef]

51. McIntosh, T.J. Short-range interactions between lipid bilayers measured by X-ray diffraction. *Curr. Opin. Struct. Biol.* **2000**, *10*, 481–485. [CrossRef]

52. Milhaud, J. New insights into water-phospholipid model membrane interactions. *Biochim. Biophys. Acta* **2004**, *1663*, 19–51. [CrossRef] [PubMed]

53. Berkowitz, M.L.; Vácha, R. Aqueous solutions at the interface with phospholipid bilayers. *Acc. Chem. Res.* **2012**, *45*, 74–82. [CrossRef] [PubMed]

54. Bergenstaahl, B.A.; Stenius, P. Phase diagrams of dioleoylphosphatidylcholine with formamide, methylformamide and dimethylformamide. *J. Phys. Chem.* **1987**, *91*, 5944–5948. [CrossRef]

55. Cheng, J.-X.; Pautot, S.; Weitz, D.A.; Xie, X.S. Ordering of water molecules between phospholipid bilayers visualized by coherent anti-Stokes Raman scattering microscopy. *Proc. Natl. Acad. Sci. USA* **2003**, *100*, 9826–9830. [CrossRef]

56. Tayebi, L.; Ma, Y.; Vashaee, D.; Chen, G.; Sinha, S.K.; Parikh, A.N. Long-range interlayer alignment of intralayer domains in stacked lipid bilayers. *Nat. Mater.* **2012**, *11*, 1074–1080. [CrossRef]

57. Marrink, S.J.; Berkowitz, M.; Berendsen, H. Molecular dynamics simulation of a membrane/water interface: The ordering of water and its relation to the hydration force. *Langmuir* **1993**, *9*, 3122–3131. [CrossRef]

58. Kjellander, R.; Marčelja, S. Perturbation of hydrogen bonding in water near polar surfaces. *Chem. Phys. Lett.* **1985**, 393–396. [CrossRef]

59. Marčelja, S.; Radić, N. Repulsion of interfaces due to boundary water. *Chem. Phys. Lett.* **1976**, *42*, 129–130. [CrossRef]

60. Attard, P.; Batchelor, M.T. A mechanism for the hydration force demonstrated in a model system. *Chem. Phys. Lett.* **1988**, *149*, 206–211. [CrossRef]

61. Tokarev, A.V.; Bondarenko, G.N.; Yampol'skii, Y.P. Chain structure and stiffness of Teflon AF glassy amorphous fluoropolymers. *Polym. Sci. Ser. A* **2007**, *49*, 909–920. [CrossRef]

62. Parra, R.D.; Zeng, X.C. Staggered and eclipsed conformations of C2F6: A systematic ab initio study. *J. Fluor. Chem.* **1997**, *83*, 51–60. [CrossRef]

63. Salerno, K.M.; Bernstein, N. Persistence Length, End-to-End Distance, and Structure of Coarse-Grained Polymers. *J. Chem. Theory Comput.* **2018**, *14*, 2219–2229. [CrossRef] [PubMed]

64. Teoh, M.M.; Chung, T.-S.; Pramoda, K.P. Thin-film polymerization and Metropolis Monte Carlo simulation of thermotropic liquid crystalline poly(ester-amide)s. *Synth. Met.* **2004**, *147*, 191–197. [CrossRef]

65. Koestner, R.; Roiter, Y.; Kozhinova, I.; Minko, S. AFM imaging of adsorbed Nafion polymer on mica and graphite at molecular level. *Langmuir* **2011**, *27*, 10157–10166. [CrossRef] [PubMed]

66. Giesselmann, F.; Zugenmaier, P. Mean-Field Coefficients and the Electroclinic Effect of a Ferroelectric Liquid-Crystal. *Phys. Rev. E* **1995**, *52*, 1762–1772. [CrossRef]

67. Jakli, A.; Harden, J.; Notz, C.; Bailey, C. Piezoelectricity of Phospholipids: A Possible Mechanism for Mechanoreception and Magnetoreception in Biology. *Liquid Cryst.* **2008**, *35*, 395–400. [CrossRef]

68. Harden, J.; Diorio, N.; Petrov, A.G.; Jakli, A. Chirality of Lipids Makes Fluid Lamellar Phases Piezoelectric. *Phys. Rev. E* **2009**, *79*. [CrossRef]
69. Harjung, M.D.; Giesselmann, F. Electroclinic effect in the chiral lamellar α phase of a lyotropic liquid crystal. *Phys. Re. E* **2018**, *97*, 032705. [CrossRef]
70. Clark, N.A.; Lagerwall, S.T. *Applications of Ferroelectric Liquid Crystals in Ferroelectric Liquid Crystals—Principles, Properties and Applications*; Taylor, G.W., Shuvalov, L.A., Eds.; Gordon and Breach Science Publishers: Philadelphia, PA, USA, 1991; ISBN 2-88124-282-0.

Review

From Equilibrium Liquid Crystal Formation and Kinetic Arrest to Photonic Bandgap Films Using Suspensions of Cellulose Nanocrystals

Christina Schütz [1,†], Johanna R. Bruckner [2,†], Camila Honorato-Rios [1], Zornitza Tosheva [1], Manos Anyfantakis [1,*] and Jan P. F. Lagerwall [1,*]

[1] Department of Physics & Materials Science, University of Luxembourg, 162a, Avenue de la faiencerie, Grand Duchy of Luxembourg, 1511 Luxembourg, Luxembourg; christina.schuetz@googlemail.com (C.S.); kmilai@gmail.com (C.H.-R.); zornitza.tosheva@uni.lu (Z.T.)

[2] Institute for Physical Chemistry, University of Stuttgart, Pfaffenwaldring 55, 70569 Stuttgart, Germany; johanna.bruckner@ipc.uni-stuttgart.de

*   Correspondence: emmanouil.anyfantakis@uni.lu (M.A.); jan.lagerwall@lcsoftmatter.com (J.P.F.L.)

†   These authors contributed equally to this work.

Received: 20 February 2020; Accepted: 10 March 2020; Published: 13 March 2020

Abstract: The lyotropic cholesteric liquid crystal phase developed by suspensions of cellulose nanocrystals (CNCs) has come increasingly into focus from numerous directions over the last few years. In part, this is because CNC suspensions are sustainably produced aqueous suspensions of a fully bio-derived nanomaterial with attractive properties. Equally important is the interesting and useful behavior exhibited by solid CNC films, created by drying a cholesteric-forming suspension. However, the pathway along which these films are realized, starting from a CNC suspension that may have low enough concentration to be fully isotropic, is more complex than often appreciated, leading to reproducibility problems and confusion. Addressing a broad audience of physicists, chemists, materials scientists and engineers, this Review focuses primarily on the physics and physical chemistry of CNC suspensions and the process of drying them. The ambition is to explain rather than to repeat, hence we spend more time than usual on the meanings and relevance of the key colloid and liquid crystal science concepts that must be mastered in order to understand the behavior of CNC suspensions, and we present some interesting analyses, arguments and data for the first time. We go through the development of cholesteric nuclei (tactoids) from the isotropic phase and their potential impact on the final dry films; the spontaneous CNC fractionation that takes place in the phase coexistence window; the kinetic arrest that sets in when the CNC mass fraction reaches $\sim$10 wt.%, preserving the cholesteric helical order until the film has dried; the 'coffee-ring effect' active prior to kinetic arrest, often ruining the uniformity in the produced films; and the compression of the helix during the final water evaporation, giving rise to visible structural color in the films.

Keywords: cellulose nanocrystals; cholesteric liquid crystals; colloidal suspensions; kinetic arrest; gelation; glass formation; coffee-ring effect; bragg reflection

---

1. Introduction

While cellulose-based liquid crystals are not novel [1], there is without doubt a strong current trend of growing interest in cellulose nanocrystals (CNCs) and the cholesteric liquid crystal phases formed by suspensions of these particles. The interest can be traced back, first, to the general current focus in society on identifying useful renewable functional materials, preferably from natural resources. CNCs—nanorods of cellulose derived from plants or other biological sources—can easily be dispersed in water, forming a cholesteric liquid crystal phase already at quite a low mass fraction [2], that spontaneously organizes the rods in a helical arrangement. This can be retained during evaporation of the water until a solid cellulose film arises that displays vivid structural colors when observed under ambient white light. This important observation [3] was warmly welcomed, because the potential of an abundantly available bioresource for the development of materials with optical functionality (photonic bandgap) was recognized, suggesting a plethora of applications, from biobased iridescent pigments that might be suitable for food and cosmetic applications [4,5] to templates for inorganic materials with complex internal structure [6], covert encryption [7], sensors of humidity [8,9] or pressure [10] and much more. Beyond the beauty of these films and their potential use in various photonic devices, the helically arranged CNCs are also considered as a basis for high-performance composite materials that show excellent mechanical properties coupled with low weight [11], mimicking the helically modulated structures of nature's top performers, as in crustacean shells or exoskeletons of beetles [12,13] (where the structure is formed by chitin, the 'relative' of cellulose in the animal kingdom). While the intensive research in the field has led to significant progress, it has also shown that the control of the physical properties of dry CNC films or of the suspensions from which they are made is far from trivial [11,14].

This brings us to the second reason for the interest, namely that liquid crystalline CNC suspensions give rise to a number of intriguing questions regarding the physics and chemistry of particle-based lyotropic cholesterics. What governs the helical self-organization and the transfer of chirality from the molecular to the mesoscopic scale? How does the phase separation into coexisting isotropic liquid and orientationally ordered liquid crystal take place in a strongly disperse rod suspension? What kind of transition takes us from an equilibrium colloidal liquid crystal phase to a kinetically arrested state, and is that state a gel or a glass? Thus, CNC suspensions are also a wonderful playground for soft matter researchers from a fundamental science point of view. As CNCs can now be acquired commercially in large volumes with high degree of reproducibility at reasonable price [15], they constitute an ideal choice for broad-scale systematic research in colloidal liquid crystals.

The first reports of CNC extraction, by sulfuric acid hydrolysis from cotton and wood, were published by Nickerson and Habrle in 1947 [16] and Rånby and Ribi in 1950 [17]. A decade later, Marchessault et al. [18] reported on the birefringent properties of aqueous CNC suspensions, connecting them for the first time to liquid crystal formation. However, it was only in 1992 that Gray and co-workers concluded that the equilibrium liquid crystal phase is of cholesteric type and that the CNCs organize in a helically modulated fashion as a result [2]. Six years later his group demonstrated that films dried from CNC suspensions can show striking iridescent colors [3]. With these (and other) seminal works, Gray set a milestone in the investigation and development of CNCs as a novel and sustainable nanomaterial. His group has constantly remained at the forefront of CNC research and the development of our understanding of these intriguing liquid crystals, and is still so today. By inspiring many others to join in, Gray has turned CNCs into one of the most interesting modern liquid crystal formers known, with great potential for the future.

This review breaks down the key scientific questions and challenges in current CNC research thematically, discussing the relevant issues, whether motivated by scientific curiosity or applied goals, connecting them to the many related other fields of colloid and liquid crystal research. The article is organized as follows: We first give readers not acquainted with CNCs a brief introduction to the material and how it is made in Section 2. In Section 3, we provide a refresher of the key concepts of electrostatically stabilized colloids that are needed to understand the phase behavior and non-equilibrium phenomena of CNC suspensions. Readers well acquainted with this topic may prefer to skip this section entirely or in part. In this context we should also point out that we adhere to the IUPAC recommendations for colloid terminology in this paper, referring to the colloidal *suspension*, not the particle (as is sometimes done), as the colloid. In the same vein, we use *disperse* and *non-disperse* rather than the IUPAC-deprecated terms polydisperse (redundant/tautological) and monodisperse (self-contradictory).

After the general colloid basics, we are then ready to go through the required basic concepts of colloidal liquid crystal research, starting in Section 4.1 with the fundamental question of what we actually mean with the chirality-influenced nematic order that is at the heart of the cholesteric phase, as well as which types of structures we can expect. We review the classic Onsager argument and its implications in Section 4.3, explaining the spontaneous appearance of long-range orientational order, and thus the formation of a (chiral) nematic liquid crystal phase, if we have well-dispersed nanorods beyond a threshold concentration that depends on rod aspect ratio. In the next subsection we look at the interesting dynamics when the isotropic and liquid crystalline phases are in coexistence, with the appearance of so-called *tactoids* of cholesteric phase in an isotropic surrounding, allowing us to fractionate a disperse CNC sample according to length, with very significant outcome.

In Section 5, we look at the non-equilibrium behavior that takes over once the CNC mass fraction passes a second threshold that is also comparatively low, typically 11–12 wt.%, and how this actually helps to preserve, within a kinetically arrested state, the order emerging from the cholesteric liquid crystalline self organization [19]. We discuss which parameters influence the onset of kinetic arrest, with particular focus on the role of ions in solution, as well as the characteristics of the CNCs themselves. The impact of the solvent characteristics, considering non-aqueous solvents as well as solvents with different added solutes, is discussed in Section 6.

We approach the applied research that aims at using CNCs in photonic bandgap films with visible selective reflection, produced by drying sessile droplets of CNC suspension on a solid substrate, in Section 7. Because structural color is at the heart of this section, we briefly review in Section 7.1 the reason why structural color may arise in cholesteric liquid crystals, and which conditions must be fulfilled for us to see it. In Section 7.2, we discuss the non-equilibrium part of the droplet drying process, during which the helical pitch is reduced by the mechanical compression of the whole film arising from water evaporation, with the consequent appearance of visible color. Here, we also discuss the strong significance of tactoids, should they remain (as is often the case) once kinetic arrest sets in. We introduce the important phenomenon dubbed coffee-ring effect in Section 7.3, allowing us to explain why CNC films typically end up rather non-uniform in thickness as well as in color. The section finishes with a discussion of means to tune the appearance of the films, including some overlooked parameters that we believe to be important.

While the main focus of our review is on CNC, we will, when instructive, compare the behavior with other nanorod colloids forming (chiral or non-chiral) nematic phases, such as carbon nanotubes, amyloid fibrils, viruses or collagen. Our ambition is not to provide an exhaustive review of all research published in the field, thus many fine papers may be left out, but rather give a comprehensive overview of our current understanding of how CNC suspensions behave, with selected illustrative examples from the literature. In today's fast-paced world with high expectations on researchers to publish rapidly and frequently—probably beyond the level which is healthy, for science as well as for people—we rarely have the opportunity to reflect on the meaning of a certain equation or a certain parameter entering our work. While writing this review, we experienced considerable joy in doing such reflections, and it is our intention with this review to share this joy with the reader. Therefore, rather than merely repeating the standard statements describing a helical director modulation in cholesterics, nematic elasticity, or tactoids nucleating from the isotropic precursor, we allow ourselves to dwell a bit longer than usual on these and other issues that are central to the review topic.

2. What Are Cellulose Nanocrystals and How Are They Made?

Cellulose, one of the world's most abundant biopolymers, is a polysaccharide responsible for the hierarchical structure in all green plants [20]. Since the dawn of mankind, wood and plant fibers (containing mainly cellulose) served as energy sources, as building materials and as clothing. Furthermore, cellulose played a very important role in the intellectual development of humanity in the form of paper. The oldest known archeological fragments of a direct precursor to modern paper were found in China and date back to the 2nd century BC. Cellulose remained an important raw material also in the early days of industrialization; aside from the fact that the cotton at the heart of the booming textile industry is basically pure cellulose, the polysaccharide was also refined into derived polymers with very different uses. Cellophane became a commodity transparent plastic film and Eastman Kodak utilized partially nitrated cellulose together with camphor as a plasticizer to produce the first flexible photographic films. We find the same combination still today, in dissolved form, in ordinary nail polish.

The recently developed ability to isolate the nanosized constituents of cellulose on relatively large scale has opened up new possibilities for the fabrication of novel cellulose-based high performance materials [11,20–23]. The attractive properties of this nanocellulose (NC), such as a high surface area and conveniently reactive hydroxyl groups at the surface, low density, low thermal expansion coefficient as well as excellent mechanical performance, offer an extensive toolbox for novel bioderived materials design. This development, together with the increasing need to replace fossil fuels not only as a source of energy but also as the raw material for the vast majority of organic materials synthesis, has resulted in revitalized research in cellulose chemistry and engineering. Interestingly, the resulting new opportunities for advanced materials development give also the physics of NC critical importance, a fact that we and co-authors highlighted in an earlier review of CNCs [11], now about six years old. In this updated review we maintain the focus on the comparatively short and rigid CNCs, comprising only crystalline cellulose, while we point out that the general term nanocellulose includes also the longer and more flexible cellulose nanofibers (CNFs), which also contain amorphous sections. While they are also fascinating materials, CNFs tend to form gels prior to developing long-range order, hence they are less interesting for liquid crystal science.

Cellulose consists of repeating cellobiose units of two anhydroglucose rings connected via a β-1,4-glycosidic bond that form a linear polysaccharide. The multiple hydroxyl groups on each monomer result in omnipresent hydrogen bonds between the cellulose chains, explaining the high stability and mechanical strength of crystalline cellulose [23] (and thus of CNCs). In the native crystal form, called cellulose I, the chains are oriented parallel to each other. Cellulose I encompasses two polymorphs [24], cellulose Iα with a triclinic and Iβ with a monoclinic structure. Different plants contain both polymorphs in ratios that are specific to the plant species [25]. Cellulose Iα is dominant in most algae [26] and bacteria [27], while Iβ is dominant in higher plant cell wall structures and in tunicate [28]. The basic building block in plants and bacteria typically consists of 20–40 cellulose polymer chains structured into highly ordered crystalline and disordered regions, which form a cellulose nanofiber.

CNCs are the smallest extractable nanoparticles from a raw cellulose source. They are nanorods with a diameter in the order of 5 nm and length from about 100 nm to 1 μm, consisting of pure cellulose in a highly crystalline state. A typical production process is depicted in Figure 1. Regardless of the preparation method, CNCs are disperse in width and, in particular, in length (30% to 50% dispersity is typical). The morphology of the cross section of a CNC is not very well defined and the small size renders it challenging to establish with certainty, but there seems to be significant variations within one and the same sample, in fact even within one and the same rod [29,30]. For practical reasons the shape is often approximated by that of a cylindrical rod but this is certainly a simplification. Further morphological variety is introduced by using different sources [31,32] and the distribution in size also varies considerably depending on the source, yielding CNCs with quite different colloidal properties. The numerous hydroxyl groups are of pivotal importance not only for the strength, but also for the surface chemistry of CNCs, in particular for attaching charged groups (see below).

Methods to isolate NC (CNCs as well as CNFs) from plants and bacteria range from bottom-up methods where bacteria are used to produce glucose cellulose nanofibrils, to top-down methods including chemical/physical and enzymatic methodologies for the isolation of NC from green plants and algae [33]. The isolation is dependent of the source of cellulose but always involves purification as well as homogenization. For instance, if wood is the source, the two other main constituents, lignin and hemicellulose, must be removed. The homogenization and separation can be done by simple mechanical treatment, a pretreatment such as chemical modification [34] or enzymatic hydrolysis reducing the required energy [35]. These processes yield microfibrillated cellulose, which is then further processed with inorganic acids in order to obtain CNCs. Purified cotton can also serve as starting material for the production of CNCs. For both routes, see Figure 1.

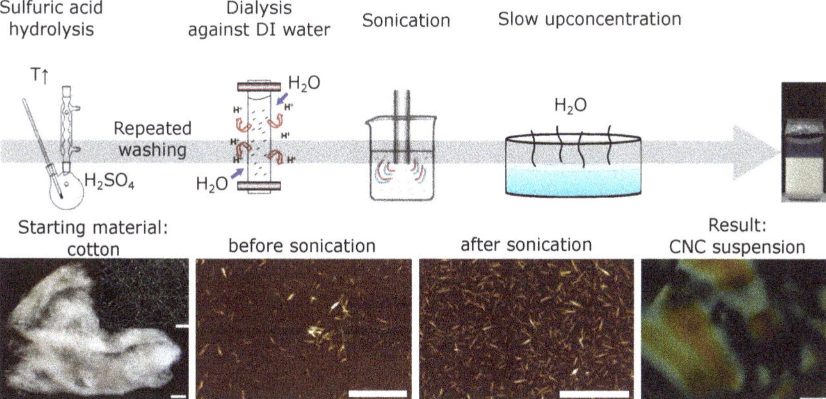

Figure 1. Upper line: simplified schematic of a typical CNC production scheme, here starting from a slurry of sulfuric acid and cotton; CNF or bacterial cellulose can be used instead. The sulfuric acid attacks the weaker regions in the cellulose fibrils and removes them, while at the same time providing the remaining CNCs with charged $-SO_3^-$ groups on their surface. The hydrolysis is terminated by dilution with excess water and the remaining acid is removed by repeated washing and dialysis. The highly diluted product is then ultra-sonicated to realize well dispersed CNCs. In a final step the diluted suspension is often concentrated (sometimes under reduced pressure) to give a maximum mass fraction of CNCs that still forms a stable suspension in water. This stock suspension can then be diluted again to achieve lower mass fractions. Lower line, from left to right: cotton as example of an untreated cellulose source (macroscopic and microscopic, as imaged by polarized light microscopy, scale bars are 1 cm and 500 μm, respectively); AFM images of CNCs before and after sonication (scale bar 1 μm), produced by sulfuric acid hydrolysis, with H^+ counter ions, both samples prepared by depositing a suspension with a mass fraction of $w = 0.002$ wt.% on the substrate used for imaging; polarized optical micrograph of the cholesteric phase formed by a CNC suspension (scale bar 100 μm).

The choice of acid for this hydrolysis is very important as it defines which surface groups are attached, thereby dictating the colloidal stability of the resulting CNCs. The most common choice is to use sulfuric acid, yielding anionic (negatively charged) CNCs in aqueous suspension, with sulfate half ester groups ($-SO_3^-$) attached to the surface, via the cellulose hydroxyl groups, and H^+ counter ions. The surface charge of CNCs makes their dispersion in water easy, giving us an electrostatically stabilized colloid. Because some $-SO_3^-$ groups, even in water, remain associated with their H^+ counter ions, the raw CNCs from sulfuric acid treatment are often subjected to an ion exchange process, in which H^+ is replaced by another counter ion, typically Na^+. For these CNCs we can generally assume that all counter ions are in solution, regardless of the pH of the suspension, making them significantly easier to handle and to analyze. Commercially available CNCs are often sold with Na^+ counter ions.

3. A Brief Reminder of the Factors Governing Electrostatic Stabilization of Colloids: the Debye Screening Length and the Ionic Strength

The classic theory for describing electrostatic stabilization (and destabilization) of colloids is due to Derjaguin and Landau, and, independently, to Vervei and Overbeek. It is therefore today referred to as the DLVO theory. The theory analyzes the interaction energy between suspended particles (Figure 2), taking, first, the Pauli exclusion principle into account, which places an absolute limit to how close the particles can approach. Their electron orbitals cannot occupy the same space, hence an infinite potential barrier appears if the distance between the particle centers would go below $x = 2r$, where r is the characteristic radius of the particle. This repulsive barrier is often referred to as hard-sphere repulsion or Born repulsion, after Max Born who was the first to draw this conclusion from the Pauli principle. Next, when x is just barely greater than $2r$, the behavior is instead dominated by the very strong attraction provided by the van der Waals interactions. Their energies scale as $1/x^6$, hence the attraction gets extremely strong at such close encounters. In fact, if x is allowed to reach the low values when van der Waals attraction dominates, the particles are at their global energy minimum separation and they aggregate irreversibly, rendering the colloid unstable. They can be separated again if enough energy is supplied (for instance mechanical energy in the form of ultrasound), but they will never separate spontaneously.

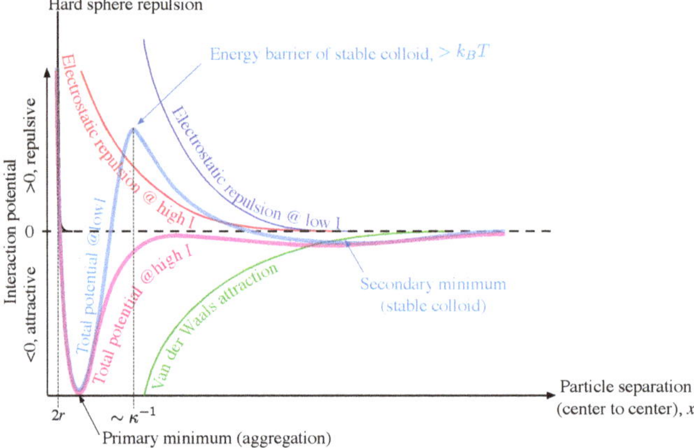

Figure 2. Qualitative sketch of the interaction potential between electrostatically stabilized colloidal particles, as a function of particle separation x, at low (blue) and high (pink) ionic strength I, respectively, according to DLVO theory. Each main contribution (thin curves) to the total energy is sketched in addition to the total interaction potential (thick curves). At low I, an electrostatic energy barrier at a distance $\sim \kappa^{-1}$ (the Debye screening length) from the particle surface, prevents the particles from reaching the primary minimum that corresponds to particle aggregation and loss of colloid stability. The reduction of the electrostatic repulsion range upon increasing I eventually reduces the energy barrier to such an extent that thermal fluctuations will bring the particles to the primary minimum, at a separation distance shorter than κ^{-1}, and the colloid loses its stability.

To keep the colloid stable, such close encounters must thus be prevented, and this is why each particle carries a certain electrostatic surface charge. As the charge has the same sign for every particle, it provides a repulsive potential more potent than Born repulsion, because it acts at a significantly larger distance. However, as can be seen in Figure 2, it can only dominate over the van der Waals attraction if the particle distance is greater than a minimum value, hence an electrostatically stabilized colloid requires that particles are separated by at least this distance to begin with. To this end, we often treat CNC suspensions gently with ultrasound, in order to break up aggregates. The colloid will then become kinetically stabilized, provided that the electrostatic repulsion provides an energy barrier that is substantially greater than the thermal energy $k_B T$, where k_B is the Boltzmann constant and T is the absolute temperature.

Often a shallower secondary energy minimum is found at somewhat larger separation. However, as its depth is usually much lower than $k_B T$, there are considerable fluctuations towards larger distances and the average particle separation is defined more by the particle volume fraction than by the location of this minimum. Only if it is made significantly deeper than $k_B T$ by introducing additional attractive forces, for instance through depletion attraction [36–40] or if the particles can develop hydrogen bond networks with each other, will particles tend to stay in this minimum. This then gives rise to loose particle aggregation called flocculation. Flocculation is reversible, in the sense that the aggregates ('flocs') dissolve spontaneously if the secondary minimum is made shallower again, such that thermal motion can move the particles out of the secondary minimum. Depletion attraction has been explored relatively rarely in the context of CNC suspensions, the only example known to us employing ionic depletants [38–40]. As a result, in this case the addition of depletant also affected the electrostatic forces in the system (see below), thus making it more difficult to draw clear-cut conclusions about the effect of depletion attraction alone. The contribution of hydrogen bonding is also not much explored, but it could become significant in certain situations, since cellulose is a very strong hydrogen bonder due to its great density of hydroxyl groups. When CNCs are dispersed in water, the extremely rich hydrogen bonding capacity of water may reduce the impact, but in non-aqueous solvents with less hydrogen bonding capacity, the hydrogen bonding of the cellulose may play a more prominent role, as will be discussed further in Section 6.

The height of the energy barrier depends not only on the particle surface charge, but also on the relative dielectric permittivity ϵ_r of the solvent, being polarized by the surface charge, and the amount of ions in solution, as ions have the effect of screening the charge, thus reducing the electrostatic repulsion. The more ions in solution, the lower the potential barrier, and eventually it will be on the order of $k_B T$, allowing particles to approach so close that van der Waals attraction dominates. This triggers particle aggregation into the primary energy minimum and loss of colloidal stability. The situation is illustrated in Figure 2 with the blue curve for low ion concentration and a stable colloid, versus the pink curve for high ion concentration and loss of colloidal stability. The ions do not need to be introduced as additives, but they are, first of all, the counter ions of the charged groups on the particles, for CNCs thus H^+ (directly after acid hydrolysis) or Na^+ (after ion exchange).

The counter ions are, of course, attracted to the oppositely charged particle surface, as described by an electrostatic potential ψ, with a maximum value ψ_0 at the particle surface. Some counter ions are fairly strongly localized the particle in what is called a Stern layer. Entropy however favors dissolution of the counter ions and the result, beyond the very thin Stern layer, is an exponential decrease in counter ion concentration with increasing distance from the surface. Because the counter ions screen the surface charge, ψ decreases with distance from the surface, and we define the distance κ^{-1} where the potential has reduced to $\psi = \psi_0/e$ as the *Debye screening length* or the *electrostatic screening length*. Quantitatively, κ^{-1} is typically on the order of 1–10 nm [41]. It is written as an inverse, since κ appears as the *Debye parameter*, multiplied by x, in the exponent of ψ, $\psi(x) = \psi_0 e^{-\kappa x}$. The primary energy minimum is typically at a distance below κ^{-1} from the particle surface, whereas the energy barrier peaks at a distance around κ^{-1} [41]. The location of the secondary minimum (if it exists) depends significantly on the parameters, but it is often on the order of $10\kappa^{-1}$.

We define the combination of the charged particle surface and the locally increased concentration of counter ions near the particle, up to κ^{-1} from the surface, as the *diffuse electric double layer*. The term was originally borrowed from the concept of a charged capacitor, with the attribute 'diffuse' emphasizing the gradual decrease in counter ion concentration as we leave the charged surface. As the Debye length κ^{-1} is the approximate location of the electrostatic repulsion peak in Figure 2, it is often used as an estimate of the range of electrostatic repulsion. One thus sometimes says that repulsion becomes significant if the particles approach each other sufficiently that the electric double layers start overlapping.

Let us now quantify these findings somewhat. First, within the so-called Poisson–Boltzmann approximation, the Debye length can be written as [41]:

$$\kappa^{-1} = \sqrt{\frac{\epsilon_0 \epsilon_r k_B T}{2q^2 I}} \tag{1}$$

where we introduced ϵ_0 for the vacuum permittivity, q for the elementary charge, and I for the *ionic strength* of the solution. The ionic strength, a measure of the free ion content of the solution and the character of these ions, is of key importance for us, since it determines the Debye length and thereby the range of electrostatic repulsion. In many classic derivations of the Poisson–Boltzmann approximation, one finds that the definition applies to 'infinite' distances from any charged surface. In colloids, because we have plenty of charged surfaces dispersed throughout the volume, the practical meaning is that I is defined for points far outside any electric double layer [42], where the distance to all surrounding (identical) charged surfaces is equal such that the ions here experience no effective potential [41]. In the most general case, the ionic strength is defined as:

$$I = \frac{1}{2} \sum_i c_i z_i^2.$$ (2)

where the index i enumerates all ion species in solution, co- as well as counter ions, and c_i and z_i are the concentration and valence of each ion, respectively. Many textbooks illustrate the use of ionic strength for rather trivial situations, such as a simple solution of NaCl in water. In this case, any point in the solution is equivalent to any other, and there is no distinction between co- and counter ions, as the Na^+ and Cl^- are equally dissolved in the water. The situation with colloids is more intricate, and here it is important to remember that I should only count *free* ions, uninfluenced by any charged surfaces. For the case of pure CNC suspensions, without other added salts, this means that native co-ions should not be counted at all, since they are all covalently bound to the CNC surface in the form of $-SO_3^-$ groups. In a first approximation, we can thus take all counter ions as contributing to the ionic strength, obtaining an effective $I_c = \frac{1}{2}c_c z_c^2$ where we set c_c and z_c the concentration and valence, respectively, of the counter ions. As we typically have monovalent counter ions (unless a deliberate attempt to destabilize the CNC suspension has been made) like Na^+ or H^+, we can set $z_c = 1$ and get $I_c = \frac{1}{2}c_c$.

Today, in the general field of colloid and interface science, it is quite common to find an alternative definition of the Debye length, valid under the assumption of only monovalent free ions with a concentration n_i:

$$\kappa^{-1} = \frac{1}{2\sqrt{\pi \lambda_B n_i}}.$$ (3)

Here, λ_B is the distance between two unit point charges at which the electrostatic interaction energy is equal to the thermal energy, dubbed the Bjerrum length:

$$\lambda_B = \frac{q^2}{4\pi\epsilon_0\epsilon_r k_B T}.$$ (4)

If we wish to better compare the two expressions for the screening length, we can insert (4) into (3), obtaining:

$$\kappa^{-1} = \sqrt{\frac{\epsilon_0\epsilon_r k_B T}{q^2 n_i}}$$ (5)

In other words, if $n_i = 2I$, the two expressions are identical. Note that n_i is the concentration of *free* ions, i.e., it does not count the co-ions bound to colloidal particles, thus we cannot use Equation (2) with all ions in the system, but we must rather use $I_c = \frac{1}{2}c_c$. We then see that $n_i = 2I_c = c_c$, i.e., the concentration of counter ions. Within the approximation that all counter ions can be considered free and that there are no free co-ions (the latter aspect no longer holds if we add simple salts like NaCl), the two expressions are thus identical.

For commercial CNC suspensions, we know the average number of $-SO_3^-$ groups per unit mass of CNC, as this information is provided by the manufacturer. Through the requirement of electroneutrality we can thus calculate the *total* concentration of counter ions c_c, which is equal to the total concentration of $-SO_3^-$ groups, often referred to as the sulfur content, c_S. However, the value n_i ($= 2I_c$) should only take the counter ions far outside the electric double layer into account, and this is more difficult to estimate. Because counter ions are adsorbed stronger to the particles the closer they are, trapping a certain fraction strongly in the Stern layer and some less strongly further out in the hydration shell (the shell of water and ions, of a thickness on the order of κ^{-1}, that moves with the particle as it translates in the solvent), n_i (and $2I_c$) will be a bit lower than c_S. Nevertheless, setting $n_i = 2I_c \approx c_S$ is a good first approximation for a CNC suspension without any other electrolytes added, but one should be aware that it overestimates the counter ion content because it neglects the excess counter ions residing near the particles.

Readers interested in learning about these issues in more details, which are far from trivial, are referred to classic treatises on colloids such as that by Israelachvili [41].

4. Liquid Crystal Formation in CNC Suspensions and Other Colloidal Nanorod Suspensions

4.1. Long-Range Orientational, Short-Range Positional Order: The (Chiral) Nematic Phase

Before going into the details of CNC-based liquid crystals, let us first introduce and explain the key liquid crystal concepts needed. The term 'cholesteric' comes from the historical fact that the first chiral nematic liquid crystal (indeed the first liquid crystal recognized as such) was seen in a cholesterol derivative [43]. The defining feature of the nematic as well as of the cholesteric phase is long-range *orientational* order—the rods tend to orient along a common direction—but no long-range *positional* order; the density or, in case of a colloid, particle volume fraction is constant throughout the system. The preferred orientation of the rods is described by the *director*, **n**. It is a signless ($\mathbf{n} = -\mathbf{n}$) pseudovector of unit length that does not change significantly over length scales that are at least 2–3 orders of magnitude greater than the characteristic size of the rods. The long-range orientational order renders many physical properties anisotropic, the director defining the symmetry axis. The most notable effect is probably that the liquid crystal phase is birefringent, i.e., it exhibits two extreme refractive indices, $n_\perp$ for light that is linearly polarized perpendicular to **n** and $n_{||}$ if the polarization is along **n**. Optically, the director **n** is thus equivalent to the optic axis of the nematic. For liquid crystals formed by rods, like CNCs, we usually have positive birefringence, $\Delta n = n_{||} - n_\perp > 0$.

The orientational order in a nematic phase is not perfect [44]. We quantify the orientational order with the order parameter S, where $S = 0$ describes an isotropic state with no long-range orientational order and $S = 1$ would be a hypothetical perfectly ordered state, where every rod points exactly along **n**. While we are not aware of any direct experimental determinations of S in CNC suspensions, values for cholesterics typically range from $S \approx 0.4$–0.9, depending on the type of liquid crystal. Colloidal liquid crystals, to which liquid crystalline CNC suspensions belong, generally have rather high orientational order, $S \approx 0.8$–0.9. Even such a high value corresponds to non-negligible orientational deviations from **n** of individual rods, typically on the order of 10–20° [45]. De France et al. [46] measured the order parameter for cholesteric CNC suspensions with respect to the *helix axis* **m** (to be defined in a moment), finding almost perfect alignment of the rods perpendicular to **m** once the helix had been sufficiently aligned in a magnetic field. Unfortunately, this is not the usual order parameter with respect to the director that we are interested in here, but the very strongly perpendicular correlation with respect to **m** suggests that also S, the orientational order with respect to **n**, should be very high.

The appearance of long-range orientational order at a transition from an isotropic liquid to a nematic phase breaks the continuous rotational symmetry that prevails in the isotropic phase. In contrast, the continuous translational symmetry remains unbroken in the non-chiral nematic phase: the positions of the rods are as random as in an isotropic liquid. When we introduce chirality in cholesterics, we break another very interesting symmetry, namely mirror symmetry. The example giving name to the concept is our hands ('hand' in ancient Greek is χείρ): a left hand is distinctly different from its mirror image, which is the right hand. Chiral molecules that are mirror images of each other are said to be enantiomers. Applied to nematic ordering, as when the molecules or objects making up the chiral nematic phase exhibit chiral interactions between each other (normally the case if their chemical structure is chiral and one enantiomer is present at higher concentration than the other), this has a strong impact. While the positional ordering is unaffected, density and particle concentration remaining constant throughout the phase, the orientational order, and rotational as well as translational symmetries, are affected in subtle ways. These are not trivial to appreciate in full, and they make the cholesteric phase so deeply fascinating. Because cellulose is chiral (more specifically, it is unichiral, i.e., only one enantiomer exists naturally), CNCs are chiral, and the nematic phase formed in CNC suspensions is cholesteric. The typical consequence of chirality added to nematic ordering, readily seen in bulk samples of CNC suspensions, is that the director rotates in space around an axis **m** (mentioned just above) that is perpendicular to **n**, as we move along **m**. This is illustrated in Figure 3. For simplicity, we will in this section define a Cartesian coordinate system with $\hat{z} = \mathbf{m}$, locating the director in the xy-plane. We point out, however, that we will need to define a different, external, Cartesian coordinate system for describing the macroscopic sample geometry when discussing film drying in Section 7.2, where **m** may point in arbitrary directions.

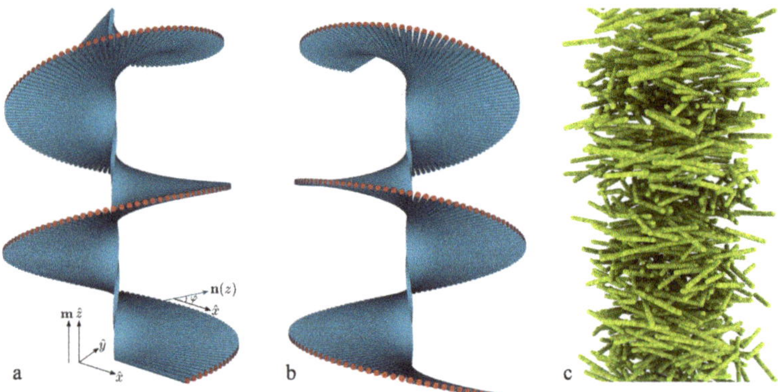

Figure 3. (**a**,**b**) Illustrations of the helix that is traced by the tip (red) of the director **n** (blue) as it continuously changes its orientation within the xy-plane when we move up the helix axis **m**, here considered along the $\hat{z}$-axis, for the case of right-handed (**a**) and left-handed (**b**) helical director modulations. Note that these drawings do not show any CNC rods, only the variation through space of **n**. A representative 'snapshot' of rod distributions, as influenced by the helical director modulation and the imperfect orientational order ($S \approx 0.8$) is instead shown in part (**c**). The corresponding director helix is left-handed, as in (**b**) and as in CNC suspensions.

The rotation angle per unit distance along **m** is constant, i.e., $\frac{d\varphi}{dz}$ has a fixed value, where we define φ as the angle between the director at any point and the $\hat{x}$ axis, see Figure 3a. The steady director rotation creates a periodic structure, since a 2π rotation around $\hat{z}$ of **n** gives us back the original director orientation. We call the period,

$$p = 2\pi / \left| \frac{d\varphi}{dz} \right|, \tag{6}$$

the pitch of the structure (the absolute value of the twist is necessary, since the distance p cannot be negative, but the twist can be right- or left-handed, with opposite signs of $\frac{d\varphi}{dz}$). If we draw the director for several consecutive points along $\hat{z}$ (thus along **m**), always setting its starting point at $x = y = 0$ for every z-value, we see that the tip of the director traces out a helix, see Figure 3a–b where the tip is highlighted in red. Indeed, we say that the director field in a cholesteric phase is helically modulated, a terminology we will come back to in a moment. For CNC suspensions, the helix is always left-handed, as confirmed empirically (the single handedness is due to the unichirality of the cellulose within the CNCs). Note that—just like the director—the helix axis is sign invariant, i.e., **m**=−**m**. A left-handed helix stays left-handed no matter along which direction you look at it.

With the helix axis $\mathbf{m}$ along $\hat{z}$, the pitch designated p, and the origin defined such that $\mathbf{n} = \hat{x}$ at $z = 0$ (equivalent to $\varphi(z = 0) = 0$), we can parametrize the left-handed helical director field of a cholesteric CNC suspension using:

$$\mathbf{n} = \left(\cos 2\pi\frac{z}{p}, \ -\sin 2\pi\frac{z}{p}, \ 0\right) \tag{7}$$

This parametrization (more commonly done for a right-handed helix, in which case all terms appear with positive sign), and the corresponding analysis of the cholesteric structure and the consequent optical properties (see Section 7.1), was first presented by Carl Wilhelm Oseen in 1928 [47]. The significance in the difference between Equation (7) and the standard equation for a left-handed helix (with $\mathbf{r}$ the space coordinate, t the independent variable, a the radius and $2\pi b = p$):

$$\mathbf{r} = (a\cos t, \ -a\sin t, \ bt) \tag{8}$$

is rarely noticed: the z-component is not zero in Equation (8) in contrast to Equation (7). The fact that $n_z = 0$ for the cholesteric director field is actually quite important. It means that $\mathbf{n}$ is everywhere perfectly perpendicular to the helix axis, $\mathbf{n} \perp \mathbf{m}$. This renders the common analogy of a screw or spiral for illustrating cholesterics (we have frequently used this analogy ourselves) potentially misleading. In contrast to the steel wire in a spiral spring or the winding of a screw, which are both *inclined* with respect to the symmetry axis and thus truly follow a helix as described by Equation (8), the cholesteric director has no component along the helix; $n_z = 0$. It is really just the *tip* of the director in a construction as in Figure 3a,b that traces out an actual helix. If you travel along $\mathbf{n}$ in a helically modulated cholesteric phase you will never move up or down the helix. You must travel *perpendicular* to $\mathbf{n}$ to do so, just like you must move perpendicular to the steps of a spiral staircase in order to climb it.

In this respect, the arrangement of steps in a spiral staircase or of the base pairs in the DNA double helix, may be a better analogy to the cholesteric helix. However, also this analogy must be adopted with care in order to avoid misunderstandings. A quite common misconception is that a cholesteric phase would be discretely layered, in the sense that the molecules or rods would be localized to layers, with a well-defined thickness and inter-layer distance. Such a discrete structure with positional order along the helix axis is present in a spiral staircase and between the base pairs in the DNA double helix, but it is not the case for cholesteric liquid crystals. The critical issue is to remember that the steps in the staircase analogy correspond to the *director*, not to rods or molecules. Furthemore, as the director has no extension in space, and the director field is continuous, the "staircase step" would be infinitely thin and there would be infinitely many "steps" in the staircase.

Rather than layers, it is better to compare the spiral staircase steps to *planes*, perpendicular to $\mathbf{m}$. A plane is by its mathematical definition two-dimensional, i.e., it has no thickness, just like the planes of constant $\mathbf{n}$ that are perpendicular to $\mathbf{m}$ in a cholesteric phase. Each plane is an imaginary 2D slice of the director field. There is an infinite number of planes of this type and we are free to draw them wherever we wish along $\mathbf{m}$, without any need for a periodicity. The periodicity is to be found only in the distance between two successive planes that have *identical* $\mathbf{n}$. This bring us to one of the most subtle aspects of cholesteric order: while the phase lacks order in the rod positions, just like in the nematic phase, the continuous *translational symmetry* is broken, in contrast to the situation in the nematic phase. A symmetry operation must bring *all* aspects of the phase back to the starting point, including orientational aspects, but a translation along $\mathbf{m}$ by a distance that is not equal to Np, where N is an integer, yields a change in orientation of $\mathbf{n}$. It is therefore not a symmetry operation.

However, if we combine the translation Δz along $\mathbf{m}$ with a rotation $\Delta z\frac{d\varphi}{dz}$, this is a symmetry operation. Thus, the continuous translational symmetry of a non-chiral nematic phase is replaced by a continuous symmetry in *coupled* translation and rotation in the cholesteric phase. This has the remarkable consequence that the periodicity of a cholesteric phase, being defined simply by the twisting strength, $\frac{d\varphi}{dz}$, is present at any sample size, even below p. The optical consequences of this have even been demonstrated experimentally, with selective reflection being seen (albeit weakly) from cholesteric samples thinner than a single pitch length [48]. This renders cholesterics quite different from discontinuous periodic structures, like truly layered arrangements, which must have an extension of at least twice the period for the periodicity to be detectable. The idea of finite-sized layers in a cholesteric thus bereaves the phase of one of its most remarkable characteristics, while also leading our thoughts astray.

There are now many examples of beautiful scanning electron microscopy (SEM) images of surfaces of fractured films that were produced by drying cholesteric CNC suspensions, which at low resolution might give the impression that there would in fact be a layered organization of the rod-like particles themselves. While many reports indeed refer to such images as evidence of a layered structure in CNC cholesterics, the explanation provided by Majoinen et al. [49] for their exquisite high-resolution images, reproduced in Figure 4, clearly shows that no layers are needed. When assessing these images it is important to remember that we are no longer looking at a liquid crystal phase, but a solid that has been formed by compressing a kinetically arrested cholesteric structure (see Section 7.2), bringing the CNC rods into direct contact. This causes some clustering and probably increases the degree of order. Such close contacts and clustering are certainly not the case in the liquid crystalline state, in which electrostatic repulsion keeps the rods individualized and thermal fluctuations keep them in constant motion.

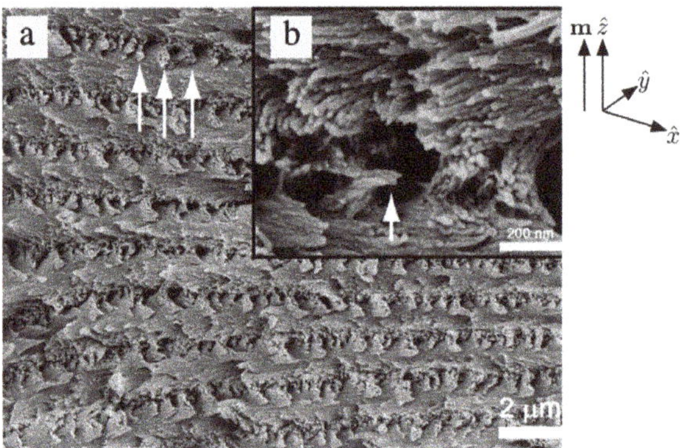

Figure 4. (**a**) Scanning Electron Microscopy image with exceptional detail of the surface obtained by fracturing a film created by drying a cholesteric CNC suspension, the axis **m** of the helical director modulation being nearly in the film plane and vertical in the image. White arrows highlight regions where CNC rods protrude from the fracture surface. The inset (**b**) shows individual CNCs near such a region, where **n** is perpendicular to the fracture plane. Next to the protruding CNCs are holes, left by the removal of the corresponding rods that ended up stuck in the opposite fracture section. Note that rods are packed an order of magnitude closer in this solid film than in an equilibrium cholesteric liquid crystal phase, and their degree of order is probably greater as well. Reprinted by permission from Springer Nature Cellulose [49], Copyright 2012.

Once the fracture is done, the orientation of the main plane of the fracture (which in the best case contains the helix axis, **m**) selects two orientations of **n** that represent opposite extremes in terms of mechanical properties. Let us assume that the helix is indeed in the fracture plane and let us for simplicity retain $\mathbf{m} = \hat{z}$ and choose $\hat{x}$ to be in the same plane, i.e., the fracture plane is an xz plane, with $\hat{y}$ its normal. Let us make the definition complete by setting $y = 0$ in the fracture plane. Wherever **n** is along $\hat{x}$, i.e., it is in the fracture plane, only the comparatively weak van der Waals forces and inter-CNC hydrogen bonds resist against the fracture. Even though hydrogen bonds should not be considered weak, it is still reasonable to expect the film to break quite cleanly along $y = 0$ at these points. In contrast, when **n** is along $\hat{y}$, perpendicular to the fracture plane, covalent bonds within the crystalline CNCs would have to be broken to obtain a clean cut at $y = 0$. Precisely because the cholesteric phase does *not* have any positional order of the rods, the majority of CNCs in regions with $\mathbf{n} \approx \hat{y}$ will be crossing right over the fracture plane, beginning at $y < 0$ and ending at $y > 0$, no matter where we choose our $y = 0$ fracture plane. Because the individual CNCs are unlikely to break, they will instead stick out from the fracture plane, see Figure 4, remaining partially embedded in either of the two surfaces created during the fracturing process (one with $y < 0$ and one with $y > 0$), at all locations where $\mathbf{n} \approx \hat{y}$. Because this happens twice per helix period (another manifestation of $\mathbf{n}=-\mathbf{n}$), the SEM image shows CNCs sticking out of the fracture plane twice per helix period. It also shows holes next to the protrusions, left when the CNCs that ended up sticking out from the other fracture section were removed. The resulting periodicity of clean versus rugged cut may at low resolution give an impression of a layered structure. The high-resolution images presented by Majoinen et al. (Figure 4) show very well that this is not the case; there is no positional order for the individual rods. Such images are very instructive, but unusual.

Another significant aspect that is not always acknowledged is that the helical pitch in an equilibrium cholesteric has nothing to do with the average particle separation distance of CNCs (or other colloidal nanorods) in the phase. As the (chiral) nematic phase has no long-range positional order, the CNCs are uniformly distributed throughout the phase and the average particle–particle distance is set by the volume fraction of the particles. If the helix of the cholesteric phase shortens at constant CNC volume fraction, only the director field rotation $\frac{d\varphi}{dz}$ gets more rapid; there is no closer approach of the CNCs. Two CNC suspensions with the same volume fraction of particles but different values of p will thus have identical average distance between the nanorods. If the ionic strength I is increased, allowing closer encounter of the rods, this may have an impact on the helix pitch, but the average particle separation is not reduced; assuming that the colloid is still stable and no aggregation or flocculation takes place, it cannot be reduced since otherwise the rods would leave unoccupied space in the system. However, the higher I, and consequent reduction in κ^{-1}, allows closer encounters as a result of fluctuations, which can be expected to strengthen the chiral interactions between rods and thereby increase the twisting $\frac{d\varphi}{dz}$. This would explain a reduction in helix pitch upon increased I without any change in the average rod separation distance.

We hope it is now clear that the cholesteric helical modulation of the director field is something quite unique. It is scientifically very interesting, but also rich in potential misinterpretation pitfalls. Let us end this section with one final example. It is a quite common misconception that a helix would be *required* for the phase to be chiral. The helix is an expression of chirality of the phase, but it is not a prerequisite. Whether a helix forms or not, the cholesteric phase remains chiral as long as the chiral interactions are active between the rods (or molecules) building up the phase. Fundamentally, the fact that the building blocks lack mirror symmetry, and that one enantiomer is present at higher concentration than the opposite, renders the phase chiral. With no impact on the chirality of the phase, a cholesteric helix can unwind to the extent that p reaches infinity. In rare occurrences, this can happen spontaneously at a certain temperature, composition or pressure of a cholesteric phase that under other circumstances forms a director helix with finite pitch. More commonly, the phenomenon can arise as a result of strong confinement or application of an external field. Regardless of its origin, such helix unwinding should not be described as "dechiralization", "cholesteric→nematic transition" or similar. A system with constant composition cannot undergo a transition between nematic and cholesteric.

For the majority of cholesterics (including all cholesteric CNC suspensions studied experimentally), which do not exhibit a twist inversion, the existence of an equilibrium, unconstrained helix pitch, p_0, could be taken as the decisive parameter regarding whether we should refer to the phase as nematic or cholesteric. If we can define a $p_0 < \infty$, then the phase is cholesteric. Thus even if we prevent helix formation by confining the phase in a small droplet or applying an external field, imposing an *effective* pitch $p \to \infty$, the fact that the phase still exhibits a well-defined natural pitch $p_0 < \infty$ makes a very clear distinction from a non-chiral nematic phase. It does not matter if p_0 enters only in the free energy calculations or if it is allowed to express itself as a helix that we can detect experimentally; the phase remains cholesteric. We will see in Section 4.4 that this is not just a matter of semantics, but there is a real impact on the physics.

A significant fact in this context is that the equilibrium pitch p_0 of CNC-based cholesterics is always much larger than the scale of a rod forming the liquid crystal phase. The pitch typically ranges from about 3 to 100 µm, thus three to five orders of magnitude greater than a CNC diameter. What this means is that chirality effects are secondary compared to the local nematic ordering of the nanorods, hence need not be considered when elucidating the origin of this ordering, as described in Section 4.3. In this review we will preferentially use the term "nematic" when we discuss local phenomena, such as the steric interactions at the heart of liquid crystal formation, that are not related to chirality or the helical modulation. To emphasize that an issue applies equally to non-chiral and chiral nematics, we sometimes write "(chiral) nematic".

4.2. Nematic Elasticity

The long-range orientational order has the remarkable consequence that a (chiral) nematic phase exhibits an elastic response, in distinct contrast to regular isotropic liquids. This manifests itself, not as an elastic force arising from a distortion of the volume of material, as in solid state elasticity, but instead in elastic torques responding to deformations in the director field, acting to bring it to a uniform orientation of **n**. Only if a volumetric distortion has an impact on the director field will it cause an elastic response. Conversely, if confinement within a certain volume leads to deformation of the director field, the elastic response of the liquid crystal can—if the interfacial tension is sufficiently low—be strong enough to distort the volume even if no external force acts to deform it, as will be discussed at length in Section 4.4. The development of our understanding of nematic elasticity [47] can be traced back to early work by Oseen and Zocher, with later refinement by Frank (who coined the modern terminology) and by Saupe and Ericksen [50].

We analyze the nematic elasticity in terms of the increase in free energy density that a deformation gives rise to, using a set of elastic constants with the dimension of force (N). Each elastic constant quantifies the cost of a particular elementary deformation of the director field. The three elementary bulk deformations splay, twist and bend, which are the ones we need to consider in this review, are illustrated in Figure 5. The modern formulation of the elastic energy density (per unit volume) due to deformations of the (chiral) nematic director field can be written, for the case of a cholesteric phase:

$$f_e = \frac{1}{2}K_1 \left(\nabla \cdot \mathbf{n}\right)^2 + \frac{1}{2}K_2 \left[\mathbf{n} \cdot \left(\nabla \times \mathbf{n}\right) - q_0\right]^2 + \frac{1}{2}K_3 \left[\mathbf{n} \times \left(\nabla \times \mathbf{n}\right)\right]^2 - \frac{1}{2}K_{24}\nabla \cdot \left(\mathbf{n}\nabla \cdot \mathbf{n} + \mathbf{n} \times \nabla \times \mathbf{n}\right) \tag{9}$$

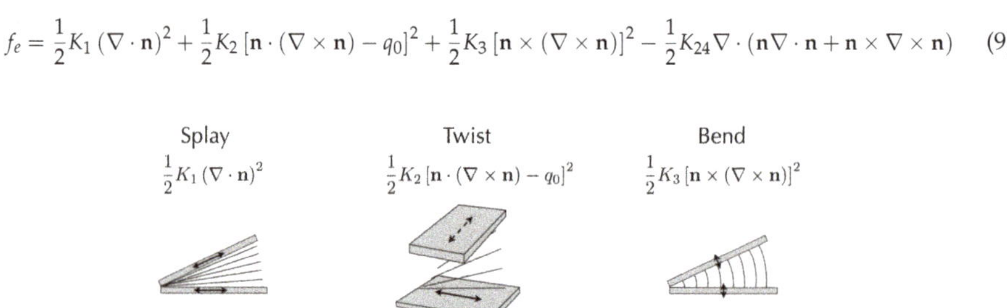

Figure 5. Illustrations of the three elemental director field deformations splay, twist and bend.

While this expression may look daunting to some, it is actually quite straightforward to interpret, given the right explanations. The energy density is a sum of four terms, each a product of an elastic constant—K_1 for splay, K_2 for twist, K_3 for bend and K_{24} for saddle-splay—and the derivative of the director field that describes the corresponding deformation. For the second term, describing twist, the derivative is compared to the natural twisting of cholesterics; we will come back to this modification in a moment. The elastic constants, with a magnitude on the order of pN, tell us how much a certain deformation 'costs' in terms of increased free energy density of the (chiral) nematic. The derivatives, in turn, tell us how strong the deformation of each type actually is in a particular director field configuration. The first three terms, corresponding to splay, twist and bend, are all positive, since K_1, K_2 and K_3 are all greater than zero and as each director field derivative is squared. This means that any such deformation increases the energy density. The fourth term, describing so-called saddle-splay deformation (the director field has negative Gaussian curvature, as in a horse saddle), can be negative or positive, hence this term can actually promote spontaneous distortion of a director field, as has been observed in lyotropic liquid crystals in strong curved confinement [51–53]. However, for this review we will ignore the fourth term, as its influence is negligible in bulk nematics or cholesterics, and in the situations of confinement that we will consider here it has been proposed that its effect can be represented by a renormalization of the splay elastic contribution [54].

An unusually large value of an elastic constant tells us that the corresponding deformation is unlikely to occur with significant strength, as it greatly increases the free energy. Conversely, if an elastic constant is unusually low, the deformation costs almost no energy, and we may expect thermal fluctuations to induce significant deformations of this type. If it is plausible to assume that a certain deformation will not occur (for instance due to an excessively large elastic constant or due to confinement that prevents the deformation) we may drop the corresponding term from Equation (9), as we can then assume that the derivative is zero. Note that the bulk deformation elastic constants, K_1, K_2 and K_3, cannot be zero, or approximated as such, as this would correspond to the unphysical situation that the deformation costs no energy, thereby allowing infinitely strong deformations.

The second term deserves special attention for this review. It describes twist deformation, a term that is straightforward to understand with respect to the uniform director field of a non-chiral nematic, where any twist amounts to a deformation. However, for a cholesteric phase, a twisted director field is the natural state, the ground state, because the helical modulation that develops spontaneously is equivalent to constant twist. Thus, a twist *deformation* in a cholesteric phase is not a twist per se, but any *deviation* from the natural twist $\left(\frac{d\varphi}{dz}\right)_0$ (again assuming $\mathbf{m} = \hat{z}$), corresponding to the equilibrium helix pitch p_0, or its corresponding wave vector magnitude,

$$q_0 = \pm 2\pi/p_0, \tag{10}$$

which enters the second term in Equation (9). This has the interesting consequence that a uniform director field, from the point of view of a cholesteric, exhibits a twist deformation! The director field has, in this case, been deformed by *untwisting* from its natural, twisted, state, which comes at a cost that is described by the second term in Equation (9). This is why it is so important not to refer to a cholesteric phase that has been prevented from forming a helix as if it were no longer cholesteric. A non-chiral nematic is in its ground state without a helix ($q_0 = 0$), but a cholesteric in which the helix formation is suppressed is in a frustrated high-energy state ($q_0 \neq 0$). A non-helical cholesteric and a non-chiral nematic may *appear* similar, but in terms of their free energies they are very different. We may note by comparison with Equation (6) that, numerically (also including the sign), $q_0 = \left(\frac{d\varphi}{dz}\right)_0$.

Note that, in contrast to the pitch p (and its equilibrium value p_0), the helix wave vector can be positive or negative-signed (hence the $\pm$ sign in Equation (10)), the first corresponding to right-handed and the second corresponding to left-handed twist. Thus also the choice of handedness given by chirality is taken care of by Equation (9): a twist with pitch p_0 but with the wrong handedness, giving the derivative in the second term the value $-q_0$, raises the energy by $2K_2 q_0^2$. The twist deformation term of Equation (9) can of course describe a non-chiral nematic just as well: we achieve this by setting $q_0 = 0$.

Equation (9) gives us the elastic energy *density*, which means that we must integrate over the volume of our (chiral) nematic in order to obtain the actual elastic contribution to the free energy of the phase. Since each derivative is a second order spatial derivative, each bulk term in Equation (9) is proportional to $1/r^2$, where r is a characteristic length describing the deformation. For twist, the characteristic length is the pitch of the twisting and is thus independent of location, whereas for splay and bend, the characteristic length is the distance from the 'source' of splay or the 'center' of bending. If splay or bend is induced by confinement of the liquid crystal in a small volume, the characteristic length of splay or bend will be on the order of the volume radius. Therefore, when we calculate the elastic energy contribution by integrating the elastic energy density over the volume, effectively multiplying by r^3, we see that the elastic free energies of splay and bend scale as r. This is an important observation for the analyses to follow in Section 4.4. However, let us first look into a more fundamental question.

4.3. The Origin of Nematic Phase Formation According to Onsager

What is the reason for the long-range orientational order in nematic nanorod suspensions? This question was first answered by Lars Onsager in the late 1940s [55], in a landmark theoretical paper that was inspired by experimental observations of nematic ordering in suspensions of tobacco mosaic virus (TMV), rod-like viruses with dimensions not too different from those of CNCs. The outcome of Onsager's analysis is that the nematic ordering is of entropic origin: the free energy is reduced by developing long-range orientational order, because the loss in orientational entropy is more than compensated by the gain in translational entropy. By aligning along a common direction, the rods are free to translate much longer distances without being blocked by each other, compared to the situation of randomly oriented rods, which will cross each other's paths much more frequently. The argument can also be reformulated in terms of excluded volume, the space not accessible for a rod due to the presence of other rods, restricting the freedom of motion and thus reducing the entropy: by lining up along a common direction, the overall excluded volume is minimized, raising the translational freedom and thus increasing the entropy, thereby reducing the free energy.

However, the gain in entropy from ordering happens only if the rods are sufficiently anisometric (anisotropic in shape) for the reduction in excluded volume per rod to be significant, and only if the volume fraction of rods is high enough for this to have any effect on the system as a whole. Intuitively, the higher the aspect ratio, the lower the required volume fraction for ordering to take place. Onsager quantified this, finding that the effective rod volume fraction ϕ, upon increasing from zero, at which the isotropic phase becomes unstable is:

$$\phi_0 = 3.3 d^e / L^e \tag{11}$$

where d^e is the effective rod diameter and L^e is its effective length (we will come back to the meaning of 'effective' in a moment). If $\phi > \phi_0$ in an Onsager suspension of nanorods, the nematic phase starts to nucleate. The phase transition is of first order and there is a significant coexistence regime, requiring a distinct difference between the two phases. For the non-disperse system that Onsager considered, where all rods are identical, the difference lies not only in the degree of order in each phase, but also in the volume fraction of rods. Where the nematic phase nucleates, it does so with a greater effective rod volume fraction:

$$\phi_1 = 4.5 d^e / L^e \tag{12}$$

The increased rod volume fraction can be easily rationalized by considering the greater packing efficiency provided by the long-range orientational order. The value ϕ_1 can be considered the stability limit of the nematic phase, upon decreasing from a volume fraction $\phi > \phi_1$.

For a *global* rod volume fraction ϕ, where $\phi_0 < \phi < \phi_1$, we have an isotropic phase with local $\phi^{\text{iso}} = \phi_0$ coexisting with a nematic phase with local $\phi^{\text{nem}} = \phi_1$. The relative volume fraction of nematic phase increases from zero to one as the global ϕ increases from ϕ_0 to ϕ_1. The higher rod volume fraction of the nematic phase, together with the fact that the rod material is almost never density matched with the solvent, means that a sample in the phase coexistence region, with $\phi_0 < \phi < \phi_1$, will separate macroscopically in a gravitational field, normally with the denser nematic phase sinking to the bottom and the less dense isotropic phase floating to the top. Should the rod material be less dense than the solvent, the opposite behavior would result, but we are not aware of any such system. This gravity-driven macroscopic separation of nematic from isotropic phase is the standard way of mapping out the coexistence region of the phase diagram: because the nematic phase is birefringent, due to the long-range orientational order, it appears bright between crossed polarizers, in contrast to the isotropic phase, which appears dark. An example is shown in Figure 6. When doing this type of experiment, it is imperative to wait long enough for gravity to do its work properly, in order to get correct results from such an investigation. The time required can be considerable, on the order of months. Indeed, beyond a certain CNC mass fraction, it will be infinite, as discussed in Section 5.

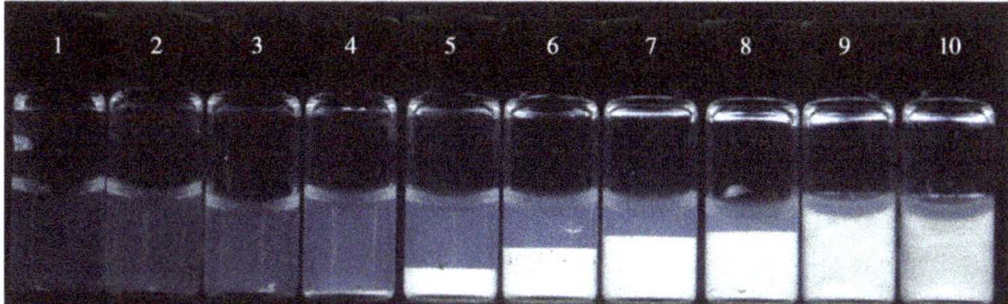

Figure 6. A series of CNC suspensions with increasing mass fraction w from left to right (the values, in wt.%, are indicated at the top of each vial), viewed between crossed polarizers after Earth gravity has separated the isotropic (dark, top) and cholesteric (bright, bottom) phases. Prior to taking the photo, all samples had been left undisturbed for 6 months. By quantifying the volume fraction of cholesteric to isotropic phase at each w, the coexistence regime of the phase diagram is mapped out.

An important complication in practical research compared to Onsager's theory is that the effective rod volume fraction ϕ is typically not experimentally accessible. Moreover, the ratio between mass and volume fraction of rods generally changes with the amount of rods. The reason is that the relevant volume is not the volume of each *physical* rod, approximately described by the diameter d and length L, but the *total* excluded volume, amounting to the volume of the actual rod plus the repulsion zone around it, approximately quantified by κ^{-1} (see Section 3). The effective volume of a rod is thus given by $\pi(d/2 + \kappa^{-1})^2(L + 2\kappa^{-1}) \approx \pi(d/2 + \kappa^{-1})^2 L$. In other words, the effective rod diameter d^e is significantly affected by the electric double layer, as κ^{-1} is on the order of the CNC diameter, whereas the effective length L^e can be approximated with the physical length L, since L is at least two orders of magnitude greater than κ^{-1}. A further complication is that CNCs are not perfect rods, they are highly disperse with strong variations in L and d, and d can vary even within one and the same CNC.

Because κ^{-1} decreases with increasing ionic strength I, and because each suspended charged rod contributes to I as its counter ions are dissociated and dissolved in the water, κ^{-1} decreases with increasing rod content. In other words, the effective diameter, and thereby the effective volume, of a rod decreases as more rods are added to the suspension. This means that the ratio between mass and volume fraction decreases with increasing CNC mass fraction. At the same time, decreasing κ^{-1} increases the effective rod aspect ratio, thus decreasing ϕ_1. Therefore, while Onsager predicts a linearly increasing volume fraction v_N of nematic phase as the rod volume fraction increases from ϕ_0 to ϕ_1 according to:

$$v_N = \frac{\phi - \phi_0}{\phi_1 - \phi_0} \tag{13}$$

the increase in v_N as a function of mass fraction w (with a w_0 corresponding to ϕ_0 and a w_1 to ϕ_1) is more complex. This is because the ratio w_0/ϕ_0 is different from w_1/ϕ_1, because I increases when w is raised from w_0 to w_1, affecting both ϕ and ϕ_1.

In an elegant demonstration of the significance of I for the phase diagram of colloidal liquid crystals [56], Dong et al. prepared a series of CNC suspensions with H^+ counter ions to which varying amounts of HCl had been added. The amount was selected such that all samples had identical ionic strength, i.e., the less CNC in suspension, the more acid was added. When plotting the phase diagram of this constant-I suspension series, a linear behavior as in Equation (13) was seen also versus mass fraction. In a follow-up study [57], Dong and Gray carried out a systematic exchange of the counter ion associated to the CNC surface, finding significant impact. We will not discuss the results with organic counter ions, which were amphiphilic and thus a rather special case, but it is interesting to look at the study of monovalent inorganic counter ions. The authors found that the onset of liquid crystal formation increases with size of the counter ion, such that $w_0^{H^+} < w_0^{Na^+} < w_0^{K^+} < w_0^{Cs^+}$. Interestingly, the plot of cholesteric phase fraction versus CNC mass fraction for different counter ions suggests that the linearity also increases with size of counter ion, to the extent that the mass fraction when the full sample became liquid crystalline was almost independent of the counter ion type. Dong and Gray also studied the effect of counter ions with higher valence than 1, e.g., Ca^{2+}. As one might expect, such ions rapidly destabilize CNC suspensions, an effect that may be attributed to the stronger screening effect (Equation (1)) and specific adsorption onto the CNC surface. Another important factor is that divalent counter ions act as 'crosslinking agents', a single counter ion binding together two CNCs. This effect was studied in the formation of hydrogels [58].

Abitbol et al. [32] wished to investigate the impact of surface charge, preparing CNCs with approximately the same aspect ratio, varying between 18 and 20, but the surface charge was altered between 0.27–0.89% sulfur (the exact meaning of the percentage was not specified). The raw material was the same, and only the acid hydrolysis conditions were varied. All suspensions formed a cholesteric phase, but the onset w_0 shifted to higher concentration with increasing surface charge. This outcome might seem surprising considering that stronger surface charge would be expected to give stronger repulsion. However, it also means a greater release of counter ions with every CNC rod. Note that the Debye length according to Equations (1)–(5) is dependent only on I, not on the surface charge. Thus, as the authors concluded, the outcome is due to the increase in I and the consequent reduction in κ^{-1} that comes with increased surface charge.

Several extensions and improvements of Onsager's original theory have been developed over the years, coming to slightly different quantitative predictions but retaining the qualitative conclusions. Most significantly, various more recent treatments [59–64] have taken the effect of dispersity into account, a parameter that is very much a characteristic of CNCs, and which will play a significant role in Section 4.5. We should also emphasize that the Onsager theory was developed for the liquid crystal class called *lyotropic*, to which colloidal nematics and cholesterics belong [65]. The liquid crystals found in many modern display devices belong to the class of *thermotropic* liquid crystals. The hallmark of lyotropics is that the liquid crystalline order develops for particles or molecule aggregates suspended in an isotropic solvent, such as water in most CNC suspensions. In contrast, thermotropic liquid crystals have no solvent; their molecules form the liquid crystal phase themselves. The entity forming the liquid crystal phase, whether the particles of colloidal liquid crystals or the molecules of thermotropics, are often referred to as *mesogens*. They differ quite strongly between lyotropics and thermotropics, a typical thermotropic molecule having a length on the order of 3 nm and width about 0.5 nm, thus very much smaller and with much smaller aspect ratio than lyotropic mesogens such as CNCs. The Onsager theory would fail to explain liquid crystalline order in thermotropics, for which the corresponding theory was developed by Maier and Saupe [66]. We will not discuss thermotropic liquid crystals in this review, apart from occasional comparisons with CNC suspensions and other colloidal lyotropics when this is of interest.

4.4. Isotropic–Nematic Phase Coexistence and the Nucleation of Tactoids

If the global rod volume fraction of an initially isotropic CNC suspension is increased to $\phi > \phi_0$, for instance in a dilute CNC suspension drop that is left to dry to form a colorful film, the cholesteric phase nucleates at various points of the isotropic background as tiny droplets, which grow in size as ϕ increases. These nuclei are called *tactoids*, a term originally introduced by Zocher [67] for the unusually shaped, spindle-like nuclei of nematic phase common in colloidal nematics. He had been the first to observe and report these nuclei a few years earlier, in suspensions of needle-like crystals of vanadium pentoxide [68]. The motivation for the term can be found in the Ancient Greek word τάξις (táxis) which can be translated as "order", hence it is a very fitting term for a nucleus of a phase with nematic-like order. It has since been used to describe nematic or cholesteric nuclei, spindle-shaped or not, in many systems [69–73]. A good historical overview of the development of the empirical understanding of tactoids has been published by Sonin [74]. If instead of a nematic or cholesteric phase, a smectic liquid crystal develops directly from the isotropic phase, the corresponding nuclei are called *bâtonnets*, reflecting their strongly elongated shape promoted by smectic order [75–77].

The first attempt to explain tactoid formation was presented by Langmuir in 1938 [78]. Being more than a decade before Onsager's explanation of nematic ordering in nanorod colloids, this was in many respects premature. In order to understand the tactoid shape, there are a few peculiarities of the nuclei of nematic phase in a nanorod suspension that we must take into account. The first is that the tactoid and its surrounding are very close in composition, thus the interfacial tension γ (equal to the energy per unit interface area) is extremely low, at least provided that $\mathbf{n}$ adopts its energy-minimizing orientation at the boundary of the nematic tactoid. Which this orientation is, and how strongly it is imposed, is dictated by a dimensionless parameter W called the *anchoring strength* at the boundary. With the interfacial energy density for an arbitrary director orientation at the tactoid boundary written $f_i = \gamma(1 + W(\mathbf{c} \cdot \mathbf{n})^2)$, where $\mathbf{c}$ is the boundary normal, we see that W effectively quantifies the difference in interfacial energy density for the unfavored versus the favored alignment. Note that f_i has the unit N/m, as it is an energy density per unit area. With this formulation of f_i, a value $W > 0$ favors tangential alignment ($(\mathbf{c} \cdot \mathbf{n})^2 = 0$ gives the minimum interfacial energy penalty) whereas $W < 0$ favors orthogonal alignment ($(\mathbf{c} \cdot \mathbf{n})^2 = 1$ gives the maximum interfacial energy reduction). For tactoids of nematic or cholesteric phase in colloidal nanorod suspensions, $W > 0$, i.e., the favored director alignment is tangential to the tactoid boundary. This can be rationalized by considering the impact of the very high aspect ratio of the mesogens: orthogonal orientation would come with significant translational entropy penalty for any mesogens approaching the boundary, which they cannot escape, thus raising the free energy at the boundary considerably [54]. In fact, it appears that this entropic penalty is so large that W is quite significant in magnitude, playing an important role in determining the characteristics of tactoids of small size.

We are only aware of one attempt to measure the interfacial tension γ between the isotropic and cholesteric phases in a CNC suspension, due to Chen and Gray [79], and the reported values were on the order of μN/m (for the equilibrium configuration with $\mathbf{n}$ tangent to the tactoid boundary, $(\mathbf{c} \cdot \mathbf{n})^2 = 0$). To put this into context, the interfacial tension of oil droplets in water stabilized by surfactants is on the order of 30 mN/m. The same comparatively high values apply for many studies of thermotropic liquid crystal droplets dispersed in chemically different continuous phases like water or glycerol, the interface often being stabilized by surfactants or polymers. With the extremely low value of γ for the isotropic-nematic interface of tactoids in CNC suspensions it is not difficult to understand that tactoids can adopt non-spherical shapes much more readily than other liquid–in–liquid emulsions. Especially when comparing the liquid crystal behavior in tactoids with that observed in liquid crystal droplets dispersed in chemically different isotropic phases this difference must be taken into account.

The next peculiarity, at least in comparing with thermotropic liquid crystals, is related to the very different mesogen shape and size. The much higher aspect ratio and orders of magnitude greater length of colloidal mesogens mean that bend elastic distortions to the director field are substantially more costly in liquid crystals formed by CNC suspensions or other colloidal nanorods. In other words, K_3 is significantly greater than for most thermotropic liquid crystals. Considering the importance of translational entropy in colloidal nematics, one may expect also the splay elastic constant in CNC suspensions to be larger than that of standard thermotropics. The elastic response to bend and splay deformations can thus be expected to have a stronger impact in liquid crystalline CNC suspensions than in thermotropic analogs, for which most measurements of elastic constants were done.

Considering these three peculiarities, large values of K_1 and K_3, very low value of the equilibrium interfacial tension γ, and strong anchoring W favoring tangential director field orientation at the tactoid boundary, we can begin to understand the behavior of tactoids. Let us, however, start by looking at what is known for droplets of liquid crystals in general, because this allows us to better highlight how tactoids of CNC suspensions (or similar chiral colloidal liquid crystals) are unusual compared with thermotropic liquid crystal droplets.

The question of how the (chiral) nematic phase behaves when confined strongly in small droplets was studied experimentally in quite some detail during the 1980s and 1990s by, among others, Lavrentovich and co-workers [80–83], by Bouligand and Livolant [84], by Crooker and co-workers [85,86] and by Drzaic [87], with theoretical contributions from Zumer and others [88] (this list is far from exhaustive). For cholesteric droplets with tangential director alignment at the boundary, much work was built on the seminal model dubbed the "Frank–Pryce model", presented by Robinson et al. [89] in 1958 (F.C. Frank and M.H.L. Pryce were not co-authors, but the appendix of the article refers to private communications with them). With **n** aligned along the boundary of the spherical droplet the helix would develop radially, thus with a defect in helix orientation at the core of the droplet. However, the tangential director field itself must contain a topological defect charge of +2 at every concentric spherical plane, a topological requirement expressed in what today is called the Poincaré–Hopf theorem [90,91]. The Frank–Pryce model thus suggests that a line of +2 defects develops along one radius of the droplet. Indeed, many experiments found such a structure in large tangentially aligned droplets of thermotropic cholesteric, provided that the equilibrium cholesteric pitch p_0 was significantly shorter than the droplet radius r [84,85,90,92]. In the regime $r \approx p_0$, the helix was found to develop gradually and with much distortion, as illustrated very well in Figure 3 in the review by Lopez-Leon and Fernandez-Nieves [90]. This is very different from the case of cholesteric colloidal suspensions, where tactoids with size on the order of the pitch generally show no trace of a helix, and then, as the tactoid grows, there is a sudden transition directly to a well-developed helix, but only along a single direction [72].

To understand this difference, we must first remember that the droplets considered in most thermotropic cholesteric studies were not in thermodynamic equilibrium with the surrounding liquid, which was not the isotropic phase of the same system but rather typically a water or glycerol phase. Here we use the term "droplet" for describing this situation, while we reserve the term "tactoid" for the equilibrium situation of liquid crystalline domains coexisting with their isotropic counterpart. When studying droplets, the interfacial tension γ is considerable, as mentioned above, ensuring that droplets are always spherical, regardless of size. In a seminal paper by Bouligand and Livolant [84], cholesteric droplets *and* tactoids in equilibrium with their isotropic phase were investigated, for thermotropics as well as for lyotropics. They found examples of tactoids of non-spherical shape as well as with non-radial helix orientations, even for thermotropic mixtures. Three recent works of large significance, focusing specifically on the deviation from spherical shape in colloidal tactoids, are due to Prinsen and van der Schoot [54] (theory), Jamali et al. [93] (nematic carbon nanotube suspensions, primarily experimental) and Nyström, Arcari and Mezzenga [72] (cholesteric suspensions of shortened amyloid fibrils, primarily experimental).

All analyses of how the liquid crystalline order develops within a droplet or tactoid are based on a comparison of the interfacial and nematic elasticity contributions to the free energy. Depending on the size of a cholesteric droplet or tactoid, represented by a characteristic length r (which is the radius in case of a sphere), we can distinguish three regimes, with distinctly different behaviors, each having a different term dominating the free energy. Because the energy due to splay or bend deformations of the nematic director field scales only as r (see Section 4.2), this dominates for the smallest tactoids or droplets, to the extent that the liquid crystal violates the anchoring imposed by the boundary and adopts a uniform, deformation-free director field [80]. The phase remains spherical in case of a thermotropic nematic droplet [80], where interfacial tension cannot be neglected, but for a tactoid of a (chiral) nematic colloidal nanorod suspension, significant elongation along **n** into the characteristic spindle-like tactoid shape is generally observed [68,74,93], see Figure 7a.

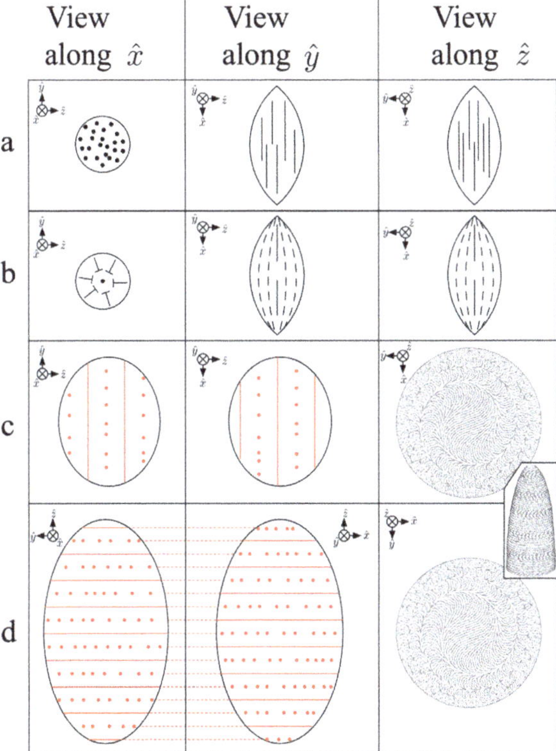

Figure 7. A schematic overview of tactoids of increasing size, from top to bottom, developing in a cholesteric-forming colloidal nanorod suspension. Each tactoid type (one per row) is sketched as it is viewed along $\hat{x}$, $\hat{y}$ and $\hat{z}$ (see local coordinate systems for guidance), where we define $\hat{z} = \mathbf{m}$ as the orientation of the helix when it develops. The drawings do not represent cross sections, but the projection into the paper plane of the outer tactoid surface. We sketch the director field in black and for the helix we draw side views in the first two columns, indicating $\mathbf{n}$ in the figure plane with red lines and $\mathbf{n}$ perpendicular to this plane with red dots, thus with a rotation of $\mathbf{n}$ by $90°$ in-between. Row (**a**) depicts the smallest tactoid (rarely seen in CNC suspensions), in which the director field is uniform, violating the anchoring conditions. The next row shows a tactoid of slightly larger size, where the anchoring condition is respected, a director field with bend and splay deformations developing within the tactoid. In the view along $\hat{x}$ we use 'nails' to illustrate the director field tilting out of the plane as a result of the surface curvature, the 'head' of the nail pointing towards us, the 'tip' pointing away. In row (**c**), a tactoid just large enough for the helix to develop is shown, the twisting developing from a non-helical director field that was originally pointing along $\hat{x}$ (as in rows **a**,**b**). The final row shows the most commonly observed tactoids in CNC suspensions, where the helix has developed with many pitches, still along a single direction throughout the tactoid (orthogonal to the corresponding direction in **c**), still violating the anchoring conditions. We draw the situation for an ellipsoidal shape, elongated along $\mathbf{m}$, which is common once the tactoids grow to sizes of several pitch lengths. The curving director pattern along $\hat{z}$ in rows (**c**) and (**d**) is the projection of a hemispherical cap viewed along the helix. It has been created using the construction of Bouligand and Livolant [84], stacking 50 slices taken at constant z distances on top of each other, from the base to the top of the hemisphere, as illustrated in the inset between rows (**c**) and (**d**). We used $20°$ rotation of $\mathbf{n}$ from one slice to the next. Note that the views along $\hat{x}$, $\hat{y}$ and $\hat{z}$ are correlated, to better illustrate the helical variation of the director field. Red dotted lines between the sketches emphasize this aspect in row (**d**).

While the extremely low interfacial tension makes it understandable that a non-spherical shape is allowed, it is not at first obvious *why* the tactoid elongates, considering that the director field is undistorted; the elongation in this regime can thus not be driven by nematic elasticity. As argued by Prinsen and van der Schoot [54], the reason is the anchoring strength W, which is significant despite the extremely low equilibrium value of γ. The cost of violating the anchoring condition, as required to achieve a uniform director field in a finite volume, is therefore enough to drive elongation into the spindle-like shape with two sharp points. The spindle shape maximizes the interface with **n** almost parallel to the boundary and it even removes the regions (as compared to a spherical boundary) where the director would be perpendicular or near perpendicular. Tactoids in this smallest regime can be recognized in the polarizing microscope by their spindle-like shape and the fact that they give complete extinction when the tactoid is rotated with its long axis along one of the polarizers, as the undistorted director field means that the optic axis is exactly along the tactoid long axis. However, as they cannot grow very large without changing character, then entering the next regime, they may be difficult to detect.

As the tactoid grows, a new regime is entered, where the interfacial energy ($\propto r^2$) contribution starts dominating over that of elastic distortion ($\propto r$). Therefore the tactoid realigns the director field in order to adhere to the anchoring conditions, thus with tangential **n** along the full boundary, at the cost of developing a distorted director field with slight splay and bend, still with **n** primarily along the spindle axis. This configuration, depicted in Figure 7b, minimizes the director field distortion as well as interfacial tension. We call this tactoid configuration 'bipolar', reflecting the fact that the director field converges into two +1 defects at each 'pole' of the tactoid (or slightly outside the tactoid, the defects then being virtual [54]). The transition to the bipolar regime is recognized in the polarizing microscope by the tactoids no longer showing extinction at any orientation, because the optic axis is no longer uniform, hence some parts of the tactoid will always show birefringence. There may be a twist distortion as well, promoted by K_{24} [94,95], but this has not yet been investigated for colloidal tactoids to the best of our knowledge.

In tactoids of CNC suspensions, it is relatively rare to see pronounced spindle shape, with the two characteristic cusps, most likely because of a lower aspect ratio of most CNCs than of the vanadium pentoxide rods studied by Zocher [68], the carbon nanotubes studied by Jamali et al. [93] or the amyloid fibrils studied by Nyström et al. [72], giving rise to less extreme values of K_3 and of W. Nevertheless, the nucleating tactoids are always anisotropic in shape, but often smoothly ellipsoidal rather than spindle-like. A rare case where clearly spindle-shaped tactoids were reported is the study of bacterial CNC of quite extreme length, about 1–2 μm, by Araki and Kuga [96]. Tactoids from both the smallest regimes, with uniform director field (uniformly black when the tactoid is aligned along a polarizer) or a bipolar one (full extinction impossible), can be seen in their micrographs, but no extensive analysis of the tactoids was carried out.

It is not surprising that the cholesteric helix, with its non-uniform director field, cannot develop in the smallest tactoids. If the whole tactoid size is on the order of the helix pitch, it is less costly to pay the twist elastic deformation cost of not developing the helix and maintaining a uniform director field. Indeed, estimates of the elastic energy contribution due to untwisting the helix by Nyström et al. [72] came to the conclusion that this term scales as r^3. This explains why the helix is suppressed in very small tactoids, dominated by splay/bend elastic energy, as well as in intermediate tactoids, where the interfacial energy dominates, but eventually a tactoid grows large enough for the contribution $\propto r^3$ to dominate, leading to development of the helix. Considering the comparatively strong anchoring favoring tangential director at the tactoid boundary, one might then expect the helix to develop radially, in order to maintain a tangential director everywhere along the tactoid boundary. However, this does not happen for CNC suspension tactoids. Instead, the cholesteric singles out one direction along which the helix forms (usually along a short axis of a tactoid, which follows from the fact that the twisting must start from a director that was initially along the tactoid long axis), thus again violating the anchoring conditions strongly, see Figure 7c–d. The reason is most likely the large value of the bend constant for colloidal liquid crystals: a radial helix entails a continuous bend, very strong at the core of the tactoid. The elastic energy cost of this bend must be much larger than the periodic violation of the anchoring conditions along the tactoid boundary that a uniform helix orientation requires.

The threshold for the helix development transition obviously depends on the twist elastic constant K_2, telling how large the cost is for not developing the natural helix with wave vector q_0, as well as the wave vector itself, which quantifies how large the twist distortion is in a non-helical cholesteric director field. Because the uniformly aligned helix axis violates anchoring, the threshold should also be related to γ and/or W. These four parameters are indeed the key components of the threshold expression established by Nyström et al. [72]:

$$\frac{V}{\alpha} \approx \left(\frac{2\gamma W}{\alpha^2 K_2 q_0^2} \right)^3 . \tag{14}$$

Here, V is the minimum tactoid volume for helix development and α is the tactoid aspect ratio, $\alpha = R/r$, with R the longest and r the shortest radius.

Also the transition to helical tactoids is easily evidenced by polarizing microscopy (Figure 8), revealing periodic dark and bright 'fingerprint' stripes (see Section 7.1) perpendicular to **m**. The tactoid appears dark where **n** orients along the viewing direction, as we are then looking along the optic axis, thus not seeing the effect of birefringence, and it is maximally bright when **n** is in the sample plane, giving maximum effect of the birefringence. Note that **m** must not be aligned parallel to either polarizer to see the brightness modulation. In Figure 8e the radial variation in **m** shows this clearly, the fingerprint texture being replaced by uniform black wherever **m** is parallel to one of the polarizers. As the tactoid grows further it often (but not always) elongates along **m**, an observation that to the best of our knowledge has not been discussed in detail, and which is not entirely obvious to explain. The tangential anchoring should favor expansion of the tactoid perpendicular to **m**, in order to maximize the interface with the favored director alignment. A possible explanation may be kinetic, the likelihood of incorporation of CNCs from the isotropic surrounding being higher at the bottom and top of the helix, since it is easier to adapt to the uniform director orientation at these points than the continuously changing orientations along the helix sides. Eventually, the tactoid may get so large ($r \approx 10p_0$) that the helix orients radially (Figure 8e) and we then see a texture similar to that of thermotropic cholesteric droplets with tangential alignment, although we never saw evidence of the Frank–Pryce defect line. We will come back to why this may be in a moment.

Spherically symmetric tactoids of large size are actually quite rare, because in most cases tactoids meet and coalesce before growing large on their own. Tactoid coalescence was studied early on by Zocher and Jacobsohn [67] and more recently by Kaznacheev et al. for vanadium pentoxide [70], by Lettinga et al. for fd virus suspensions [97] and by Moser et al. for collagen-based cholesterics [73], showing a common evolution of merging tactoids. For CNC suspensions it has recently been studied extensively by the MacLachlan group, under the targeted concept of 'tactoid annealing' [98], an issue we will come back to when discussing the challenge of obtaining optically uniform CNC films in Section 7. The coalescence process generally causes topological defects in the director field in the resulting larger tactoid. When multiple tactoids have coalesced a hole may even be left in the middle. Rearranging the director field can compensate for these defects but this typically takes a long time. The low value of γ means that very irregular shapes arise and remain stable for long time. Furthermore, due to the slightly higher density of the tactoids compared to the surrounding isotropic phase, because of their higher nanorod density (Equation (12)), they will gradually settle to the bottom of the suspension where they coalesce into a continuous and macroscopically visible liquid crystal phase.

The study of collagen-based cholesterics by Mosser et al. [73] is worth mentioning also for their experimental approach to probe a large region of the phase diagram in a single sample. Because the main thermodynamic control variable in lyotropic liquid crystals is the concentration of mesogen, we need to vary the concentration in a controlled way in order to map out the phase diagram. Typically this is done in a rather tedious manner by preparing individual samples with varying mesogen concentration and investigating each one individually (Figure 6), but Mosser et al. instead developed two ingenious set-ups for evaporating the water from the original collagen solutions in such a way that a controlled concentration gradient developed along the sample. This experimental approach may be useful to explore also in CNC suspension research.

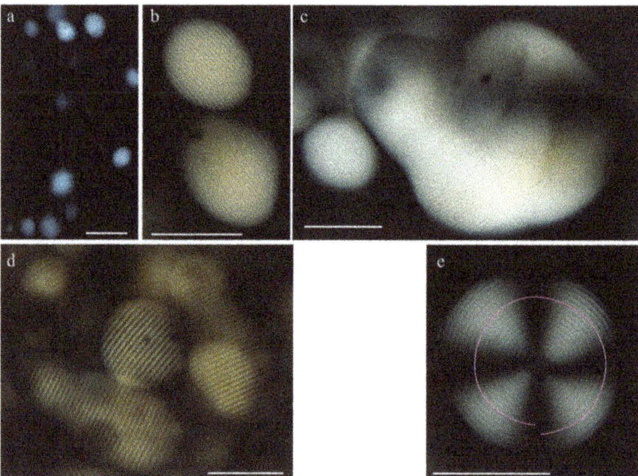

Figure 8. Representative polarizing microscopy micrographs (crossed polarizers) of tactoids of cholesteric CNC suspensions, soon after 'birth' (**a**) and later in the growth process (**b–d**), eventually reaching sizes where radial helix orientation occurs (**e**). Even in the smallest tactoids in (**a**), the helix has probably developed, although the focus may render it difficult to see (see discussion in main text). In the last panel one 'dark ring' has been emphasized, to highlight the non-closing spiral-like character of the stripes in the 'fingerprint' texture decorating the tactoids. The distance between two consecutive stripes is $p/2$ (see Section 7.1). Scale bars correspond to 50 μm.

Some representative polarizing micrographs of CNC suspension tactoids are shown in Figure 8. In the smallest tactoids (panel a), fingerprint stripes due to the helical director modulation can be seen in some but not all tactoids. Those where no stripes are seen are often out of focus. Most likely, these are suspended in the bulk of the isotropic phase whereas the tactoids that are sharply imaged have reached the bottom of the capillary in which the sample is studied, providing a clean cut through the tactoid that is easier to image optically. The absence of visible fingerprint texture in the other tactoids thus does not mean that these tactoids do not exhibit a helical director field, only that the imaging does not allow its detection. In addition to the tactoid being out of focus, another reason for not seeing the stripes may be that the tactoid is oriented with the helix axis **m** more or less along the viewing direction. The breaking of the symmetry that the glass substrate of the capillary provides for those tactoids that have settled at the bottom may provide an aligning influence, guiding **m** into the plane, thereby making it easier to visualize the stripes. Indeed, the fact that the stripe distance is relatively constant from tactoid to tactoid suggests that this is the case; if **m** were randomly oriented in space, we should see the fingerprint texture with great variation in periodicity, depending on how **m** is oriented within the tactoid with respect to the viewing direction.

As the tactoids grow, a variety of tactoid shapes can be seen, some clearly being remnants of tactoid merging (c–d). Panel (e) shows a relatively rare case of a nearly spherical tactoid with radial helix orientation, showing a texture similar to those seen in many studies of thermotropic cholesteric droplets. However, no trace of the Frank–Pryce defect line can be seen, which could indicate that the line is directed along the viewing direction. However, as mentioned above, tactoids of this size, that can be imaged with sharp pitch bands, are very likely sessile on the capillary bottom, i.e., the tactoid is only a hemisphere. This would allow the director field to develop without defects except at the central core, hence the Frank–Pryce defect line would not be needed. Interestingly, the tactoid looks quite dark around the center, suggesting that the central core is not cholesteric but isotropic, an energetically favorable way of removing the defect, as demonstrated by Kumacheva and co-workers for fully spherical CNC suspension droplets [99]. Yet, the center is not as dark as the surrounding isotropic phase, which we can explain by the fact that, in the middle, we are looking through the cholesteric exterior of the tactoid, and here we are observing it along **m**; the long pitch then gives rise to optical rotation of linearly polarized light [100,101], letting some light through the crossed polarizers. Another interesting observation is that the rings are not circles but spirals, as emphasized by highlighting one such spiral in Figure 8e. When the highlighted spiral passes its origin, it has skipped a full helix pitch. This indicates a certain asymmetry within the tactoid.

For long time, the only method for studying tactoids experimentally was polarizing microscopy, giving images such as those in Figure 8. Confocal microscopy was also used in the study of amyloid fibril suspensions by Nyström et al. [72] and of virus suspensions by Lettinga et al. [97]. The limited resolution of these optical methods, applied to objects like tactoids which can be tens or even hundreds of microns in size, restricts the detail with which tactoid formation, growth and merging can be studied. This changed thanks to the work of MacLachlan and coworkers [98,102,103], who managed to capture multiple stages of tactoid development in CNC suspensions by 'freezing-in' the momentary structure using UV-induced photopolymerization of a reactive monomer dissolved in the water. The solid objects with structures templated by the liquid crystalline tactoids, as well as their isotropic surrounding, could then be beautifully investigated with high resolution by conducting scanning electron microscopy on the surfaces arising by fracturing the samples.

4.5. CNC Fractionation by Liquid Crystal Phase Separation and the Impact of Aspect Ratio on Phase Sequence

Tactoids, and the early stage of isotropic–nematic phase separation at which they occur, are of great interest for a disperse system such as CNCs. The reason is to be found in Equation (11): because ϕ_0 decreases with increasing rod aspect ratio, the very first tactoids that form in a disperse CNC suspension as its mass fraction w is increased will contain the longest rods. Likewise, the last remaining isotropic phase as we approach the CNC fraction where the full sample is cholesteric (or the first isotropic droplets (sometimes called *atactoids*) emerging if we dilute a fully cholesteric CNC suspension) will contain the shortest rods. Indeed, in a study by Dong et al. on filter paper-derived CNCs [104], later confirmed for bacterial cellulose by Hirai et al. [105], longer rods were found to predominantly populate the cholesteric phase, and shorter rods predominantly the isotropic phase, consistent with theoretical considerations by Lekkerkerker [60]. Interestingly, the experimental studies reported higher dispersity in rod aspect ratio in the cholesteric than in the isotropic fraction.

Already Onsager hinted at the possibility of fractionating length disperse colloidal suspensions but he did not develop a theory that takes dispersity into account. The baton was first picked up by Flory and Abe [59], then by Lekkerkerker and co-workers [60–62] and later by Bates and Frenkel [63] and others. A very recent contribution, discussing also CNCs as example, was published in this journal by Wensink [64]. These analyses look into how the biphasic coexistence regime is affected by the dispersity, a broadening being a significant consequence, and to what extent fractionation should be possible. The work of Wensink also looks into issues like the expected helical pitch development.

In a recent study of our group [106] we put this size-selective sorting provided by liquid crystal phase separation to use, repeatedly separating physically the cholesteric phase developing in a CNC suspension in the biphasic regime ($w_0 < w < w_1$), and subjecting each fraction, the cholesteric and the isotropic, to another phase separation of the same type. The cholesteric phase was diluted until it again was roughly in the middle of the coexistence regime while the isotropic phase was concentrated until the same point. By repeating the separation three times, a final cholesteric fraction with significantly increased average rod length and a final isotropic fraction with significantly reduced average rod length resulted. In [106] we reported roughly unchanged dispersity for the cholesteric fraction, but we subsequently found that this was an artifact due to the difficulty in physically separating the phases;

we have improved the method and confirmed that both fractions have reduced dispersity, as will be published soon.

The beauty of this procedure is that the change in rod character and reduction in dispersity has a strong impact on the liquid crystal phase sequence. The pristine disperse CNC suspension had an average rod length of about 0.2 µm and diameter about 4 nm, whereas after fractionation the average length in the cholesteric-derived sample was 0.23 µm but only 0.17 µm in the sample derived from the isotropic phases. There was less change in the diameter, and the average aspect ratios (L/d) in the final fractions were 61 and 46, respectively (as determined by atomic force microscopy, AFM). When establishing the phase diagrams for the two fractionated samples and comparing it to the original one, we found significant and—in the case of the long-rod fraction—very beneficial changes. The onset mass fraction of liquid crystal formation had reduced from the original $w_0^{\mathrm{init}} \approx 5$ wt.% to $w_0^{\mathrm{long}} \approx 3$ wt.%, and the stability limit of the cholesteric phase shifted from $w_1^{\mathrm{init}} \approx 11.8$ wt.% to $w_1^{\mathrm{long}} \approx 6$ wt.%. In other words, not only had the onset of liquid crystal formation been pushed down to much lower CNC mass fractions, but also the phase coexistence window had been made much narrower by the reduction in dispersity. The reductions in w_0, w_1 and $w_1 - w_0$ are highly beneficial, as they lead to an expanded equilibrium range of a completely liquid crystalline phase that can be studied experimentally. This is because the phenomenon of kinetic arrest, ending the equilibrium behavior, was unaffected by the fractionation. We will discuss the reason for this unexpected and very interesting finding in Section 5.

The liquid crystal-based fractionation is so powerful because all fractions originate from the same CNC suspension. In other words, *only* the rod aspect ratio changes, while surface charge, type of stabilization and all other properties remain the same (assuming that there is no systematic variation with rod length of these properties in the mother suspension). This is a significant difference from previous studies trying to assess the impact of rod length, carried out on CNCs, and often also CNFs, that were produced using different procedures, from different raw materials. Therefore, rod lengths are varied at the same time as many other properties, making it difficult to draw clearcut conclusions. Nevertheless, similar trends were reported in these works. Dispersions of wood-derived CNC showed a decrease in w_0 from $\sim$5 wt.% to $\sim$3 wt.% as the aspect ratio L/d increased, again connected to an increase of the average length L from 100 nm to 150 nm [107,108]. However, the rods with higher L/d also showed lower surface charge densities, complicating the analysis. The authors proposed that the rod dimensions and geometry influence w_0 to a greater extent than the surface charge. CNCs derived from bacterial cellulose [96,105,109] and tunicin [109] have significantly larger aspect ratios, in the range of 50–150, and very low values of $w_0 \approx 0.5$ wt.%. In case of CNCs obtained from filter paper (thus originally wood-derived) with L/d in the range of 13–50, the biphasic region was found in the range between 5 and 13 wt.% [56]. This can be compared to CNCs derived directly from wood with aspect ratios in the range of 13–20 and biphasic region spanning 3–7 wt.% [108].

5. The End of Equilibrium: Kinetic Arrest, Percolation, Gel Formation and Glass Transitions

When we concentrate a CNC suspension to a mass fraction on the order of $w = 10$ wt.%, we leave the equilibrium phase diagram and the sample enters a kinetically arrested gel-like state. From a practical point of view, the kinetic arrest is on the one hand a blessing, because it locks the cholesteric order in place even as the water is evaporated and the particle fraction is increased all the way to 100 wt.% [19]. Were it not for the kinetic arrest, other structures would replace the cholesteric order, such as smectic states or solid crystalline states, and the helical structure would be lost. It is thus the kinetic arrest of the helical structure that allows us to form the colorful cellulose films from CNC suspensions that has created such a strong applied interest in the material, simply by evaporating the water. The loss of equilibrium is a mixed blessing, however, because gelation often sets in at such low CNC mass fraction that the sample has not yet reached liquid crystalline order throughout the sample. This limits us, from a fundamental science point of view, because we cannot study the liquid crystal behavior of many CNC suspensions without the presence of isotropic droplets coexisting with the liquid crystal. From a practical point of view, it means that we have plenty of internal boundaries between tactoids that have not coalesced, and even coalesced tactoids are unable to unify their helix orientations. The helix orientation in the final film can thus be largely uncorrelated from place to place, giving rise to optical properties that are highly non-uniform. This will be further discussed in Section 7.

Do we today understand what is going on at this stage? We believe the answer is largely yes, but in order to provide the explanation, we need to complement our understanding of the behavior of electrostatically stabilized colloids with a digression into non-equilibrium phenomena such as gelation and glass transition.

5.1. Percolation and Kinetic Arrest in CNC Suspensions

A high particle aspect ratio, as for nanorods like CNCs, promotes not only liquid crystalline ordering, but also the chance of close approaches between particles. For a certain overall mass or volume fraction, the likelihood of close approaches scales with the effective aspect ratio, thus $\propto L/(d + 2\kappa^{-1})$. At sufficient concentration of particles with sufficient aspect ratio, we may eventually reach a state of potential particle connectivity extending throughout the system, a phenomenon that physicists call *geometric percolation* or *connectivity percolation*. It is important to note that almost all research on percolation has been done for isotropic systems, i.e., where there is no long-range order in the rod orientation. As we will come back to in a moment, the orientational order of (chiral) nematics actually has a profound impact on percolation [110], an aspect that needs to be considered in order to fully understand the behavior of kinetic arrest in cholesteric CNC suspensions.

Percolation can have multiple consequences for the behavior of the system, which ones depending on the nature of the percolating particles and the interactions between them. If the particles that came into contact remain connected, to the extent that the percolating network can resist mechanical stresses, we get what is called *rigidity percolation*: once a stable continuous network of particles exists throughout the system, the whole system stops flowing. We can also call this state of long-term connectivity *gelation*, and the colloidal suspension has transitioned from a *sol* (unconnected particles suspended in a fluid matrix) to a *gel*. This is a non-equilibrium state, stabilized kinetically: the system is generally not in a state of minimum free energy, but nevertheless retains its structure because it is unable to rearrange into the energy-minimizing structure. An energy barrier much greater than thermal energy must be overcome in order to leave the state. We say that the rod arrangement is *kinetically arrested*. Rigidity percolation is highly relevant to CNC suspensions, since they all enter a kinetically arrested gel-like state above a threshold mass fraction w_k, that shows many of the hallmarks of rigidity percolation.

The stable connectivity of rigidity percolation requires attractive interactions that maintain the connection despite the impact of thermal fluctuations and, at least to some extent, sustain also externally imposed stresses (a soft gel may easily be broken). As discussed in Section 3, this can be provided by the van der Waals attraction, which is very strong when particles are in close contact. However, because we are dealing with colloids, in which particles should generally not aggregate, the CNCs have been prepared with the surface charge that prevents such close contact. On the other hand, we have seen in Section 3, illustrated in Figure 2, that an increased ionic strength, provided, e.g., by the additional counter ions as we add more CNCs into suspension, will weaken the electrostatic repulsion, eventually to the extent that no barrier exists. Considering other attractive interactions, e.g., hydrogen bonding in case of CNCs, we might also have a situation where a secondary minimum gets sufficiently deep upon reduction of electrostatic repulsion to trigger flocculation of particles into loose aggregates, without inducing permanent aggregation into the primary minimum. By increasing the CNC mass fraction w, we might thus expect to reach the threshold for geometric percolation as well as the limit of colloidal stability, which would then give rise to rigidity percolation, since long-term stable bonds between CNCs would be formed. Importantly, the CNC mass fraction required to loose colloidal stability, in the sense that the electrostatic stabilization no longer dominates over van der Waals attraction, is apparently above the geometric percolation threshold. If this were not the case, the loss of colloidal stability would simply lead to local rod aggregation and sedimentation rather than gelation. Further contributions may well be expected from hydrophobic interactions, still not very well investigated in the context of CNC suspensions.

The mass or volume fraction of rods required for percolation is called the percolation threshold. Because of the dependence on aspect ratio of the likelihood of encounters, the percolation threshold is proportional to $(d + 2\kappa^{-1})/L$, just like the threshold for liquid crystal formation. This makes percolation a particularly delicate phenomenon in case of CNC suspensions: one would expect an increased rod aspect ratio to promote liquid crystal formation as well as gelation, but as will be discussed below, the impact is more complex. In contrast to liquid crystal formation, percolation does not require the rod to be very stiff or straight. Considering this, it is not difficult to understand that CNFs, much longer than CNCs but also more flexible, kinked and potentially even branched, tend to percolate and form gels before any long-range orientational order develops. We are not aware of any study reporting liquid crystal formation in CNF suspensions, even if some long-range order may appear together with kinetic arrest.

5.2. Kinetic Arrest by Glass Transition or Gelation?

In contrast to the percolation threshold, the reduction of κ^{-1} does not depend on the rod aspect ratio, only on the ionic strength I, which in turn is proportional to the total amount of CNCs in suspension, assuming that no further ions are added ($I = I_c$). Therefore, a very high rod aspect ratio could reduce the percolation threshold to such an extent that we reach geometric percolation long before the electrostatic repulsion is sufficiently reduced to create lasting bonds between rods. Would this mean that no kinetic arrest can take place? Not necessarily. There is another possible origin of kinetic arrest, that would occur in the opposite extreme of *low* ionic strength. This is not strictly a percolation transition, because it is not triggered by the appearance of continuous networks throughout the system, but it is also related to the volume fraction of rods, and their aspect ratio. The kinetic arrest would in this case be more of a glass transition, in the sense that the rods become trapped from rearrangement, not by forming networks but rather by being stuck in a 'cage'. This 'electrostatic caging' is promoted by large κ^{-1}—hence the requirement of low ionic strength—as well as sufficient effective volume fraction of rods to form the cage. The mobility of any individual probe rod is restricted by the presence of all the surrounding rods through the excluded volume that they cause for the probe rod, given by the volume of the actual rods and the addition provided by κ^{-1}. Above the threshold for kinetic arrest in this scenario, a rod could not leave its 'cage' of excluded volume formed by surrounding rods, although it is not part of a network.

A complication is that the surface charge and ionic strength are not independent, as we always have electroneutrality. The higher the surface charge, the more counter ions are released when a rod is suspended, and thus the higher the ionic strength I. This scenario thus requires a delicate balance of surface charge and particle aspect ratio, but it has been proposed to occur, e.g., for fd virus suspensions [111]. These nanorods have many similarities to CNCs, but there are also significant differences. While CNCs are highly disperse, viruses are, by virtue of their biological origin, non-disperse, i.e., every virus particle has the same length and diameter. They also have a significantly greater aspect ratio than most CNCs (bacterial CNCs being a possible exception), with $d = 6.6$ nm and $L = 0.88$ μm [112] for fd. Finally, and very significantly for this discussion, their surface charge is substantially greater than for CNCs. Rod-like viruses, like fd, consist of a single helically packed nucleic acid chain surrounded by a 'capsid', a cylindrical coating of proteins. The coating of fd consists of about 2700 proteins, each with the same mix of charged amino acids [113]. At neutral pH, the charge amounts to about $-3.4q$ per protein [112], where q is the elementary charge. Therefore, each fd virus nanorod carries a negative surface charge of more than $9000q$!

How does this compare with typical CNC qualities? For the commercial CNC that we studied in Ref. [106], finding clear evidence of counter ion-induced gelation (see below), the supplier gave the information that the particles have a sulfur content of 0.95 wt.%. Our AFM investigations of the

CNC showed that the pristine sample had an average physical rod length of about 0.2 μm and an average physical rod radius of about 2 nm, yielding a volume of an average rod of cellulose that is $V_{1CNC} = \pi(2 \times 10^{-7}\text{cm})^2 \times 0.2 \times 10^{-4}\text{cm} \approx 2.5 \times 10^{-18}$ cm^3. With a cellulose density of 1.5 g/cm^3, this gives us a mass of an average CNC of 3.8×10^{-18} g. The manufacturer's information then tells us that an average rod carries 3.6×10^{-20} g of sulfur. To turn this into a count of the $-SO_3^-$ groups on the CNC, we divide by the molar mass of sulfur and multiply by Avogadro's constant, obtaining:

$$N_S = \frac{3.6 \times 10^{-20}\text{g}}{32.065\text{g/mol}} \times 6.022 \times 10^{23}\text{mol}^{-1} \approx 700. \tag{15}$$

In other words, a typical CNC rod has a surface charge of about $700q$, an order of magnitude lower than fd virus. In addition, the CNC does not have nearly the same aspect ratio, so in order to compare the two systems, we should normalize to particle mass or length (the diameters of CNC and fd are similar enough to ignore the difference), the latter being simpler. With a CNC having roughly a quarter of the length of an fd rod, we thus find that the length-normalized surface charge is about three times greater for fd than for CNC.

As discussed above, an increase in surface charge comes with greater I_c for the same amount of rods, hence the rod mass fraction must be low in order not to loose colloidal stability when the surface charge is as high as in fd. Because of the high L/d, this is the case for cholesteric fd virus suspensions, where the caging-based kinetic arrest was proposed. Indeed, Kang and Dhont reported remarkably low $w_0 \approx 0.15$ wt.% and $w_1 \approx 0.34$ wt.%. Furthermore, w_k, the threshold mass fraction for kinetic arrest, is much lower, $w_k \approx 1.2$ wt.%, a factor 10 lower than w_k in the CNC suspensions studied in [106] (details below). As the length-normalized surface charge of fd is only about a factor 3 greater, the counter ion concentration at comparable mass fractions should also be three times greater, meaning that w_k for fd corresponds to about three times lower I_c than in CNC suspensions. This could indeed suggest that the reasons for kinetic arrest in the fd virus and in the wood-derived CNC suspensions are different.

With bacterially derived CNC, we may actually reach situations not too different from fd virus suspensions, and one interesting report suggests that we may then see kinetic arrest by glass transition also here. This is the study by Araki and Kuga [96] of the phase behavior of bacterial CNC suspensions discussed in the context of tactoid formation in Section 4.4. As mentioned there, these CNCs were exceptionally long, 1–2 μm, and they had a board-like shape, with a typical cross section of 10 nm by 50 nm. The surface charge per unit area was fairly normal for CNC, 0.02 charges per square nanometer surface area, which computes to a total CNC charge of $2400q$ assuming 1 μm length. Normalized by the length, this is not too different from the $700q$ per rod for the wood-derived CNC just mentioned. Because of the very long rods, and a higher than usual aspect ratio, the onset of liquid crystallinity was very low, $w_0 \approx 0.1$ wt.%, comparable to the case of fd virus. In contrast, the mass fraction for achieving a completely liquid crystalline sample was much higher than for fd, $w_1 \approx 1.7$ wt.%, a difference most likely due to the high dispersity of the bacterial CNC, the presence of short CNCs raising w_1 (see Equation (12)).

The onset of kinetic arrest was detected earlier than for most CNC suspensions, with $w_k \approx 3$ wt.%, but compared to the fd suspension, this is still a factor 3 greater. This compares well with the fact that the fd had about three times greater surface charge. Araki and Kuga did not see textures indicative of helix formation in the pristine cholesteric CNC suspensions, an observation that is probably related to the high viscosity of these suspensions (phase separation between isotropic and cholesteric phases was reported to be slow) and the fact that kinetic arrest sets in at such low CNC mass fractions. Surprisingly, the viscosity decreased upon addition of NaCl salt (the effect of salt addition in CNC suspensions will be discussed in more detail in Section 6.3) and now a periodically striped texture indicative of helix formation was seen with 0.1 mM NaCl added to a 1.58 wt.% CNC suspension. This amount of salt also increased the liquid crystal onset to $w_0^{0.1\text{ mM NaCl}} \approx 0.4$ wt.%. The similarities with the fd virus suspensions and the fact that salt addition counteracted the kinetic arrest seen in the pristine system renders it plausible that these bacterial CNC suspensions may indeed be experiencing the electrostatic caging-induced glass transition in the pristine state. The addition of NaCl would reduce the Debye length to such an extent that the caging does not occur, the viscosity decreases, and the helix then has a chance to develop.

Nordenström et al. [114] set out to determine experimentally whether the kinetic arrest in various CNC and CNF suspensions is of gelation or of glass transition type, and also if the rod volume fraction required for kinetic arrest scaled as d/L or not. To this end they compared several different types of CNC as well as CNF, thereby spanning a broad range of aspect ratios, from 29 for the shortest CNC type to about 1000 for the longest CNF. Their experiments suggest that the threshold indeed follows classical percolation behavior within this range. While they did not mention if their samples formed liquid crystal phases or not, the d/L scaling for the percolation threshold, expected for isotropic but not for liquid crystalline suspensions [110] (see below), suggests that their samples were predominantly isotropic. In order to test for gelation versus glass formation, they re-diluted the sample, in order to check if the sample started flowing again. As they found that this was always the case, their conclusion was that the kinetic arrest was of the glass transition type.

There are, however, a few significant aspects that the study did not take into account. First, a recovery of flow with complete redispersion of the particles could be expected even after rigidity percolation if the connectivity triggering the network formation is due to CNCs being stuck in a deep *secondary* minimum in Figure 2 rather than reaching the primary minimum. As mentioned in Section 3, the strong hydrogen bonding capacity of cellulose might contribute to such a situation, although it needs to be investigated further. Second, even if rigidity percolation is driven by particles reaching the primary energy minimum separation, dilution of the gel might still cause the sample to start flowing again, because now the particle fraction is below the percolation threshold. The expansion upon dilution could break the volume-spanning network, creating sections of aggregated particle networks suspended in the solvent. We may thus get back to a flowing system, but it is not necessarily identical at the individual particle scale to the system prior to kinetic arrest. As will become clear in a moment, we had this situation in our study of kinetically arrested CNC suspensions. Finally, by using so drastically different sample types for varying the rod aspect ratio, many other parameters are varied at the same time, not least surface charge, stiffness and linearity when comparing CNF with CNC. This leaves the study open in the sense that we do not know the impact of the other parameters, the variations of which were not negligible.

In our recent systematic study of kinetic arrest in CNC suspensions [106] we did not use different samples prepared by different production methods, but instead we compared different fractions separated from one and the same original CNC suspension, with Na^+ counter ions, using the liquid crystal-based fractionation scheme described in Section 4.5. Recall that our fractionation changed the average rod length from 0.2 µm in the pristine suspension to 0.23 µm in the cholesteric-derived long-rod fraction and 0.17 µm in the isotropic-derived short-rod fraction, corresponding to average L/d values of 61 and 46, respectively. At first sight, these differences may not appear to be so dramatic, but as described in Section 4.5, the liquid crystal phase diagram was much enhanced, moving w_0 and w_1 to significantly lower values, narrowing the coexistence range, and leaving a large range of equilibrium cholesteric phase without coexisting isotropic droplets. Yet there was no change to w_k, the onset mass fraction for kinetic arrest. For the pristine suspension as well as for the short- and long-rod fractions, $w_k \approx 12$ wt.%. Based on this fractionation-independent w_k we conjectured that kinetic arrest is triggered by the system reaching a critical ionic strength I^k or, more specifically (see below), a critical counter ion concentration c_c^k. This gives a strong hint that the gelation is related to loss of colloidal stability, enabling rigidity percolation.

To further test this hypothesis, we added NaCl salt to several samples of CNC suspensions, from the long- as well as from the short-rod fractions, with CNC mass fractions from $w = 1$ wt.% up to $w = 11$ wt.% and in each sample we tuned the NaCl concentration such that the Na^+ counter ion concentration c_c reached the critical value c_c^k, the same as in a salt-free sample at w_k. Indeed, now *all* samples gelled, down to $w = 3$ wt.%! This is a very clear indication that the kinetic arrest in this study is a gelation phenomenon and that the increase of c_c beyond c_c^k, apparently the limit of colloidal stability, is the trigger. Further support of this conclusion was found in rheological experiments. Freely flowing samples showed a greater loss modulus than storage modulus ($G'' > G'$), as typical for viscoelastic liquids. This changed for the kinetically arrested samples, which all had $G'' < G'$, typical of a gel. We note (see Equation (2)) that the addition of NaCl up to $c_c = c_c^k$ leads to an ionic strength I that is greater than I^k of a salt-free sample at w_k, since I is also affected by the added free Cl^- co-ions. The loss of colloidal stability thus appears not to be a simple consequence of κ^{-1} being reduced below a critical value, as κ^{-1} fundamentally is a function of I, not of c_c. Possibly, ion-specific processes within the double layers surrounding each CNC rod, where Na^+ ions are much more prominent than Cl^- ions [41], need to be taken into account explicitly.

When we studied suspensions with mass fractions near w_k by AFM, another curious observation was that rods tend to connect in chains, more or less end-to-end, effectively forming much longer 'supra rods', albeit also more flexible. Based on this, we conjectured that the kinetically arrested state is an example of rigidity percolation where the percolating entities are not the individual CNCs, but rather the supra rods formed by this peculiar aggregation. The reason that the fractionation had no effect on w_k could then be easily understood: the aspect ratio even of the longest CNCs could be so low that the percolation threshold for *individual* rods ends up at mass fractions far beyond the mass fractions where rods start connecting. The connection drastically increases the effective rod length (with no difference depending on the starting sample), bringing down the percolation threshold and thus rapidly triggering rigidity percolation and gelation. The AFM measurements are illustrative also in terms of distinguishing aggregation by particles reaching the primary or secondary minimum in the free energy curve. As discussed above and as probed by Nordenström et al. [114], dilution should lead to redispersion of rods that have aggregated only by flocculation. In contrast, truly coagulated rods will stay connected, as energy is required to move them out of the deep primary minimum in which they are stuck. When preparing samples for AFM measurements, the suspension is diluted by many orders of magnitude, as we need sparse deposition on the substrate for proper imaging. Despite this very strong dilution (which certainly led to very easily flowing samples), we clearly saw chains of CNCs in the AFM images. This gives clear evidence that the kinetic arrest in this case was indeed due to rigidity percolation through permanently connected CNCs, thus to gelation.

Since we are discussing liquid crystal phases, a key question that nevertheless is almost always overlooked in the context of CNC suspensions is what impact the orientational order has on percolation. The vast majority of studies of percolation, experimental, theoretical and simulations, have been carried out on isotropic systems, with random-aligned rods. An exception is the team of van der Schoot and Schilling, who have studied percolation in liquid crystalline systems for some time. A recent work from the team, by Finner et al. [110], has critical consequences for the understanding of kinetic arrest in CNC suspensions. While their study focused on electrical percolation, i.e., where finite-sized conductive rods eventually connect to form a macroscopically conductive system, their conclusions can be transferred to the case of rigidity percolation with some additional considerations. Importantly, the systems considered in [110] deal with a colloid of only one type of rods, which may undergo an isotropic–nematic transition and/or percolation as the rod volume/mass fraction is increased. In this respect, their system is perfectly analogous to the CNC suspensions considered in this review.

The take-home messages in [110] are that (1) rod *connectivity* is the most critical component for percolation when nematic ordering is considered, and (2) that much of what we know about percolation in isotropic systems is turned on its head when the effect of long-range orientational order is taken into account. How is then the critical parameter 'connectivity' defined? Finner et al. define it by introducing a critical rod-to-rod distance d_c, such that two rods are considered connected if they approach each other by less than d_c. They find four regions of very different behavior depending on the connectivity, i.e., on the magnitude of d_c. For low d_c, i.e., weak connectivity (the rods need to get very close to be considered connected), percolation never takes place; while in the isotropic phase the likelihood of two rods getting close to each other is greater, due to the random orientation, the low value of d_c means that nematic ordering takes over before percolation of isotropically arranged rods can be seen. Once the nematic phase forms, the essential outcome is that d_c must pass a certain threshold value for having percolation but, very interestingly, percolation then happens for *all* rod fractions and *any* rod aspect ratio, as long as the phase stays nematic.

This reminds very much of our system, where gelation was seen beyond a critical value of the counter ion concentration, c_c^k, provided that we can correlate c_c^k with d_c. Indeed, such a correlation is plausible, because while a high d_c in the work of Finner et al. means that two rods are considered connected even if they are not very close, in our case $c_c = c_c^k$ means that connectivity in terms of attraction—by van der Waals interactions, hydrogen bonding or the hydrophobic effect—dominating over electrostatic repulsion will happen although it requires close encounters, because the high counter ion content enables such close encounters. Thus, while we without doubt have quasi-linear chain aggregation in CNC suspensions as we approach $c_c = c_c^k$, and thus effective rods with much greater aspect ratio, this may not be the most important point. Rather, the critical issue is that the gelation threshold value of c_c^k corresponds to a threshold connectivity, which brings the cholesteric CNC suspension into the regime where percolation takes place regardless of aspect ratio.

Nevertheless, a remaining important question is what the origin of the peculiar end-to-end aggregation might be. Finner et al. noticed in their simulations that their percolating clusters were extended along **n**, attributing this to the nematic order, but a detailed explanation of why we do not see mainly side-to-side aggregation, which should also be promoted by orientational order, was not given. In [106], we proposed a potentially generic explanation based on the different effects of counter ion fluctuations near a rod end and near a rod side. However, we were later made aware of a significant aspect of the chemical properties of CNCs that may be even more relevant. Lokanathan et al. [115] demonstrated that the ends of CNCs can be selectively functionalized by reductive amination without any alterations at the sides of the CNCs. This can only be explained with the presence of aldehyde groups and proves that the terminal glucosyl moiety of each CNC rod is able to undergo spontaneous isomerization, in analogy to the mutarotation known from cyclic sugars. In Figure 9, the isomerization process of CNCs in suspension is depicted, showing how the terminal glucosyl moiety can transform from cyclic β-D-glucopyranosyl via the open-chained D-glucosyl, which incorporates the aldehyde functionality, to α-D-glucopyranosyl, again cyclic. The isomerization is catalyzed, for example, by acids, which are often still present in CNC suspensions due to the acidic hydrolysis which is applied during production.

Looking at the different isomers, one can imagine that at least the open-chained D-glucosyl as well as the α-D-glucopyranosyl moieties possess polarizabilities differing from the remaining $\beta(1 \rightarrow 4)$ glycosidic linked D-glucosyl moieties, and thus may also lead to different surface densities of -SO_3^- groups during acid hydrolysis. These variations in the surface charge and/or the van der Waals attractions between the sides and the terminal ends of the CNCs might explain the end-to-end aggregation. Furthermore, we may speculate about the formation of new $\beta(1 \rightarrow 4)$ glycosidic bonds if the hydroxyl group of a starting D-glucosyl moiety, marked in green on the left in Figure 9, would attack the aldehyde group of the terminal D-glucosyl moiety of another rod. This would then lead to an end-to-end condensation of the rods. While these ideas would need to be corroborated with further experiments, they seem quite plausible, and they would provide a rather straight-forward explanation to the end-to-end aggregation. At $w \approx w_k$ (or by adding NaCl until reaching $c_c \approx c_c^k$), the colloidal stability is still just barely retained on the sides but not at the ends, where the surface charge is presumably lower. As CNCs fluctuate in the sample, they thus build the chains seen in AFM, probably contributing to triggering rigidity percolation.

It is interesting to note that also fd virus has different proteins coating the ends compared to the cylindrical side [116], thus also here the side-to-side interaction may be different from the end-to-end interaction. In fact, Petrova et al. found linear chain formation of fd virus at high ionic strength, very similar to our situation, when the viruses were adsorbed on a weakly charged cationic lipid membrane [116]. While the authors attributed the linear chain formation to the interactions with the membrane, the fact that linear chain formation was observed only at high I, and considering the contrast between sides and ends of the fd virus proteins, a further investigation of localized variations in colloidal stability would certainly be motivated. Possibly, the influence of ordering the virus rods into a plane provided by the lipid membrane might be comparable to the influence of liquid crystalline order in a cholesteric nanorod suspension. There are clearly interesting questions to investigate here in the future, where comparisons between CNC and fd virus suspensions may be very worthwhile.

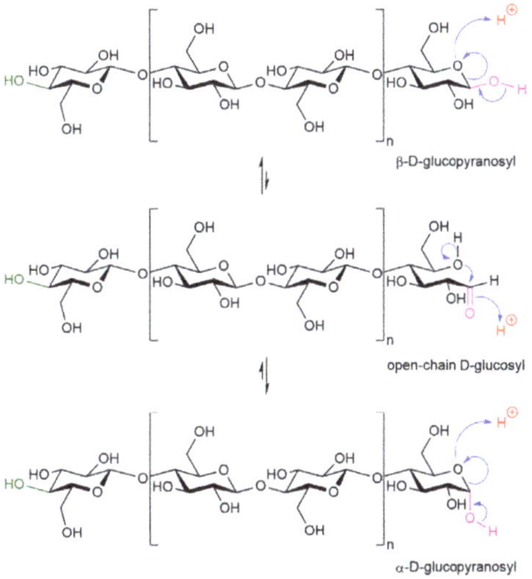

Figure 9. CNCs are composed of several thousands of $\beta(1 \rightarrow 4)$ linked D-glucosyl moieties. The terminal glucosyl moiety undergoes acid catalyzed isomerization. Thereby the functionality of the group which is highlighted in pink changes from hydroxyl to aldehyde. The blue arrows indicate the rearrangement of the chemical bonds. Next to the two shown six-membered cyclic pyranosyl isomers, two five-membered cyclic furanosyl isomers (not shown) could form additionally. Isomerization of the starting glucosyl moiety is not possible.

5.3. How Do We Detect Kinetic Arrest Experimentally?

A simple, approximate experimental method of detecting kinetic arrest is to gently turn a sample vial upside down and observe the evolution with time. If after 24 h the suspension did not sink to the bottom, we tend to consider the sample to be kinetically arrested [114,117], although it can, righteously, be considered too simplistic for an absolute confirmation [118]. Nordenström et al. [114] used this method combined with dynamic light scattering to investigate the kinetic arrest in their study (see above). Phan-Xuan et al. [119] measured the critical aggregation concentration of a CNC system with addition of mono-, di- and trivalent ions using the appearance of turbidity as a criterion. They also studied the system with small angle X-ray scattering and supported the experiments with a simple coarse-grain model in Monte Carlo simulations. They showed that, at low ionic strength, the suspensions with monovalent ions are stable, the critical aggregation concentration depending on ion type along the order $Li^+ > Na^+ > K^+ > Cs^+$. For di- or trivalent ions the critical concentrations are significantly lower, as expected from the squared dependence on z_i in Equation (2).

Rheology is another key tool for identifying the gelation threshold and analyzing the nature of the formed network. A differentiation can be made between a viscoelastic liquid and a solid-like gel by evaluating the development of the two moduli, G' and G'', as a function of frequency, where for a viscoelastic liquid $G' < G''$ and a solid-like gel has $G' > G''$. The first reported systematic investigation of the rheological behavior of CNC suspensions that we are aware of was done in 2011 by

Ureña-Benavides et al. [120]. Shafiei-Sabat et al. [121] later conducted a rheological study in which they compared two wood-derived CNC suspensions with different degrees of sulfonation, in disordered and ordered states, finding that CNCs with higher surface charge are kinetically arrested at higher mass fraction than CNCs with lower surface charge. Higher sulfate group content led to a broad biphasic transition region, forming a gelled state at higher w compared to those with lower degree of sulfonation. Probably, the lower surface charge in this case decreased the electrostatic repulsion sufficiently to promote aggregation at lower w. The authors did not detect any difference in particle size between the two systems they studied, although this could have had an influence. In 2016, some of us conducted a study confirming these observations [117]. Our rheological investigation demonstrated a clear difference for CNCs with high and low surface charge in a semi-logarithmic representation of shear viscosity versus concentration. For CNCs with low surface charge the transition to kinetic arrest was distinguished by a sudden superlogarithmic increase of viscosity whereas for CNCs carrying more surface charges no such trend was observed at any shear rate in the investigated concentration regime.

Considering that cholesteric CNC suspensions tend to develop a helical structure; also, the texture as observed in polarizing microscopy can be used as a test of kinetic arrest. We used this method in a study of CNC suspensions in different non-aqueous solvents [122], interpreting the inability over several days of a birefringent bulk CNC suspension to form a fingerprint texture as a sign of kinetic arrest. Comparison with rheological data for the suspensions showed that this interpretation was correct. A general point to remember when investigating CNC suspension experimentally, concerning kinetic arrest or other aspects, is that the sample history can have significant impact, as is typical for non-ergodic systems such as gels. Differences in kinetic arrest behavior have been attributed to varying hydrolysis conditions [117,120] or differences in the sonication procedure [108].

6. Tuning the Equilibrium Behavior—And Its Range—By Modifying the Solvent

So far in this review, the continuous phase surrounding the CNCs has been discussed almost only in terms of its content of counter ions released from the CNCs when they are brought in suspension, and the assumption has generally been that the liquid used for the continuous phase is water. There is, however, also a rich body of research on mixed or entirely non-aqueous suspension hosts [123–127], sometimes with additives that can range from simple salts to reactive molecules that can be transformed into macromolecular structures like organic polymers or silica. Looking back at the definition of the Debye screening length, Equation (5), we find that both the relative permittivity of the solvent and the concentration of free ions have a major impact on κ^{-1} and thus on the behavior of colloidal suspensions. This makes it worthwhile to have a deeper look into this topic.

There are many applied reasons for modifying the continuous phase, for instance in order to make composites where helically arranged CNCs are embedded in a solid matrix, inorganic [128–133] or organic [134–141], that is templated by the CNCs (the CNCs may be removed afterwards). To make mechanically durable and flexible composites it is interesting to incorporate synthetic polymers, but as many of them (and their corresponding monomers) are not water-soluble, a switch to a non-aqueous solvent for the CNC suspension may be necessary. On the other hand—and more interesting from the perspective of this review—much can also be learnt from such studies from a basic science point of view. Our understanding of the particle–particle and particle–solvent interactions in CNC suspensions, not least how the chirality transfer mechanism at the heart of the cholesteric helix formation works, is for instance far from satisfactory, and it may benefit from seeing how other solvents affect the self-organization.

To keep this review contained we will not go into most of these rich paths of explorative CNC-based materials science, but we will focus only on two aspects, from which we can learn much about how the organization of CNCs is affected by the continuous phase. We will look at the effects of replacing water by non-aqueous solvents, and we will look into how the equilibrium phase diagram and helical pitch of aqueous cholesteric CNC suspensions are affected by addition of salt or glucose. However, first we look at a practical challenge of large significance.

6.1. The Challenge of Preparing Non-Aqueous CNC Suspensions and How to Overcome It

Suspending CNCs in non-aqueous solvents is challenging for two reasons. The first reason is that we may need alternative methods to stabilize the CNCs in the new suspension medium, since electrostatic repulsion is not necessarily a viable option in organic solvents. The second reason is that CNCs are usually prepared in acidic aqueous solutions and thus have to be transferred to another solvent without risking irreversible agglomeration of the CNCs (see Section 3). Concerning the colloidal stability of the, usually highly charged, CNCs in non-aqueous solvents, a variety of different approaches can be found in the literature. The most straightforward way is to choose other polar solvents such as the conventional organic solvents dimethyl sulfoxide (DMSO), N, N-dimethyl formamide (DMF) or tetrahydrofuran (THF) [142–147] or mixtures with ionic liquids [148–150]. Okura et al. [151] investigated the dispersibility and shear-induced birefringence of CNCs that had been produced by HCl hydrolysis in 21 different solvents. They concluded that a high relative permittivity ϵ_r as well as a high viscosity of the solvent are beneficial for stabilizing CNCs (the viscosity obviously only contributing to kinetic stabilization). Furthermore, they noted that the CNCs could only be dispersed in amphoteric solvents, i.e., solvents which are strong Lewis acids and bases at the same time, due to their ability to strongly interact with the hydrophilic CNCs.

Further methods to promote colloidal stability of CNCs in non-aqueous solvents are to change the acid used for hydrolysis [146], neutralize the CNCs with base to make them less polar [138] or alter their surface properties by covalently bonding larger chemical moieties to their surface [123,144,152–166]. As work in the latter field has been extensively documented, we will not go into detail here, but refer to two reviews by Eyley and Thielemans [124] and by Habibi [123] about surface modification of CNCs. With the right choice of modification and degree of substitution, the range of solvents stabilizing CNC suspensions can be largely expanded. For example, Yuan et al. [167] acetylated the surface with iso-octadecenyl succinic anhydride and investigated dispersion quality in 16 different solvents. While the pristine CNCs showed good suspension properties in solvents with a relative permittivity of $\epsilon_r \geq 37.8$, the modified CNCs could only be suspended in solvents with a moderate to low relative permittivity of $37.8 \geq \epsilon_r \geq 2.2$. Last but not least, it is possible to stabilize CNCs in apolar solvents such as cyclohexane or toluene by adding surfactants that adsorb to the hydrophilic surface of the CNCs and thus make them hydrophobic [125,126,161,168,169].

To transfer CNCs from one solvent to another, a common approach has been to freeze-dry aqueous suspensions of pristine [143,144,147,158,159], surface modified [153,154,157,162,165,167] or surfactant-stabilized [125,126,155,161,168,169] CNCs. Especially for pristine CNCs, freeze drying is preferable to other drying methods, since the strong capability of the CNCs to form hydrogen bonds may otherwise lead to irreversible agglomeration. This effect can be avoided by introducing stabilizers, such as surfactants or surface modifications, before drying. Furthermore, two groups showed that stabilized CNCs can be dried in vacuum at 45 °C to 50 °C without aggregation [158,164]. A second method that can be used when the old and new solvent are miscible is to strongly centrifuge the original CNC suspension to bring the CNCs to the bottom of the vial, remove the supernatant liquid and replace it with the desired new solvent; this process is then repeated until almost all water is gone [163]. Tian et al. [164] compared the two methods using pristine CNCs and concluded that the second one is preferable, because they found in their study that CNCs agglomerated stronger during freeze-drying, to the extent that it might be impossible to re-suspend them again. A third method is to mix the aqueous CNC suspension with the other solvent, followed by evaporation of the water under atmospheric [144] or reduced pressure [122] conditions. Obviously, this method can only be applied to solvents with boiling point considerably higher than that of water and which do not form a eutectic mixture with water. The new solvent must also either be able to disperse the CNCs on its own or it must host the stabilizer required for stable dispersion. Furthermore, elevated temperatures encountered in the procedure might lead to hydrolysis of sulfate groups, altering the properties of the CNCs.

6.2. Liquid Crystallinity and Kinetic Arrest of Non-Aqueous CNC Suspensions

Up to now, only few studies investigated liquid crystalline organization of CNC suspensions in non-aqueous media in detail. Often the sole evidence published for anisotropy is the observation of birefringence in larger vials [144,157,167,169] or flow-induced birefringence [144,147,151,154,162]. Some researchers went one step further by conducting polarizing microscopy investigations of the textures of the non-aqueous suspensions. Two publications describe the formation of tactoids in toluene [168] or DMF [138]. Several more report the observation of fingerprint textures [122,125,126,138,155,158,168,170], which proves that the helix formation is preserved in a variety of solvents. On the other hand, some authors only find uncharacteristic textures without any evidence for the formation of macroscopically chiral structures in suspensions of surface-modified [153,154,159] or surfactant-stabilized [169] CNCs in toluene or chlorobenzene. Xu and coworkers [159] concluded from the lack of evidence of helical modulation that surface grafting makes "the twisty rods smooth and straight". Indeed, the chiral interactions between rods may well be affected by the process. However, as discussed in Section 4.1, absence of a detectable helix is not a sign of the phase not being cholesteric; there are many factors that may prevent or delay the development of the helical director field modulation so we may still have $q_0 \neq 0$. At least three [153,154,169] of the four publications which report uncharacteristic textures additionally mention a high viscosity of the suspensions. This leads us to propose that the suspensions may have been in or in the proximity of a kinetically arrested state at the investigated concentrations. The inability to develop a helix would then be due to a loss of equilibrium, rather than loss of chirality. We have seen this type of kinetic arrest-induced loss of helix-related textures in our own studies of non-aqueous CNC suspensions [122].

Regarding the pitch of the helix, when it develops in non-aqueous solvents, quite varying values have been published: Yi et al. [158] found a pitch of 1–2 µm for polystyrene-grafted CNCs in DMF compared to 2–3 µm for ungrafted CNCs in water. For surfactant-stabilized CNCs Heux et al. [125] measured a pitch of 4 µm in cyclohexane, while Zhou et al. [168] reported a value of 17 µm in toluene. It is difficult to draw conclusions and improve our understanding of the physics behind the helix formation from such singled-out values, especially if we consider that these values cannot be compared directly due to potentially very different CNC mass fractions for each reported value, and differing stabilization and production methods employed. Consequently, in the following we will focus on only a few publications which investigate liquid crystallinity of CNC suspensions in non-aqueous media in a systematic and comparative manner.

Frka-Petesic et al. studied the orientation and unwinding of the cholesteric helix by an applied electric field [155]. To overcome certain disadvantages of water, e.g., the high electrical conductivity, they used toluene as a solvent, using surfactants for stabilizing the CNCs after transfer to this nonpolar host. In this context, they also investigated the phase behavior and equilibrium pitch p_0 of the system. They found a two-phase coexistence region roughly between 20 and 25 wt.%, where the mass fraction refers to combined CNC and surfactant in relation to the total mass, and they did not observe kinetic arrest up to the maximum investigated CNC content of 38 wt.%. Although the mass fraction of the pure CNC is somewhat lower, as the surfactant is also considered, we note that the rod mass fractions are significantly higher here than in aqueous CNC suspensions. With increasing mass fraction, the authors found p_0 to decrease from roughly 4 to 2 µm. Interestingly, they mentioned that the chiral interactions which lead to the helical precession of the CNCs cannot be transmitted from one CNC to the next by electrostatic interactions, because ion dissociation is hindered by the low relative permittivity of toluene ($\epsilon_r = 2.38$) and thus cannot contribute to the chirality transfer through the solvent.

A similar system of surfactant-stabilized CNCs in cyclohexane was investigated by Elazzouzi-Hafraoui et al. [126]. They changed not only the concentration range but also the rod length of the CNCs by altering the temperature during treatment with sulfuric acid. They found a similar dependence of aspect ratio for the phase diagram as predicted by Onsager theory and known from aqueous CNC suspensions (see Section 4.3 or Section 4.5, respectively): w_0 shifts from 16 to 21 wt.% and w_1 from 32 to 41 wt.% when the aspect ratio was reduced. Next to the aspect ratio, an accordingly altered surface charge due to the changed treatment time with sulfuric acid might play a role, too. However, this is not discussed in the publication. For very long rods (and thus a comparatively low surface charge) they found a kinetically arrested state, then above 22 wt.%, immediately after the end of the isotropic phase. Considering the theoretical results of Finner et al. [110] (see Section 5.1), we suspect that percolation and ordering, with appearance of birefringence, probably happened simultaneously in that case. We have seen the exact same phenomenon in CNC suspensions in the low-permittivity solvent DMF [122].

For the helical pitch Elazzouzi-Hafraoui et al. found a decrease from roughly 4.6 to 2.5 µm for high aspect ratios and from 6.2 to 2.6 µm for low aspect ratios when increasing the mass fraction. Note that, since the aspect ratio influences w_0 and w_1, this does not necessarily mean that the aspect ratio has a direct impact on the pitch, but it can rather be related to the differences in CNC mass fraction at which the helix pitch was measured for the different aspect ratios. The authors compared their values to the ones found in an aqueous CNC system [56] which was produced with a similar method, showing pitch values one order of magnitude larger than in the cyclohexane system. Elazzouzi-Hafraoui et al. attribute the difference to stronger chiral interactions of CNCs in solvents with a low relative permittivity such as cyclohexane ($\epsilon_r = 2.02$) compared to water ($\epsilon_r = 80.20$). While this may be true, it is important to note that neither the lack of surfactants nor the substantially smaller diameter of the CNCs in the pristine aqueous system were considered, nor the roughly one order of magnitude greater mass fraction of CNC in the non-aqueous system. All these differences need to be taken into account, hence it is unclear whether or not the differences in pitch can be traced back to the solvent properties alone.

The only way to learn more about the specific effect of the solvent on CNC suspensions, is to investigate the same type of CNCs, i.e., with no differences in the CNC source, production method, surface charge and particle size, in varying solvents. We did this in a study from 2016 [122] for the solvents water ($\epsilon_r = 80.20$), formamide ($\epsilon_r = 111.0$), N-methyl formamide (NMF, $\epsilon_r = 189.0$) and N,N-dimethyl formamide (DMF, $\epsilon_r = 38.25$) and found striking correlations between the relative permittivity and the properties of the suspension. This set of solvents is interesting also from another perspective, and this is the ability to form hydrogen bonds. While formamide resembles water in the sense that both solvents form 3D networks of hydrogen bonds, NMF can only form 1D hydrogen bond chains, and DMF can act only as hydrogen bond acceptor, hence there are no solvent–solvent hydrogen bonds in DMF.

Starting with this latter solvent, we found that kinetic arrest coincides with the occurrence of birefringence at the low value $w_k \approx 2$ wt.%. This can be explained by the poor stabilization of the charged rods in DMF due to its low permittivity, reducing κ^{-1} according to Equation (1) and thus allowing closer encounters between rods. An additional effect may be the greater prominence of inter-CNC hydrogen bonding in a solvent that is such a poor hydrogen bond partner, deepening the secondary energy minimum in Figure 2 and thus increasing the chance of long-lasting connections between rods that meet. Percolation and gelation at low w would be the result. Note that the situation is not comparable to that of Frka-Petesic et al. [155] and Elazzouzi-Hafraoui et al. [126], where the solvents toluene and cyclohexane have much lower values of ϵ_r still, because the CNCs were stabilized by surfactant in those studies. In contrast, we studied the behavior of pristine CNCs prepared by hydrolysis with sulfuric acid, thus electrostatically stabilized with H^+ counter ions, after transfer to DMF.

For the other three solvents, we initially reported that all phase diagrams were shifted to slightly lower mass fractions as ϵ_r increased. A more careful look at the data, however, reveals that the onset CNC mass fraction w_0 is increased slightly as the permittivity of the solvent is increased, while w_1 shifts in the opposite direction. For water, $w_0 \approx 2.5$ wt.%, increasing to $w_0 \approx 2.6$ wt.% in NMF and $w_0 \approx 2.7$ wt.% in formamide. The variations are on the order of the error so one should be careful about drawing far-reaching conclusions here. The corresponding values for the stability limit of the cholesteric phase change more, $w_1 \approx 8$ wt.% for water, $w_1 \approx 7.5$ wt.% for formamide and $w_1 \approx 7.3$ wt.%

for NMF. In the biphasic regime, we see a cross-over behavior around $w \approx 4$ wt.%, corresponding to roughly 20% cholesteric and 80% isotropic phase. It is important to remember that the values reported here refer to the full macroscopic volume of the sample, which is highly disperse in terms of CNC rod length. Remembering the discussion in Section 4.5, this means that w_0 always refers to the longest rods of the sample nucleating the first tactoids of cholesteric phase, and w_1 always corresponds to the shortest rods, requiring the highest mass fraction to develop long-range order.

In order to make sense of these observations we note that the change in ϵ_r has a direct as well as an indirect effect on the suspension. The direct effect is an increase of the Debye screening length as we move from water to formamide to NMF, since $\kappa^{-1} \propto \sqrt{\epsilon_r}$, see Equation (1). This would thus increase the effective volume fraction of the CNC at the same mass fraction, which would tend to reduce the critical mass fractions, explaining the change in w_1, although the effect on the critical volume fraction ϕ_0 would also need to be considered. The indirect effect is more subtle. The CNCs used here still had the H^+ counter ions remaining after sulfuric acid hydrolysis, and here the degree of dissociation from the $-SO_3^-$ groups at the CNC surface can vary quite significantly depending on the solvent. A higher value of ϵ_r promotes dissociation, as seen in the definition of the Bjerrum length, Equation (4), which decreases with increasing ϵ_r. A higher ϵ_r thus makes it easier for thermal energy to keep counter ions dissociated from the particles, yielding a greater concentration c_c of free counter ions in the solvent. As Equation (1) has ϵ_r in the numerator and I in the denominator, we see that the changes in dielectric permittivity and in free counter ion concentration could compensate each other, depending on how much c_c is affected. It could be that the dissociation is very high in a high-permittivity solvent at low CNC content, while increasing CNC content gradually reduces the degree of H^+ dissociation, such that the increase in κ^{-1} provided by higher ϵ_r starts dominating. This might explain the cross-over to lower critical mass fractions for high-permittivity solvents as w increases, but more extensive investigations are needed to draw clear conclusions. The dispersity of the sample, and the different effects of a greater Debye length on short and long rods, respectively, may be another factor to consider.

As for the helical pitch and how it depends on CNC mass fraction, we detected quite a remarkable effect of the relative permittivity of the solvent. In contrast to the conclusions from Elazzouzi-Hafraoui et al., we found that the magnitude of the pitch decreases with increasing ϵ_r, a difference we attribute to our CNCs being pristine while theirs were surfactant-stabilized. However, most significantly, the variation in pitch with CNC mass fraction is greatly reduced as ϵ_r increases. With water we found that p_0 decreases over a broad range, from 48 to 16 µm as the CNC mass fraction increases from $w_0 \approx 2.5$ wt.% to $w_1 \approx 8$ wt.%. With formamide the pitch started out much shorter, decreasing from 16 to 12 µm as the CNC content changed from $w_0 \approx 2.7$ wt.% to $w_1 \approx 7.5$ wt.%. With NMF, having the highest ϵ_r, p_0 stayed more or less constant at the very low value of 3 µm regardless of CNC mass fraction, from $w_0 \approx 2.6$ wt.% to $w_1 \approx 7.3$ wt.%. Moreover, also the time to develop the helix was strongly affected by the variations in ϵ_r. Even though a linear dependence of this time on viscosity (increasing with increasing w) was found for all samples, the slope of this dependence was much greater for water than for the high-permittivity solvents formamide and NMF. The effect is significant, as illustrated by the fact that the time to reach the equilibrium p_0 at $w = 8$ wt.% decreased from 3 days in water to 14 h in formamide. It thus appears that high solvent permittivity has a dramatic effect in strengthening the chiral interactions between the CNCs, but the reason for this is not yet clear.

Concerning the mass fraction w_k at which kinetic arrest occurs, no clear connection to the relative permittivity was found. It increased slightly from $w_k \approx 8$ wt.% in water to $w_k \approx 9$ wt.% in formamide and NMF. The highest w_k value was actually found for formamide, possibly because this solvent gives a maximum dielectric permittivity while still maintaining a 3D hydrogen network between solvent molecules, thus counteracting the attractive interactions between CNCs due to inter-CNC hydrogen bonding.

Recapping this section, we want to point out that CNCs suspended in a variety of solvents are able to form the same liquid crystal structures as in conventional aqueous suspensions and that the solvent properties, especially the relative permittivity, have proved to be potent tools to alter characteristics such as the phase behavior, the helix pitch and its dependence on CNC mass fraction, as well as the kinetics of the system. However, one has to consider that alternative stabilization methods, i.e., the use of surfactants or surface grafting, have a major impact on the particle characteristics compared to suspensions with pristine CNCs. Moreover, it is important to recognize that the relative permittivity of the solvent can affect the behavior of a colloidal suspension in more than one way, directly and indirectly, and that variations in hydrogen bonding capacity of the solvent molecules can have strong impact on the interaction with the CNCs. This makes intuitive predictions about the colloid behavior difficult.

6.3. The Response of the Equilibrium Phase Diagram and Helix Pitch to Selected Solutes in the Aqueous Phase

Many papers have been published where the effects of salt addition on the characteristics of films produced by drying CNC suspensions have been investigated. The problem is that quite a few of them do not look into the equilibrium phase diagram behavior; often, the starting point is a low-concentration suspension that is fully isotropic, and the investigations are then focused on films formed by drying this suspension, while little or no investigations are done at intermediate concentrations. This renders it tremendously difficult to draw clear conclusions regarding the reasons for the characteristics of the final dry film, because—as should be clear from the previous sections of this review—there are so many factors of the equilibrium system that influence the end results, from the onset and completion of liquid crystal formation, to the helical modulation of the director field, to the tactoid dynamics, to the end of equilibrium via kinetic arrest, that we cannot know if one is more important than the other, and how they may be connected. Therefore, in this section we look exclusively at the papers that have investigated the effects of additives on the *equilibrium* behavior of aqueous CNC suspensions. As above, we point out that any mentions of w_0 and w_1 refer to the macroscopic, disperse, sample, hence the two values effectively correspond to different rod aspect ratios.

A classic paper is the already mentioned study by Dong et al. [56], in which they obtained a linear relationship between cholesteric volume fraction and CNC mass fraction by keeping the ionic strength constant through adjustment with HCl addition (see Section 4.3 in the context of Equation (13)). They also looked at how the phase diagram was shifted by salt addition, comparing HCl, NaCl and KCl. The authors found that the choice of salt did not matter, H^+, Na^+ and K^+ cations having identical effects, and that all of them shifted the phase diagram to higher CNC mass fractions. A plot of w_0 versus concentration of added NaCl suggests a dependence on salt addition that roughly follows a square root relationship, but the authors did not fit any functions to the data. Considering the reported lack of sensitivity to the effect of counter ion of the added salt, it is worthwhile in this context to remind of the follow-up paper by Dong and Gray [57] discussed in Section 4.3, where they showed that the type of ion *does* matter, at least for the onset of liquid crystal formation, if salt is not simply added but if the counter ions of the CNCs are instead exchanged.

In [56], the authors also looked at the impact on the helical pitch of salt addition, again finding that the choice of counter ion was irrelevant, all salts having a very strong reducing effect on the pitch. For simplicity they considered a constant CNC mass fraction w_{fix}, corresponding to the mid point of the phase coexistence region prior to salt addition, and they found that p_0 decreased non-linearly (again no fit was done so the exact function was not established) from $\sim$63 µm in the pristine suspension to $\sim$30 µm with the maximum possible salt addition of 2.5 mM. [171] When evaluating this effect one should note that the CNC mass fraction stayed constant, but the system moved from the mid point of the phase coexistence region to very near w_0 due to the effect of salt addition on the phase diagram. It would have been interesting to know how the pitch versus CNC mass fraction function was affected, but the reduction in p_0 as $w_{\text{fix}} \rightarrow w_0$ via salt addition suggests that the effect may be similar to that of increasing ϵ_r of the solvents, discussed above. Therefore, we may expect that increased ionic strength reduces not only the absolute value of the pitch but also its dependence on w.

In our study of two series of sulfuric acid-derived CNC with different strengths of the surface charge [117] we found that the relative effect of salt addition on w_0 is roughly the same for both CNC types. For instance, the addition of 2 mM NaCl to a suspension of the low surface charge (L) CNC increased the onset from $w_0^L \approx 3$ wt.% to about 5 wt.%, whereas the same NaCl addition to the high surface charge (H) CNC-type shifted the onset from $w_0^L \approx 3.5$ wt.% to about 5.5 wt.%, thus in both cases a shift of about 2 wt.%. Regarding the helix pitch, the salt-free suspensions showed slightly longer p_0 at identical w for low surface charge and both sample types showed a general trend of decreasing pitch upon NaCl addition, although experimental circumstances made data unreliable as well as somewhat inconsistent.

From a practical point of view in terms of tuning the parameters of cholesteric CNC suspensions and solid films made from them (discussed in Section 7.4), a problem with salt addition is that both the pitch and the phase diagram are strongly affected, making it impossible to tune one parameter without affecting also the other. In this respect, a very interesting additive is glucose, which was explored by Mu and Gray in a paper [172] investigating both the impact on the equilibrium CNC suspension properties and on the films dried from them, using acid form CNC (H^+ counter ions). We will come back to their observations regarding the latter aspect in Section 7.4, as we here focus on the equilibrium behavior. Interestingly, no impact whatsoever of glucose addition was found on the equilibrium liquid crystal phase diagram, w_0 being identical to the glucose-free suspension even up to 10% D-(+)-glucose [173]. In contrast, the equilibrium pitch was affected significantly by the glucose addition. Using a stock solution with $w = 5.2$ wt.% (in the coexistence regime, nearer w_1 than w_0), the cholesteric pitch was found to decrease from 11 μm without glucose to 7 μm with 10% glucose. In a preliminary experiment (to be published) where we tried to repeat this using a CNC suspension with Na^+ counter ions, we confirmed the absence of an effect on w_0 but the change in pitch was in the opposite direction, towards slightly longer values. While these data need to be confirmed, it could be that the type of counter ions plays an important role for the response of the helix formation even to uncharged additives like glucose. While Mu and Gray did not give quantitative data for w_1 and w_k, the discussion suggests that w_k was decreased, as glucose addition apparently promoted gelation. As we will see in Section 7.4, this had a striking effect on the color of the dried films produced by these suspensions.

7. Creating Color with CNCs by Drying Suspensions Into Solid Films: Where Intriguing Science Meets Promising Applications

We now have a quite good overview of how CNC suspensions behave in equilibrium and where the equilibrium range ends. When we move our focus to the process of making the colorful dried films that provides much of the applied motivation for CNC research, by evaporating the solvent from a sessile drop of CNC suspension, we must leave equilibrium for two reasons. First, the process of evaporating liquid from a droplet is fundamentally a non-equilibrium situation as we continuously change the composition of our system. Second, as we have seen above, the kinetic arrest that marks the end of equilibrium even in a closed system typically happens with some 90 wt.% water left in the system. This means that the path from the equilibrium CNC suspension to a selectively reflecting film covers a very extensive range of non-equilibrium behavior. Luckily, this range has recently been exquisitely analyzed by Frka-Petesic et al. [174], allowing us to largely refer to their paper for the details of the process. In this final section of our review before it is time to conclude, we will, nevertheless, summarize the key components of this important step. However, first we will explain where the selective reflection colors come from in the first place, pointing out what is needed to see the colors and how their presence should be tested experimentally. We also need to look into another non-equilibrium phenomenon that takes place in a droplet from which water is being evaporated, occurring prior to kinetic arrest, namely the coffee-ring effect. We will focus solely on aqueous suspensions of CNC in this section, as we are not aware of reports of films produced by drying non-aqueous CNC suspensions.

7.1. Bragg Reflection and Structural Color in Cholesteric Liquid Crystals

If we place a short-pitch cholesteric liquid crystal sample with vertical helix axis **m** on a black background and observe it along **m** (normal incidence), it will be difficult not to notice that the sample intensely (and quite beautifully) reflects a narrow wavelength range of light, giving it a strongly colored appearance, see Figure 10b,c. With "short-pitch", we here mean $0.25 \lesssim p \lesssim 0.5$ µm, the range boundaries to be explained below. Moreover, if you look at the sample through right- and left-handed circular polarizers (for instance the right and left eyes of a pair of 3D cinema glasses, as in Figure 10), you will notice that the reflected color is unaffected by one of the polarizers but completely blocked by the other. Now, if you incline the sample such that you are looking along an angle $\theta \neq 0$ away from **m** (oblique incidence), making sure that ambient light is sufficiently diffuse that the sample gets illuminated from the opposite side at the corresponding angle $-\theta$, you will find that the reflection color is blue-shifted, i.e., the wavelength of the reflected color is shorter.

All these effects are illustrated in Figure 10b,c for the case of a large film prepared by drying a CNC suspension as well as for a small sample of thermotropic cholesteric liquid crystal enclosed between glass substrates with aligning layers, as reference. Both have reflection largely in the orange-red region for normal incidence. While the CNC film has a left-handed helical modulation the thermotropic sample is right-handed. You will notice that the effects just described apply perfectly to the small thermotropic sample, which we know has uniform helix pitch and perfectly vertical orientation of **m**, thanks to the aligning layers. With the CNC film, in contrast, we notice in Figure 10b that the left-handed circular polarization of the reflected light quickly is lost as θ is increased from zero. Moreover, the blue-shifting of the reflection color is more dramatic for the thermotropic sample than for the CNC sample. Figure 10c shows that the reflection from the thermotropic sample, viewed under the exact same conditions, retains very strong right-handed circular polarization even for very large θ. What we can conclude from this comparison is that dry CNC films indeed exhibit much of the optical characteristics that are inherent to equilibrium short-pitch cholesteric liquid crystals, but the behavior is often less pronounced, suggesting imperfections in the structure. As we will discuss a bit further below, removing these imperfections is a major goal for applied CNC research today. However, in order to do so, we first need to be able to characterize the films properly. To this end, let us first establish what the expected optical behavior is for an ideal cholesteric liquid crystal.

As mentioned in Section 4.1 the long-range orientational order renders a nematic phase birefringent, with refractive indices $n_\perp$ and $n_{||}$ for polarization perpendicular and parallel, respectively, to **n**. In terms of optics, the helical modulation of the director in a cholesteric is thus equivalent to a helical modulation of the optic axis. For a certain linear polarization of light entering along the helix direction, this means that the refractive index varies periodically between $n_{||}$ and $n_\perp$ with a period $p/2$, see Figure 10a.

Elucidating the optical properties of a birefringent medium with a helically modulated optic axis is not trivial: one needs to solve Maxwell's equation for a medium with a continuously rotating dielectric permittivity tensor. The first to attempt this was Oseen [47], publishing the results in German in a journal of the Swedish Academy in 1928, then in English in his seminal 1933 Faraday Transactions paper on the theory of liquid crystals [175]. This allowed him to explain some of the curious optical properties of cholesterics, namely the fact that there is a wavelength band gap in which light incident along the helix axis **m** is separated into two circularly polarized eigenmodes, one of which is fully reflected and one of which is transmitted. Although his calculations were mainly restricted to the case of normal incidence, he also noted that the reflection at oblique incidence should be shifted to shorter wavelengths, i.e., the observed blueshift was qualitatively accounted for.

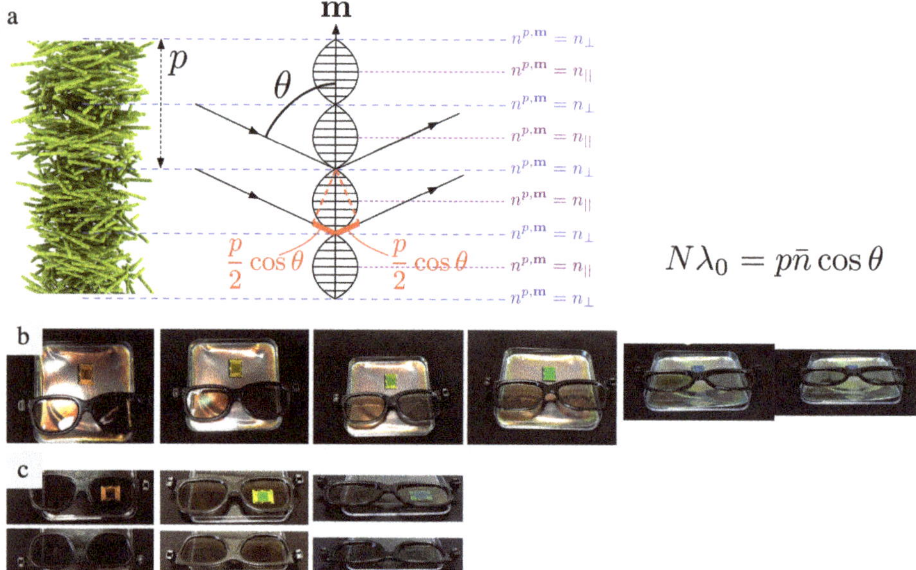

Figure 10. Illustration of visible Bragg reflection from cholesteric structures. (**a**) Schematic illustration of how the helical modulation of the director, and thereby the optic axis, gives rise to angle-dependent Bragg reflection. On the left, a realistic depiction of rods in a cholesteric arrangement is shown. Next to it a commonly used simplified representation is drawn, where a long horizontal line represents **n** in the paper plane whereas a vanishing line means that **n** is perpendicular. If we consider light incident along the helix axis **m** with polarization in the paper plane (p) it will experience a refractive index $n^{p,\mathbf{m}}$ that varies with a period of $p/2$ between $n_\perp$ and $n_{||}$, as indicated on the right. If we instead consider light entering at an oblique angle θ with respect to **m**, we can apply Bragg's simplification and consider that light is reflected every time the refractive index repeats itself, for instance being $n_\perp$ as in the figure. The two drawn light rays reflected at different levels along **m**, separated by $p/2$ from each other, will then have slightly different path lengths, as indicated in red. In order to get constructive interference, the difference must be an integer N times the wavelength, i.e., we obtain Bragg's condition $N\lambda_0 = p\bar{n}\cos\theta$, where λ_0 is the central reflection wavelength as measured in air and $\bar{n}$ is the average refractive index (see main text). (**b**) Viewing angle-dependent selective reflection of a large square dried CNC film (11.5 cm side length) and a small cell containing a thermotropic cholesteric. The CNC film is viewed partially through 3D cinema glasses, thus through **a** left-handed circular polarizer on the left and **a** right-handed one on the right. For normal incidence the reflected light is left-handed circularly polarized but with increasing viewing angle the circular polarization is lost. The color is blue-shifted, i.e., λ_0 moves to shorter values, as expected for increasing θ. (**c**) For comparison, the thermotropic sample is viewed through the right and left circular polarizers under identical illumination and imaging conditions as a function of viewing angle. The circular polarization (here right-handed) is retained even for a large θ.

Twenty years later, Hessel de Vries published the next milestone paper [101] regarding the optical properties of cholesterics (without mentioning Oseen's prior work). This was more complete and covered many other aspects, including the variation in optical rotatory power for wavelengths outside the bandgap and the effect of the sample boundaries. The language in de Vries' article is closer to modern terminology than that of Oseen's, and we can thus easily recognize many key points in this article, such as the width of the bandgap being equal to $p\Delta n$, where $\Delta n = n_{||} - n_\perp$, i.e., the magnitude of the nematic birefringence. While de Vries also restricted the calculations to normal incidence, he makes a note in the introduction (without derivation or motivation) that the selective reflection wavelength at oblique incidence approximately follows Bragg's law, noting that there appears to be repeated reflections at periodically spaced internal planes. These are, of course, as imaginary as were Lawrence Bragg's planes applied to x-ray crystallography, but they are a helpful tool. In cholesterics, the planes are defined by the locations along **m** where the refractive index repeats itself, as graphically illustrated in Figure 10a (see caption for further details). More detailed analyses of the behavior at oblique incidence were presented by Fergason [176] (this paper considers samples with arbitrary helix orientation with respect to the sample plane), Dreher and Meier [177] (here an analysis of how far the analogy with Bragg reflection holds is provided) and Berreman and Scheffer [178,179] (numerical analyses of oblique incidence properties are compared with experimental data).

As the optics of cholesterics is truly complex we will not go into further detail, but refer the interested reader to the above references. Here, we instead summarize the key outcomes of the theories, which are very important for properly understanding what to expect from cholesteric CNC suspensions and films prepared by drying them, and for analyzing samples in a proper manner.

- Light incident along the helix axis, **m**, with a wavelength λ^c, as measured inside the cholesteric, that is identical to the helix pitch p, is separated into two eigenmodes that are perfectly circularly polarized, one right-handed and one left-handed. The mode with handedness opposite to that of the helix propagates unobstructed, but the one with the same handedness cannot propagate at all; it is totally back-reflected. This means that analysis of the reflected light through circular polarizers will show the reflected color if the polarizer has the handedness of the helix, while it will show no reflection if the polarizer has the opposite handedness.

- Because the cholesteric is birefringent and thus does not have one single refractive index, the condition $\lambda^c = p$ is not uniquely defined when considering light wavelengths λ in air, prior to entering the cholesteric. Therefore, we find a band of air wavelengths that satisfy the reflection condition, from $\lambda_{min} = pn_\perp$ to $\lambda_{max} = pn_{||}$ (assuming $\Delta n > 0$). This gives us a reflection band of width $p\Delta n$ for the air wavelengths around $\lambda_0 = p\bar{n}$ that are selectively reflected, where the average refractive index is calculated as $\bar{n} = \sqrt{\frac{n_{||}^2 + n_\perp^2}{2}}$. (To calculate the average refractive index we must first calculate the average dielectric permittivity, $\epsilon_r = n^2$, , which is the proper material constant for use with Maxwell's equations, and then take the square root, $n = \sqrt{\epsilon_r}$.) Considering that visible light wavelengths in air range from about 400 nm (violet) to about 700 nm (deep red), and that the average refractive index in dry CNC films can be expected to be $\bar{n} \approx 1.6$ (based on values reported for $n_\perp$ and $n_{||}$ for crystalline cellulose [14]), this means that the helix pitch in the film should be in the range $\sim 250 < p < 440$ nm in order to see selective reflection at normal incidence.

- For oblique incidence, i.e., if the incoming light beam does not enter along **m** but at an angle θ from the helix axis, the mean selective reflection wavelength, as measured in air, varies approximately according to Bragg's law:

$$N\lambda_0 = p\bar{n}\cos\theta. \tag{16}$$

Here, N is an integer that can be greater than 1 for oblique incidence, i.e., higher-order reflections may occur under these conditions. For normal incidence, $N = 1$ strictly, because the perfectly sinusoidal modulation of the refractive index along the helix has no harmonics [177].

- The polarizations of the eigenmodes at oblique incidence, $\theta \neq 0$, are elliptical rather than circular. This means that, since any elliptical polarization different from circular can be separated into one left- and one right-handed circular component, neither of which is zero, analysis of the reflected light through circular polarizers should not give perfect extinction with the polarizer that has opposite handedness of the helix. However, for a well-aligned ideal cholesteric helix, the deviation from circular polarization is very small even up to large angles, hence the effect is practically negligible, see Figure 10c. Thus, even at oblique incidence, circular polarizers are very useful for assessing the quality of a cholesteric film.

- If the helix axis **m** is in the plane of the sample, no selective reflection will be seen. If the pitch is smaller than light wavelengths, the light averages the effect of the helix and experiences regular birefringence where **m** is the optic axis, with extraordinary component $n_e = n_\perp$ for polarization along **m** and ordinary component $n_o = \bar{n}$ for polarization perpendicular to **m**. (The refractive indices $n_\perp$ and $n_\parallel$, and through them also $\bar{n}$, still refer to orientations with respect to **n**.) Note that this means that the birefringence of the short-pitch cholesteric observed perpendicular to **m** is negative. On the other hand, if the pitch is longer than light wavelengths, light will resolve the variations of optic axis orientation, and the sample appears with the characteristic 'fingerprint' texture that we have seen in, e.g., Figure 8, where dark bands correspond to **n** normal to the sample plane, along the light incidence, and bright bands are seen for **n** in the sample plane (assuming that **m** is parallel to neither polarizer nor analyzer). This means that the distance between two consecutive dark (or bright) bands is $p/2$ (**n** $=-$**n**).

7.2. Drying a Cholesteric Gel: The Appearance of Color and Its Dependence on Drying Conditions

As we have just seen, visible selective reflection typically calls for a helix pitch below 0.5 µm, yet the equilibrium cholesteric pitch in CNC suspensions is often on the order of tens of µm, going down to ~3 µm in the shortest cases. So, how do we explain the appearance of visible color in the films? The answer must be sought in the non-equilibrium process that starts with kinetic arrest and ends with a dry colorful film, a stage of drying that was long left largely uninvestigated [11]. Frka-Petesic et al. took on the task of filling this knowledge gap, considering the process first on a more qualitative level [14], then with extraordinary detail, combining theory, simulation and experiments, in an absolutely exquisite study [174]. We believe this will become a landmark paper, as it explains almost all mysteries of the optical characteristics of films prepared by drying CNC suspensions.

The kinetic arrest marks the end of dynamic liquid crystalline organization within the sessile droplet from which dry CNC films are made. Whatever arrangement of CNCs prevails at this point is locked in place; tactoids can no longer merge or reorient and any boundaries or discontinuities in the director and/or helix orientation field will remain. However, at this point we still have 90 wt.% of the sample mass left in the form of water, which will leave the sample by evaporation (or diffusion; see below) as the kinetically arrested droplet is transformed into a dry film. If we consider a droplet that is pinned to its substrate, which is generally the case on hydrophilic substrates [14], the removal of the water will induce a significant compression along the vertical direction of the lab frame, which we now denote $\hat{z}$ for clarity. Considering the typical quite low values of $w_k \approx 10$ wt.%, the compression factor $\alpha = d_f/d_{ka}$ can be considerable, where we define the droplet thickness at kinetic arrest as d_{ka} and the final film thickness as d_f. A value of $\alpha = 0.1$ can readily be expected, the result seen nicely in Figure 4; the rods are stacked right on top of each other in the dry film, very differently from the equilibrium liquid crystal phase situation. The significance of this compression on the optical properties can be tremendous, because of the complex distortion of the helical structure that it entails, as illustrated in Figure 11, from [174].

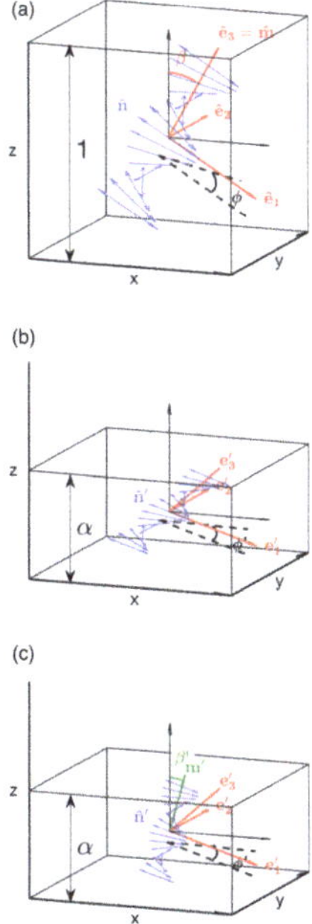

Figure 11. Illustration of the process of helix compression and axis reorientation upon drying a CNC suspension after the kinetic arrest, following Frka-Petesic et al. [174]. The uncompressed state (**a**) is compared to the state after uniaxial compression, showing representative director arrows that, before compression, were perpendicular to the original helix (**b**), as well as arrows perpendicular to the redefined helix representative of the compressed and deformed state (**c**). See main text for further explanation. Reprinted figure with permission from [174], http://dx.doi.org/.10.1103/physrevmaterials.3.045601. Copyright (2019) by the American Physical Society.

The first key observation of Frka-Petesic et al. is that the uniaxial compression will not only compress the helix pitch to a much lower value, thereby explaining why we get pitch values on the order of visible light wavelengths in the dry films although the equilibrium helix pitch never goes below $p_0 \approx 3\ \mu\mathrm{m}$, but also that the factor by which the pitch changes depends greatly on the orientation of **m** with respect to the lab frame at the point of kinetic arrest. The original helix orientation at kinetic arrest of an arbitrary domain, with final pitch p_k (which may be longer than p_0, since the time to acquire the equilibrium pitch diverges with the diverging viscosity as w_k is approached), is illustrated in Figure 11a. A local Cartesian coordinate system $(\acute{e}_1, \acute{e}_2, \acute{e}_3)$ is introduced, with the original helix axis orientation given by **m** and the director rotating in the $\acute{e}_1, \acute{e}_2$ plane. Only if $\mathbf{m} = \acute{e}_3 = \hat{z}$ would the helix pitch be maximally reduced to the minimum final value of $p_f^{\min} = \alpha p_k$. With a non-zero angle β between $\mathbf{m} = \acute{e}_3$ and $\hat{z}$ the compression will be smaller, as is qualitatively easy to see from Figure 11. Quantitatively, the helix pitch compression depends on the original helix tilt β as [174]:

$$p_f/p_k = \frac{1}{\sqrt{\sin^2 \beta + \alpha^{-2} \cos^2 \beta}} \tag{17}$$

We may quickly verify for the extreme values that $p_f/p_k(\beta = 0°) = \alpha$ and that $p_f/p_k(\beta = 90°) = 1$, as expected.

Next, the authors also realized that the compression along $-\hat{z}$ applied to a domain with inclined helix, $\mathbf{m} = \acute{e}_3 \neq \hat{z}$, will also *reorient* the helix, towards a slightly more vertical direction $\mathbf{m}'$. The reason is that the only physically meaningful definition of a helix axis is one that is perpendicular to the director, but as is clear from Figure 11b, showing the helix after compression by a factor α, the direction $\hat{e}_3$ is clearly no longer perpendicular to **n** after compression. Instead, a new helix orientation **m$'$** can be identified as in Figure 11c. This is related to the fact that the original local coordinate system $(\hat{e}_1, \hat{e}_2, \hat{e}_3)$ is no longer orthonormal, i.e., it is no longer a Cartesian coordinate system.

Finally, and very importantly, the new helix axis **m$'$** no longer describes a perfectly helicoidal modulation of the optic axis, in which the optic axis rotates by a constant amount $d\varphi/dm'$ (where we define the variable m' to indicate location along the new helix axis **m$'$**). Instead we find regions of decreased twist $(d\varphi/dm')_-$ alternating with regions of increased twist $(d\varphi/dm')_+ > (d\varphi/dm')_-$ (see [174] for details). This is significant because this deformed helix can no longer be described by a pure sinusoidal function. Its function contains higher harmonics, which means that the selective reflection is no longer circularly polarized as for an ideal cholesteric helix. Moreover, in contrast to the case of equilibrium cholesterics, with a perfectly sinusoidal helical director modulation, we can now have higher-order Bragg reflections even for observation along the helix axis ($N > 1$ in Equation (16)). Thus, domains that did not exhibit a vertical helix axis at the point of kinetic arrest, $\mathbf{m} \neq \hat{\mathbf{z}}$, will (1) have a longer final pitch p_f than vertical domains, (2) have a helix axis that is not oriented along the film normal (although it is closer to the film normal than the original helix axis was), (3) reflect light that is not perfectly circularly polarized, hence it passes both a right- *and* a left-handed circular polarizer, and (4) give rise to higher-order reflections even for observation along the helix.

With this excellent analysis by Frka-Petesic et al. of the effect of vertical compression of a kinetically arrested cholesteric CNC suspension we realize that there are considerable analogies between the way in which colorful dry CNC films are made and the 'anisotropic deswelling' method introduced by Heino Finkelmann and co-workers for aligning cholesteric liquid crystal elastomers (CLCEs) [180]. When making a CLCE, a solution of reactive mesogens in excess volatile solvent is kept in a vial and the solvent is evaporated while the reactive mesogens polymerize to form a 3D network. The network formation corresponds to the kinetic arrest in a drying CNC suspension. While the original anisotropic deswelling method employed long-term centrifugation during solvent evaporation to ensure the required unidirectionality of sample compression [180,181], thereby orienting the helix perpendicular to the resulting elastomer film, we recently demonstrated that the same effect can be achieved without centrifugation if the solution droplet is pinned to the substrate to avoid horizontal compression [182], just as for the case of drying CNC suspensions. Moreover, when a CLCE is stretched in the film plane, perpendicular to the helix axis, the perfect sinusoidal modulation of the optic axis is lost. Just like for films produced by drying CNC suspensions in which **m** is not vertical, an alternating arrangement of rapid and slow director rotations results, removing the circular polarization of the reflected light such that the color is seen almost with equal strength through left- and right-handed circular polarizers [182,183].

We may now go back to the clearly non-ideal optical behavior of the dried CNC film in Figure 10, explaining all observed discrepancies by applying the model of Frka-Petesic et al. First, the reason why the color stays red over a much greater tilting range than for the thermotropic sample is that there were domains in the kinetically arrested sample from which the film was made in which the helix axis was not vertical in the lab frame, $\mathbf{m} \neq \hat{\mathbf{z}}$. We note that the color at normal incidence observation is red, which means, according to Bragg's law, that the pitch of the domains with vertically oriented helix in the film must be on the long side of what generates visible light, i.e., $p \approx 0.5$ μm. Frka-Petesic et al. tell us that these domains have the strongest compressed helix, i.e., the shortest pitch, which means that domains with longer pitch will be reflecting in the infrared for observation along the helix. We will thus see almost only the reflection from the domains with vertically oriented helix, which retains a near-perfect helicoidal modulation and thus gives us strong circular polarization, the color disappearing when observing through a right-handed polarizer. As we start observing the film obliquely, however, domains with longer pitch will eventually start reflecting in the visible, as explained by Bragg's law. This means that, while the reflection from the original domains blue-shifts towards yellow, this reflection is mixed with a new red reflection from the tilted domains, and it is difficult to see a clear change in color. What is easy to see is the loss of circular polarization, which we can now understand to be a result of the reflection from the tilted domains, in which the helix has been distorted by the compression, losing its circularly polarized reflection.

The analysis of Frka-Petesic et al. goes much further still, also discussing, for instance, which periodic patterns may arise at the intersection between tactoids and substrate when the helix is not vertical, and how the distribution of final helix orientations should change depending on the compression factor and the original quality of alignment. We encourage the reader to discover these and other aspects by reading the original paper.

Strengthened with this insight we can conclude that in order to get as close as possible to ideal cholesteric optics from dried CNC films, it is imperative to ensure a vertical helix orientation as well as a single value of p throughout the sample prior to kinetic arrest. There are a few strategies aimed at reaching this goal, two of which we will describe now. The first approach is to apply a magnetic field in the vertical direction [3,46,184–186]. Because CNCs have negative diamagnetic anisotropy, **n** tends to align perpendicular to an applied magnetic field, which means that **m** will align along the field. The first studies of magnetic field alignment of cholesteric CNC suspensions reported rather high required field strengths [3,184,185] but more recent reports show that very good results can be achieved with reasonable fields (0.5–1.2 T), easily obtainable with commercially available permanent NdFeB magnets [186]. Second, a smart way that was introduced by the MacLachlan group is 'tactoid annealing' [98], meaning that the evaporation of the water is slowed down very much by raising the humidity of the surrounding atmosphere. Tactoids then have time to merge and adopt a uniform helix orientation, tending to vertical as the tactoids sink to the bottom and feel the planar-aligning influence of the bottom substrate. If the drying is carried out in a petri dish, for instance, once the CNC mass fraction reaches a high enough value to have a significant cholesteric fraction, but low enough to not enter kinetic arrest, a cover can be placed on it to stop evaporation for a time on the order of a day or two. This gives the tactoids that have formed time to merge and the helix to align vertically. The cover can then be removed, or a small hole can be opened, in order to complete the water evaporation to produce the dry CNC film, with improved optical properties thanks to a more uniform orientation of **m**.

The slowing-down of the droplet drying can be done in another smart way, which also has another highly beneficial effect. We will see how in the following section, after we have introduced one more critically important non-equilibrium phenomenon.

7.3. The Coffee-Ring Effect and How to Avoid It

From a general colloids and interface science perspective, studying the evaporation of colloidal suspensions and the linked process of particle deposition on the supporting substrates is a phenomenologically simple physical problem, or at least it may appear so. This also holds true for the case of CNC suspensions that are dried to yield colorful films. However, a more detailed look at the problem reveals that, regardless of the particular drying geometry under investigation (e.g., a sessile drop on a glass slide or a thick liquid film inside a Petri dish), a multitude of physicochemical phenomena are simultaneously at work. These typically include, for instance, macroscopic phenomena such as hydrodynamic flow patterns [187] and microscopic effects such as particle-particle and particle-interface interactions [188]. Additionally, phenomena that simultaneously involve different length- as well as time-scales, such as wetting and the dynamics of the three-phase contact line, are always part of the puzzle [187,188]. Apart from its scientific beauty, the problem of evaporating colloidal suspensions—not only CNC suspensions—is an important one from a practical viewpoint. Developing multifunctional surfaces requires patterning of (nano) particles in a controlled and reproducible fashion, so that their collective properties, for instance, can emerge. Intuitively, a simple strategy to pattern particles on solids is to cast a given volume of particle suspension on the surface and let the solvent evaporate, as we do when making dry CNC films. Despite its conceptual and practical simplicity, however, this method poses several challenges, as we will now explain.

In a typical experiment of drying a colloidal suspension, a sessile drop is dispensed on a partially wetting solid substrate, i.e., with a contact angle of less than $90°$ (volume on the order of µL, drop base radius on the order of mm). Very frequently, the suspension drop is pinned on the substrate, which means that its contact line is not allowed to recede (or advance) to a new position other than its initial one. Contact line pinning is a general phenomenon that is observed even in drops of pure liquids, and it is attributed to heterogeneities on the substrate surface, which can be either of geometrical (i.e., roughness) or chemical (i.e., non-uniform surface composition) nature. In particle-laden drops, the additional effect of self-pinning might occur, where the concentration of particles at the contact line can strengthen the initial pinning [189]. When the solvent in a pinned sessile drop evaporates, the mass of liquid lost near the edge of the drop is larger than that at its center. This is because the probability of escape of an evaporating solvent molecule depends on its point of departure. Escaping molecules undergo a random walk, and a molecule starting from the center of the drop has a higher probability to be re-adsorbed compared to a molecule near the contact line [189]. Surface tension dictates that the free surface of the drop (i.e., the liquid–gas interface) must always assume the shape of a spherical cap. This, in combination with the geometrical constraint of contact line pinning, and the non-uniform evaporation rate across the free surface of the drop, results in liquid being squeezed outwards to replenish the molecules lost there. The particles suspended in the drop will be dragged to the edge because of this flow, leading to a progressive increase in the local particle concentration (i.e., at the drop edge). When all solvent molecules are lost, most particles that were initially uniformly dispersed in the drop are gathered at the initial drop periphery, forming ring-shaped deposits. This phenomenon, known as the coffee-ring effect, was first explained by Deegan and co-corkers in a seminal paper published in 1997 [190].

The coffee-ring effect occurs in a plethora of situations of evaporating colloidal suspensions, regardless of the specific properties of the system (substrate, solvent and particles) and it is insensitive to numerous experimental conditions [189]. This can be understood if we consider that the main requirements for the coffee-ring effect to take place, namely (i) a non-zero contact angle, (ii) contact line pinning, (iii) enhanced evaporation at the drop edge and (iv) absence of counter-acting flows such as Marangoni effects (discussed below), are met in numerous occasions. The requirements are also fulfilled in the case of evaporating CNC suspensions, hence it is reasonable to hypothesize that the coffee-ring effect is an integral part of the process of CNC film formation. Figure 12a, showing superpositions of images acquired during the evaporation of an aqueous CNC suspension droplet containing a small amount of polystyrene microspheres that act as flow tracers, confirms this hypothesis. During the interval $0.80t_{evap} - 0.87t_{evap}$, where t_{evap} is the total evaporation time, tracers follow a radial flow with a direction from the drop center to its edge. However, the tracers cannot reach the edge of the drop because a large number of them has already accumulated in a thick band that starts from the contact line. Qualitatively, the same hydrodynamic pattern is retained until the end of the drying process, but the thickness of the band significantly increases with increasing t (see image corresponding to $0.93t_{evap} - 1.00t_{evap}$). It is reasonable to assume that the dispersed CNC particles also follow the flow pattern revealed by the tracer microparticles. This is confirmed by the characteristic ring-shaped deposition patterns that are obtained after complete water evaporation (inset in Figure 12a and top panel of Figure 12b). The ring morphology of the CNC film is quantitatively verified by measuring the height profile with a profilometer (Figure 12b, bottom panel).

The first authors that reported on the connection of the coffee-ring effect to the morphological and optical properties of dry CNC films resulting from sessile drop evaporation, were Mu and Gray five years ago [191]. These authors used profilometry to characterize the morphology of the obtained CNC deposits, and mentioned that the increased CNC concentration at the drop periphery must be due to the coffee-ring effect, which transports the CNCs toward the edge of the pinned drop. Furthermore, they attempted to connect the longer wavelength reflection colors at the outer region of their films to the enhanced tendency of the system to 'freeze-in' the cholesteric structure due to this local increase in CNC concentration. The authors concluded that mass transfer and gradients in CNC concentration during the film formation process affect the pitch and the orientation of the cholesteric helix [191]. Despite the reasonable reference to the coffee-ring effect, Mu and Gray did not provide direct evidence of its occurrence. To our knowledge, Figure 12a is the first report of flow visualization of the coffee-ring effect in drying drops of aqueous CNC suspensions.

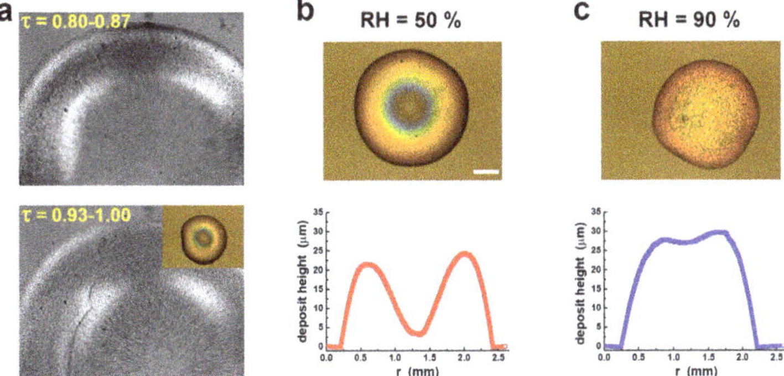

Figure 12. The coffee-ring effect in evaporating droplets of aqueous CNC suspension and its elimination via slow drying. (**a**) Superposition of transmission optical microscopy images acquired during the evaporation of a droplet (volume 1 μL) of an unsonicated aqueous CNC suspension (starting concentration $w = 4$ wt.%) to which we have added 0.1 mg/mL polystyrene spheres (diameter 1 μm). The droplet was deposited on a glass coverslip and dried at ambient laboratory conditions (relative humidity, $RH = 40\%$). The polystyrene spheres act as tracers that enable flow visualization. The variable τ is a reduced evaporation time, $\tau = t/t_{evap}$, where t_{evap} is the total evaporation time. The inset shows the colorful reflection of a fully dried film that was obtained by drying the same suspension, under very similar experimental conditions (volume 2 μL, $RH = 50\%$). (**b,c**) Reflection optical microscopy images and corresponding height profiles, obtained by profilometry, of CNC deposits from droplets (volume 2 μL) of unsonicated aqueous CNC suspensions (starting concentration $w = 5$ wt.%) deposited on glass coverslips, dried at low ($RH = 50\%$, a) and high ($RH = 90\%$, b) relative humidity; from M.Sc. thesis of M. Dupas, U. Luxembourg 2018, supervised by M. Anyfantakis.

To further understand the coffee-ring effect and its impact, we need to look into the details of the film drying procedure, this time focusing mainly on the period *before* kinetic arrest. It is worth noting that, although the coffee-ring effect occurs in sessile drops, its main component, the evaporation-induced outward flow, has a strong impact on the characteristics of deposits obtained from drying situations/configurations other than sessile drops, if a pinned contact line exists. Therefore, the following discussion is important regarding the drying of CNC suspensions in general, regardless of the specific drying geometry. Figure 12b immediately reveals the most common shortcoming of employing suspension evaporation for CNC film production. When viewed under white light illumination and at normal incidence, the film displays a radial variation of reflection colors. Applying what we have learnt above, we can conclude that there is a radial variation in the microstructure within the material, possibly with pitch as well as helix orientation variations.

Apart from its profound influence on the morphology of the final deposit, the capillary outward flow (see Figure 12a) obviously dictates the evolution of particle distribution within the drop, both in space and time during the evaporation process. The resulting particle concentration gradients are particularly important in the case of drying CNC suspension drops, because we are dealing with a lyotropic liquid crystal: the CNC mass fraction w is, after all, the main parameter which dictates the phase behavior, helix pitch, and thus the properties of the system. Any drying drop is, by definition, an out-of-equilibrium system, where at any stage insufficient time is provided for the system to reach an equilibrium configuration that corresponds to its particular particle concentration at that point in time. To further complicate matters, a given average mass fraction for the drop does not provide information about the local mass fractions at different locations within the drop. These are defined by the hydrodynamic flow patterns, which transport particles to specific locations, acting against diffusion which tends to distribute particles uniformly. If we wish to avoid the radial color variations and acquire homogeneous optical properties, we thus must counteract these flows, so we must inhibit the factors that induce them.

A crucial parameter that calls for special attention is the total evaporation time (in other words, the drop lifetime), a slowed-down drying process being the key element of the previously mentioned tactoid annealing strategy [98]. This slowing-down was achieved by increasing the relative humidity, RH, allowing the tactoids inside the droplet to rearrange, fuse and adopt a uniform helix orientation. In addition, increasing the RH of the environment in which a sessile CNC suspension drop dries dramatically weakens the strength and impact of the coffee-ring effect on the CNC deposition after the evaporation process is complete. The benefits of weakening the evaporation-induced flow can be seen in Figure 12c, where RH was significantly increased during the droplet drying. This resulted in the final CNC deposit adopting a more uniform disk-like morphology (compared to the ring-shaped pattern in Figure 12b for lower RH), as evidenced from the measured height profile across the film.

We have very recently carried out a detailed investigation on the impact of RH on the CNC distribution in the dry films and their optical uniformity [192], finding that the relative humidity strongly affects the hydrodynamics of the drying drops. In drops that dry quickly under ambient humidity, the coffee-ring effect dominates for a large portion of the drop lifetime. On the contrary, in drops that dry slowly under humid conditions, there is an antagonism between a weakened evaporative outward flow and a circulatory flow linked to the formation of tactoids. The former dominates only toward the end of the drying process, rendering the impact of the coffee-ring effect significant only for a small portion of the drop lifetime. This results in the homogeneous distribution of CNC particles in the dry film, leading also to quite uniform structural color, as the one shown in Figure 12c.

One of the best demonstrations of how useful it is to slow down the drying process can be found in the study by Zhao et al., where arrays of extremely uniform CNC microfilms were fabricated by drying multiple nanoliter sessile drops on substrates with patterned wettability [9]. Their key trick was to inhibit evaporation completely by covering the CNC suspension droplet under a thin layer of immiscible oil (hexadecane), thereby replacing water loss by evaporation with water loss by diffusion into the oil. Since this is saturated in water at very low concentrations, accepting more water only after the water has evaporated from the oil phase, the water extraction is dramatically slowed down and it takes place uniformly across the film, thus not setting the usual hydrodynamic flows in motion. This strategy led to films of a single color and a dome-shaped morphology, as verified by profilometry. Contrarily, when a droplet of CNC suspension of the same type (and the same volume) was dried in ambient conditions, the strong evaporation-driven radial flow led to strong coffee-ring-like particle deposition and films with radial color variation. The slowing-down of water loss by coverage by an oil phase was systematically studied earlier by Shimura et al. for the purpose of avoiding evaporation-driven abnormalities in aqueous protein solutions during capillary electrophoresis [193]. The data in Figure 2 of that paper are very useful for identifying the right amount of oil to a certain aqueous droplet volume for tuning the drying rate to a desired value.

Another approach was explored by Gençer et al. [194]. They introduced ethanol into the atmosphere in which a sessile CNC suspension droplet was evaporating, thereby setting up a circulatory Marangoni flow that partially compensated for the evaporative outward flow. Marangoni flows are surface flows that are caused by gradients in the interfacial tension of a liquid, themselves caused by gradients in temperature (thermal Marangoni) or concentration of a surface-active component such as surfactants (solutal Marangoni). It has been shown that both thermal [195] and solutal [196] Marangoni flows can have a dramatic impact on the distribution of colloidal particles upon drying of a sessile suspension drop. In the case of Gençer et al., [194], adsorption of ethanol at the air–water interface led to a local reduction in the surface tension and thus the emergence of a solutal Marangoni flow. While this should allow for a much faster drying than with oil coverage, the balancing of the flows is challenging. Although the results presented were promising, the films were not as uniform as when drying the CNC suspension under oil. Moreover, the ethanol will diffuse into the CNC suspension, thereby certainly affecting the liquid crystal self-organization.

7.4. To What Extent Can We Tune the Properties of CNC Films?

If we succeed in achieving uniform film color, the next desirable step would be to be able to select *which* color the film exhibits. It would also be valuable if we can work on the mechanical properties of the film, in particular making them less brittle and more flexible than a pristine CNC film. Several groups have worked along these goals for a long time, also long before the full complexity of the drying process was realized. The focus has often been only on the final film, without paying attention to the intermediate stages, hence it is not always easy to draw clearcut conclusions. Nevertheless, the experimental findings may give valuable insights, so we summarize some reports on tuning the properties of CNC films that we are aware of.

Pan et al. [197] reported an increase in the helix pitch of dry CNC films when increasing the initial concentration of the suspension used, although the range studied for the starting suspension was always in the low-concentration, fully isotropic regime. They also explored addition of NaCl up to 0.5 mM, then finding a linear decrease with salt concentration for the pitch in the solid film. Beck et al. investigated the influence of ultrasonic energy treatment of aqueous CNC suspensions that were dried in polystyrene Petri dishes to produce iridescent films [198]. Ultrasonication by sonotrode tip red-shifted the reflection wavelength band, the magnitude increasing with the energy input per gram of CNC, see Figure 13a. Suspensions treated with different energy inputs could be mixed and drying of these mixtures led to films with reflection bands corresponding to an intermediate level of ultrasonication. The authors suggested that this effect was electrostatic, involving the release of ions trapped in the bound water layer at the surface of the CNC. Parker et al. argued that the ion release cannot explain the observations on its own, and suggested that ultrasonic breaking of CNCs effectively leads to a reduction in chiral species, which could thus explain a redshift [14]. To the best of our knowledge, potential artifacts from titanium particle release from the sonotrode tip have not been investigated in this context. At least in the field of carbon nanotube research, the contamination by sonication-derived titanium particles has been confirmed to be significant [199], and it is not impossible that such particles will also influence the cholesteric liquid crystal self-organization in CNC suspensions. Nguyen et al. varied the substrate on which CNC suspensions were dried, achieving varying contact angles of the suspension drops [200]. Opaque films were formed on the most hydrophobic substrates like polytetrafluoroethylene, whereas blue films were formed on the most hydrophilic surfaces, such as aluminum. Drying on substrates with intermediate wettability, such as cellulose acetate, gave rise to yellow-red films. One should note that studying the effect of varying wettability is far from trivial, because it is always accompanied by changes in other parameters that also play important roles in different aspects of the drying (and the liquid crystal self-organization) process.

Several groups have explored the role of additives introduced in CNC suspensions on the properties of the dried films, thus connecting to the discussion in Section 6.3 but now with the focus on the final dry film rather than the equilibrium properties. The shortening of the equilibrium cholesteric pitch that is typically provided by salt addition (Section 6.3) would lead to the expectation that also the pitch of the final film is blue-shifted by salt addition. This is indeed what Revol et al. found in their seminal study focusing on achieving visible reflection from dried CNC films [3], tuning the normal incidence reflection color from the infrared to yellow, see Figure 13b. They achieved similar results by reducing the CNC surface charge by controlled desulfation.

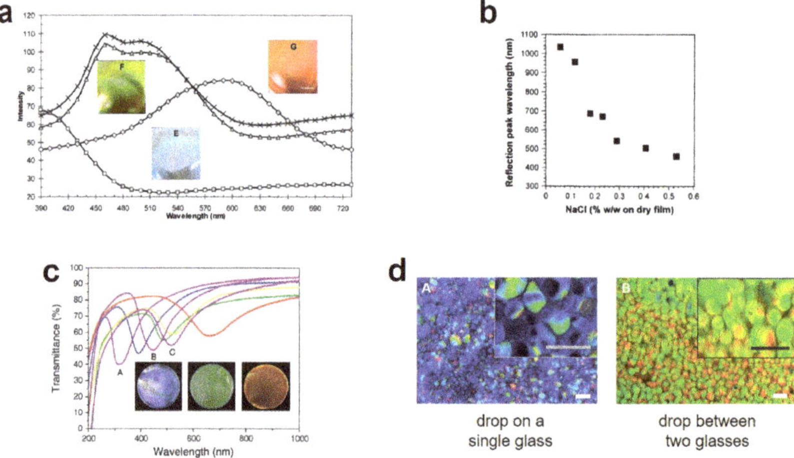

Figure 13. Examples of parameters and conditions affecting the selective reflection of films formed by drying CNC suspensions. (**a**) Impact of ultrasonication treatment. Photographs (insets) and reflection spectra (diagram) of CNC films prepared from suspensions treated with ultrasonication energies of 750 J/g CNC (E) and 2225 J/g CNC (G). Film (F) was prepared by drying a 1:1 mixture of the suspensions used to prepare films (E) and (G). The scale bar is 1 cm. Adapted with permission from Ref. [198]. Copyright 2011 American Chemical Society. (**b**) The influence of electrolyte addition. Increasing the concentration of NaCl added to an acid-form CNC suspension leads to a decrease in the wavelength of the reflection band of the dried CNC films. Reprinted with permission from Ref. [3]. Copyright 1998 Pulp and Paper Technical Association of Canada. (**c**) The effect of added polymers. Transmission spectra and reflection photographs of films of CNC/polyethylene glycol composite films (diameter 9 cm) at various CNC/polymer compositions; blue: 90/10; green: 80/20 and red: 70/30. Adapted with permission from Ref. [134]. (**d**) The effect of surface anchoring and orientation of liquid crystalline CNC suspensions. Cross-polarized images of films dried in water vapor saturated atmosphere assisted by orbital shear with a glass coverslip off (left) and on (right) during drying. The scale bars are 100 μm. Reprinted with permission from Ref. [201]. Copyright 2018 American Chemical Society.

We can now come back to the study by Mu and Gray of addition of D-(+)-glucose to CNC suspensions, discussed regarding the equilibrium behavior in Section 6.3. As mentioned there, the glucose tends to decrease the equilibrium pitch without impacting w_0, but it also promotes gelation, suggesting that w_k is reduced. When investigating the films prepared by drying glucose-enriched CNC suspensions, Mu and Gray found the surprising result that the color was red-shifted the more glucose was added. The authors proposed that the explanation is to be found in the promoted gelation: although the glucose tends to reduce the equilibrium pitch at a certain CNC mass fraction, the earlier kinetic arrest means that the pitch locked in by gelation was still longer than without glucose, leading to longer pitch in the dry film and thus a red shift of the reflection color. Another aspect is that the glucose remains in the final film, thus probably reducing the effective compression somewhat. A shift to longer reflection wavelengths was also observed by Guidetti et al. [202] when adding zwitterionic surfactants, with head groups that contain positive ammonium as well as negative sulfate groups. A model was proposed where the ammonium group was assumed to electrostatically interact with the negatively charged sulfate groups of the CNC, the hydrophobic alkyl tail of the surfactant acting as a small 'spring' between CNCs. With increasing surfactant concentration (keeping the CNC mass fraction w constant), the normal incidence reflection shifted from mainly green-yellow to red and eventually to infrared.

Apart from low molar mass additives, the influence of macromolecules on the optical properties of CNC films has been studied extensively. Bardet et al. [203] explored the effect of an anionic polyelectrolyte (sodium polyacrylate) and a neutral polymer (polyethylene glycol) on the self-assembly properties of CNC in suspension and films produced by drying. The film color was blue-shifted by increasing the amount of non-adsorbing sodium polyacrylate. This effect was attributed to a decrease in the equilibrium chiral nematic pitch, which could be either due to osmotic pressure differences or because of the highly hygroscopic character of sodium polyacrylate, resulting in enhanced adsorption of water molecules. Polyethylene glycol was found to increase the flexibility of the CNC films. In a following study, Yao et al. reported that intercalation of polyethylene glycol macromolecules between CNC led to an increase in the pitch of the composite film. As a result, increasing the relative amount of polyethylene glycol caused a continuous red-shift of the reflectance band, which enabled the authors to tune the photonic bandgap of the film by varying its composition [134], see Figure 13c.

The beauty of non-cellulosic polymer incorporation is that the films also become much less brittle, reaching sufficient flexibilities for complete bending without breaking [134,203]. An interesting alternative is to polymerize reactive monomers within the drying CNC suspension, a route explored by Wu et al. [135]. By making a sandwich structure including a thin polymer sheet acting as a $\lambda/2$ plate between two polymerized CNC sheets, they mimicked the marvelously smart structure of the *Plusiotis resplendens* beetle [12,204] that reflects all light with wavelengths in the reflection wavelength band, thus right- as well as left-handed polarizations, thanks to the handedness inversion provided by the $\lambda/2$ plate. The resulting composites were not only flexible and highly uniform in color, but they also changed their shape in response to exposure to humid air.

An interesting variation of the concept with an introduced $\lambda/2$ plate was presented by Fernandes et al. [205]. They took advantage of the cracks that frequently form within films produced by drying CNC suspensions, filling this with the commonly used thermotropic nematic liquid crystal 5CB (4′-n-pentylbiphenyl-4-carbonitrile). As it aligned with its director, and thus optic axis, in the plane of the film, and as the crack thickness was about right to give the 5CB layer a $\lambda/2$ retardation effect, the relaxed-state film reflected both circular polarizations. However, by applying an electric field they could switch the director of the 5CB layer out of plane, thereby removing the optical effect of the intermediate layer, getting the one-handed circular polarization reflectivity back. This allowed them to dynamically switch between polarized and unpolarized reflectivity. MacLachlan and co-workers explored a similar approach [136], infiltrating a dried CNC film with the thermotropic liquid crystal 8CB (4′-n-octylbiphenyl-4-carbonitrile) such that it replaced the air in the pores of the film. By heating or cooling the composite past the temperature where 8CB transitions into an isotropic liquid, they could switch the overall reflective properties thanks to the refractive index modulation in the 8CB-filled pores. At room temperature, the film reflected green light, whereas at 48°C, with 8CB in the isotropic phase, the film was colorless.

7.5. Parameters and Effects Meriting Further Investigations

Our understanding of CNC suspensions and their liquid crystal as well as non-equilibrium behavior is finally starting to mature, such that clear strategies can be laid out on how to make valuable materials from CNCs. Nevertheless, there are still significant open questions remaining and a few issues have not yet been thoroughly addressed. We end this review by highlighting some of them, as inspiration for future research thrusts.

Although there has been a strong increase in the number of groups exploring film formation from drying CNC suspensions, there remains the question about the quality and the properties of the CNC particles utilized. Some reproducibility difficulties can certainly be attributed to incomplete characterization of the CNCs, with uncontrolled parameters leading to different results during different experiments. Many groups prepare their own CNC batches, typically applying sulfuric acid hydrolysis on materials from cotton or wood sources. Even though a specific preparation protocol is frequently followed [104], small variations in the procedure are perhaps unavoidable. On the other hand, there are now several companies that produce CNCs on industrial scale. Reid et al pointed out the importance of the various sources and the preparation methods (e.g., hydrolysis conditions and post-extraction treatments) on the properties and hence the behavior of CNCs [15]. This calls for a thorough characterization of any type of CNCs prior to using it for fundamental or applied studies. They discussed in detail the methods and procedures that should be employed, for instance to characterize the dimensions and surface properties of CNCs and their colloidal stability. As we have discussed in detail in this review, these properties dictate both the equilibrium and non-equilibrium behaviors of CNC suspensions and thus the properties of CNC films resulting from drying them. Furthermore, a detailed comparison of the characteristics of CNCs prepared on the bench scale to commercial batches from various companies was done.

Some phenomena that can be expected to affect the formation of films, and thus their optical features, have either so far been generally overlooked or they have just not been investigated thoroughly yet. For instance, naturally occurring flows (i.e., not imposed by external fields) in drying aqueous CNC suspensions have not yet been sufficiently investigated. The impact of surface tension-driven Marangoni flows was so far discussed only in the previously mentioned paper by Gençer et al. [194]. The authors provided interfacial tension measurements on drops of varying CNC concentration and calculated the appropriate dimensionless groups (e.g., the Marangoni number). Despite the many hints that flow phenomena are crucial in defining the macroscopic film morphology and its microstructure, the only published flow visualization data we are aware of are those in Figure 12a. Therefore, more studies on these surface flows (mostly solutal, as thermal Marangoni flows are typically very weak for aqueous drops) are needed. Moreover, convective flows can arise from density variations inside a drying drop [206], which can be for example caused by the simultaneous presence of lighter isotropic regions and heavier anisotropic regions; the influence of such flows has not been studied yet.

The discussion about the impact of CNC alignment and anchoring on the interfaces involved in suspension drying (i.e., the various substrate/suspension and suspension/surrounding interfaces) has been very limited up to now. Saha and Davis recently investigated the effect of surface anchoring by slowly drying CNC suspensions sandwiched between a glass slide and a glass coverslip, under high relative humidity [201]. The resulting films displayed dot-shaped domains with planar helix orientation and subtle mosaic defects, see Figure 13d. On the contrary, suspension drying without the coverslip yielded films with many mosaic defects, fingerprint textures and randomly oriented domains of different colors. The authors suggested that the anchoring of CNC particles to both glass surfaces had significant influence on particle ordering during water loss. Their alignment on both surfaces should lead to a tendency for planar helix orientation (i.e., the helix axis being perpendicular to the film). The authors mentioned, however, that other effects such as the confined evaporation and the prevention of skin formation might play a vital role in this drying geometry.

Entirely different approaches to depositing the CNC suspension like dip coating [207] can be used, in particular when thin films are desired, but the analysis of the complex non-equilibrium situation here is challenging. Hoeger et al. used blade coating to deposit CNC suspensions on mica, gold and silica substrates, with and without pre-adsorbed cationic polyelectrolyte, for studying the relevance of the CNC-substrate interaction [208]. They discussed their observations in the context of shear, capillary and electrostatic forces. Although attention starts to be paid on these issues, quantitative investigations of the alignment of CNC particles on different types of substrates (e.g., hydrophilic vs. hydrophobic [200]) or at the free surface of the suspensions (that could be a suspension/air or suspension/oil interface [9]) are still largely missing.

There are still some apparently conflicting observations between different studies, for instance regarding gelation or glass formation and regarding the impact on helix pitch of certain additives. One aspect that seems not to have been sufficiently investigated in this context is the importance of the native CNC counter ions and the additives, at least regarding the onset of kinetic arrest. Many studies on the effect of ionic strength were carried out with acid-form CNCs, thus with native H^+

counter ions, but then NaCl was added as the electrolyte to tune the ionic strength. It is not given that the effect will be identical to the case where HCl is added to acid-form CNCs (to our knowledge, only investigated systematically in one study [56]) or where NaCl is added to CNCs with native Na^+ counter ions. Even the interaction with non-ionic additives like glucose or nonionic polymers may be different for acid- and salt-form CNCs.

8. Summary and Outlook

Compared to many research activities requiring advanced and often expensive machinery, perhaps even large-scale facilities setting up highly unusual experimental conditions, and/or exclusive compounds available only in minute quantities, research on CNC suspensions may seem incredibly simple and low-key: what could possibly be complicated about a drop of water containing suspended particles of cellulose, the most abundant organic polymer on Earth, evaporating on a table in a regular lab? We hope that this review has demonstrated how false such a view point is, and that the processes taking place as the water leaves that droplet, giving birth to a beautiful as well as very useful colored film, are immensely complex. In order to reach the applied goal of making cellulose films with controlled and tunable structural colors, we need to understand all aspects of each of these processes. We need to understand what governs the colloidal stability of the CNC nanorods, with the choice of solvent, surface charge and counter ions all playing vital parts. We need to understand why and how the rod suspension may organize into a long-range ordered cholesteric phase, what happens inside the tactoids and at their boundaries, and how the tactoids eventually merge and form a macroscopic ordered phase. This includes being aware of the many subtleties of the helical director modulation that makes the CNC suspension so interesting, in equilibrium as well as after kinetic arrest, when the helix is being compressed as the water evaporates. We then need to understand the optics of a birefringent medium in which the optic axis rotates along a direction that may or may not be vertical, and the rotation may or may not be constant within the film. Furthermore since we start with a droplet of fluid particle suspension from which the solvent evaporates, we need to understand the hydrodynamic flows that take place within the droplet, moving particles around, and how these are affected by the underlying substrate as well as the phase bounding the droplet at its top. Of course, one also needs to understand the chemistry going into making the CNCs in the first place.

With this review we have tried to explain and discuss most of these aspects except for the CNC production in a way that any researcher working with, or desiring to work with, CNC suspensions has a chance to grasp the full complexity of these materials, providing links to further reading when our account does not go sufficiently deep. The present time is actually particularly exciting, because we are now, for the first time, at a stage where all the processes of the complex procedure behind CNC film production are recognized, and most of them are on the way of being understood, if not already clarified at a satisfactory level. We sincerely hope that our review can be helpful as well as stimulating, that it might guide experimental work into the right directions, and thereby make our understanding of the complex class of soft matter that colloidal cholesteric liquid crystals constitute a bit richer and a bit more complete. In the long run, this will hopefully also bring the dream of making sustainable materials with controlled tunable structural color, and/or enhanced mechanical properties, from wood or other renewable cellulose sources, a little bit closer to realization.

Author Contributions: This review was coordinated by J.P.F.L. who also wrote most of the text, with all authors giving input throughout. In particular, C.S. wrote most of Section 2 and provided Figure 1 (with AFM images from C.H.-R.), J.R.B. wrote most of Section 6 and provided Figure 9, and M.A. wrote most of Section 7.3 and provided Figure 12. C.H.-R. and Z.T. contributed with components to images (Figures 1, 6, 8 and 10) and gave contributions to the text in several parts of the article. All authors have read and agreed to the published version of the manuscript.

Funding: This research was funded by The Fonds National de la Recherche (FNR), Luxembourg, grants MISONANCE (C14/MS/8331546), SSh (C17/MS/11688643) and COReLIGHT (C18/MS/12701231). C. Schütz acknowledges financial support from the Alexander von Humboldt Foundation for her stay in Luxembourg.

Conflicts of Interest: The authors declare no conflict of interest. The funders had no role in the writing of the manuscript.

References and Notes

1. Zugenmaier, P. *Crystalline Cellulose and Derivatives: Characterization and Structures*; Springer Series in Wood Science; Springer: Berlin, Germany, 2010; p. 285.
2. Revol, J.; Bradford, H.; Giasson, J.; Marchessault, R.; Gray, D. Helicoidal self-ordering of cellulose microfibrils in aqueous suspension. *Int. J. Biol. Macromol.* **1992**, *14*, 170–172. [CrossRef]
3. Revol, J.F.; Godbout, L.; Gray, D. Solid self-assembled films of cellulose with chiral nematic order and optically variable properties. *J. Pulp Pap. Sci.* **1998**, *24*, 146–149.
4. Bardet, R.; Roussel, F.; Coindeau, S.; Belgacem, N.; Bras, J. Engineered pigments based on iridescent cellulose nanocrystal films. *Carbohydr. Polym.* **2015**, *122*, 367–375. [CrossRef]
5. Frka-Petesic, B.; Vignolini, S. So much more than paper. *Nat. Photon.* **2019**, *13*, 365–367. [CrossRef]
6. Giese, M.; Blusch, L.K.; Khan, M.K.; MacLachlan, M.J. Functional Materials from Cellulose-Derived Liquid-Crystal Templates. *Angew. Chem. Int. Ed.* **2015**, *54*, 2888–2910. [CrossRef] [PubMed]
7. Zhang, Y.P.; Chodavarapu, V.P.; Kirk, A.G.; Andrews, M.P. Nanocrystalline cellulose for covert optical encryption. *SPIE Org. Photonic Mater. Devices XIV* **2012**, *8258*, 825808.
8. Zhang, Y.P.; Chodavarapu, V.P.; Kirk, A.G.; Andrews, M.P. Structured color humidity indicator from reversible pitch tuning in self-assembled nanocrystalline cellulose films. *Sens. Actuators B* **2013**, *176*, 692–697. [CrossRef]
9. Zhao, T.H.; Parker, R.M.; Williams, C.A.; Lim, K.T.P.; Frka-Petesic, B.; Vignolini, S. Printing of Responsive Photonic Cellulose Nanocrystal Microfilm Arrays. *Adv. Funct. Mater.* **2018**, *29*, 1804531. [CrossRef]
10. Giese, M.; Blusch, L.; Khan, M.; Hamad, W.; MacLachlan, M. Responsive mesoporous photonic cellulose films by supramolecular cotemplating. *Angew. Chem. Int. Ed.* **2014**, *53*, 8880–8884. [CrossRef]
11. Lagerwall, J.P.F.; Schütz, C.; Salajkova, M.; Noh, J.; Park, J.H.; Scalia, G.; Bergström, L. Cellulose nanocrystal-based materials: From liquid crystal self-assembly and glass formation to multifunctional thin films. *NPG Asia Mater.* **2014**, *6*, e80. [CrossRef]
12. Srinivasarao, M. Nano-optics in the biological world: Beetles, butterflies, birds, and moths. *Chem. Rev.* **1999**, *99*, 1935–1961. [CrossRef] [PubMed]
13. Mitov, M. Cholesteric liquid crystals in living matter. *Soft Matter* **2017**, *13*, 4176–4209. [CrossRef] [PubMed]
14. Parker, R.; Guidetti, G.; Williams, C.; Zhao, T.; Narkevicius, A.; Vignolini, S.; Frka-Petesic, B. The Self-Assembly of Cellulose Nanocrystals: Hierarchical Design of Visual Appearance. *Adv. Mater.* **2017**, *30*, 1704477. [CrossRef] [PubMed]
15. Reid, M.; Villalobos, M.; Cranston, E. Benchmarking Cellulose Nanocrystals: From the Laboratory to Industrial Production. *Langmuir* **2017**, *33*, 1583–1598. [CrossRef] [PubMed]
16. Nickerson, R.F.; Habrle, J.A. Cellulose Intercrystalline Structure. *Ind. Eng. Chem.* **1947**, *39*, 1507–1512. [CrossRef]
17. Rånby, B.; Ribi, E. Über den Feinbau der Zellulose. *Experientia* **1950**, *6*, 12–14. [CrossRef]
18. Marchessault, R.; Morehead, F.; Walter, N. Liquid crystal systems from fibrillar polysaccharides. *Nature* **1959**, *184*, 632–633. [CrossRef]
19. Liu, Q.; Campbell, M.; Evans, J.; Smalyukh, I. Orientationally ordered colloidal co-dispersions of gold nanorods and cellulose nanocrystals. *Adv. Mater.* **2014**, *26*, 7178–7184. [CrossRef]
20. Moon, R.; Martini, A.; Nairn, J.; Simonsen, J.; Youngblood, J. Cellulose nanomaterials review: Structure, properties and nanocomposites. *Chem. Soc. Rev.* **2011**, *40*, 3941–3994. [CrossRef]
21. Wicklein, B.; Salazar-Alvarez, G. Functional hybrids based on biogenic nanofibrils and inorganic nanomaterials. *J. Mater. Chem. A* **2013**, *1*, 5469–5478. [CrossRef]
22. Eichhorn, S.J. Cellulose nanowhiskers: Promising materials for advanced applications. *Soft Matter* **2011**, *7*, 303–315. [CrossRef]
23. Klemm, D.; Kramer, F.; Moritz, S.; Lindstrom, T.; Ankerfors, M.; Gray, D.; Dorris, A. Nanocelluloses: A New Family of Nature-Based Materials. *Angew. Chem. Int. Ed.* **2011**, *50*, 5438–5466. [CrossRef] [PubMed]

24. Nishiyama, Y.; Langan, P.; Chanzy, H. Crystal structure and hydrogen-bonding system in cellulose Iβ from synchrotron X-ray and neutron fiber diffraction. *J. Am. Chem. Soc.* **2002**, *124*, 9074–9082. [CrossRef] [PubMed]

25. Atalla, R.; Vanderhart, D. Native cellulose: A composite of two distinct crystalline forms. *Science* **1984**, *223*, 283–285. [CrossRef] [PubMed]

26. Eichhorn, S.; Dufresne, A.; Aranguren, M.; Marcovich, N.; Capadona, J.; Rowan, S.; Weder, C.; Thielemans, W.; Roman, M.; Renneckar, S.; et al. Review: Current international research into cellulose nanofibres and nanocomposites. *J. Mater. Sci.* **2010**, *45*, 1–33. [CrossRef]

27. Yamamoto, H.; Horii, F. CPMAS carbon-13 NMR analysis of the crystal transformation induced for Valonia cellulose by annealing at high temperatures. *Macromolecules* **1993**, *26*, 1313–1317. [CrossRef]

28. Yamamoto, H.; Horii, F.; Hirai, A. In situ crystallization of bacterial cellulose II. Influences of different polymeric additives on the formation of celluloses I_α and I_β at the early stage of incubation. *Cellulose* **1996**, *3*, 229–242. [CrossRef]

29. Majoinen, J.; Haataja, J.; Appelhans, D.; Lederer, A.; Olszewska, A.; Seitsonen, J.; Aseyev, V.; Kontturi, E.; Rosilo, H.; Österberg, M.; et al. Supracolloidal multivalent interactions and wrapping of dendronized glycopolymers on native cellulose nanocrystals. *J. Am. Chem. Soc.* **2014**, *136*, 866–869. [CrossRef]

30. Usov, I.; Nyström, G.; Adamcik, J.; Handschin, S.; Schütz, C.; Fall, A.; Bergström, L.; Mezzenga, R. Understanding nanocellulose chirality and structure-properties relationship at the single fibril level. *Nat. Commun.* **2015**, *6*, 7564. [CrossRef]

31. Belton, P.S.; Tanner, S.F.; Cartier, N.; Chanzy, H. High-resolution solid-state carbon-13 nuclear magnetic resonance spectroscopy of tunicin, an animal cellulose. *Macromolecules* **1989**, *22*, 1615–1617. [CrossRef]

32. Abitbol, T.; Kam, D.; Levi-Kalisman, Y.; Gray, D.; Shoseyov, O. Surface Charge Influence on the Phase Separation and Viscosity of Cellulose Nanocrystals. *Langmuir* **2018**, *34*, 3925–3933. [CrossRef] [PubMed]

33. Su, Y.; Burger, C.; Hsiao, B.S.; Chu, B. Characterization of TEMPO-oxidized cellulose nanofibers in aqueous suspension by small-angle X-ray scattering. *J. Appl. Crystallogr.* **2014**, *47*, 788–798. [CrossRef]

34. Schütz, C.; Agthe, M.; Fall, A.; Gordeyeva, K.; Guccini, V.; Salajková, M.; Plivelic, T.; Lagerwall, J.; Salazar-Alvarez, G.; Bergström, L. Rod Packing in Chiral Nematic Cellulose Nanocrystal Dispersions Studied by Small-Angle X-ray Scattering and Laser Diffraction. *Langmuir* **2015**, *31*, 6507–6513. [CrossRef] [PubMed]

35. Elazzouzi-Hafraoui, S.; Nishiyama, Y.; Putaux, J.; Heux, L.; Dubreuil, F.; Rochas, C. The shape and size distribution of crystalline nanoparticles prepared by acid hydrolysis of native cellulose. *Biomacromolecules* **2008**, *9*, 57–65. [CrossRef] [PubMed]

36. Zakri, C.; Poulin, P. Phase behavior of nanotube suspensions: From attraction induced percolation to liquid crystalline phases. *J. Mater. Chem.* **2006**, *16*, 4095–4098. [CrossRef]

37. Barry, E.; Dogic, Z. Entropy driven self-assembly of nonamphiphilic colloidal membranes. *Proc. Natl. Acad. Sci. USA* **2010**, *107*, 10348–10353. [CrossRef] [PubMed]

38. Edgar, C.; Gray, D. Influence of dextran on the phase Behavior of suspensions of cellulose nanocrystals. *Macromolecules* **2002**, *35*, 7400–7406. [CrossRef]

39. Beck-Candanedo, S.; Viet, D.; Gray, D. Induced phase separation in cellulose nanocrystal suspensions containing ionic dye species. *Cellulose* **2006**, *13*, 629–635. [CrossRef]

40. Beck-Candanedo, S.; Viet, D.; Gray, D. Induced phase separation in low-ionic-strength cellulose nanocrystal suspensions containing high-molecular-weight blue dextrans. *Langmuir* **2006**, *22*, 8690–8695. [CrossRef]

41. Israelachvili, J.N. *Intermolecular and Surface Forces*, 3rd ed.; Academic Press: Burlington, MA, USA, 2010.

42. Kang, K.; Wilk, A.; Patkowski, A.; Dhont, J. Diffusion of spheres in isotropic and nematic networks of rods: Electrostatic interactions and hydrodynamic screening. *J. Chem. Phys.* **2007**, *126*, 214501. [CrossRef]

43. Sluckin, T.J.; Dunmur, D.A.; Stegemeyer, H. *Crystals that Flow: Classic Papers from the History of Liquid Crystals*; Taylor and Francis: London, UK, 2004.

44. Lagerwall, J.P.; Scalia, G. Introduction. In *Liquid Crystals with Nano and Microparticles: (In 2 Volumes)*; Lagerwall, J.P., Scalia, G., Eds.; World Scientific: Singapore, 2016.

45. Lagerwall, J.P.F.; Giesselmann, F. Current topics in smectic liquid crystal research. *ChemPhysChem* **2006**, *7*, 20–45. [CrossRef] [PubMed]

46. De France, K.; Yager, K.; Hoare, T.; Cranston, E. Cooperative Ordering and Kinetics of Cellulose Nanocrystal Alignment in a Magnetic Field. *Langmuir* **2016**, *32*, 7564–7571. [CrossRef] [PubMed]

47. Lagerwall, S.T. On some important chapters in the history of liquid crystals. *Liq. Cryst.* **2013**, *40*, 1698–1729. [CrossRef]

48. Enz, E.; La Ferrara, V.; Scalia, G. Confinement-Sensitive Optical Response of Cholesteric Liquid Crystals in Electrospun Fibers. *ACS Nano* **2013**, *7*, 6627–6635. [CrossRef] [PubMed]

49. Majoinen, J.; Kontturi, E.; Ikkala, O.; Gray, D.G. SEM imaging of chiral nematic films cast from cellulose nanocrystal suspensions. *Cellulose* **2012**, *19*, 1599–1605. [CrossRef]

50. de Gennes, P.G.; Prost, J. *The Physics of Liquid Crystals*; Clarendon Press: Oxford, UK, 1993.

51. Nayani, K.; Chang, R.; Fu, J.; Ellis, P.; Fernandez-Nieves, A.; Park, J.; Srinivasarao, M. Spontaneous emergence of chirality in achiral lyotropic chromonic liquid crystals confined to cylinders. *Nat. Commun.* **2015**, *6*, 8067. [CrossRef]

52. Jeong, J.; Kang, L.; Davidson, Z.; Collings, P.; Lubensky, T.; Yodh, A. Chiral structures from achiral liquid crystals in cylindrical capillaries. *Proc. Natl. Acad. Sci. USA* **2015**, *112*, E1837–E1844. [CrossRef]

53. Davidson, Z.; Kang, L.; Jeong, J.; Still, T.; Collings, P.; Lubensky, T.; Yodh, A. Chiral structures and defects of lyotropic chromonic liquid crystals induced by saddle-splay elasticity. *Phys. Rev. E* **2015**, *91*, 050501. [CrossRef]

54. Prinsen, P.; van der Schoot, P. Shape and director-field transformation of tactoids. *Phys. Rev. E* **2003**, *68*, 021701. [CrossRef]

55. Onsager, L. The effects of shape on the interaction of colloidal particles. *Ann. N. Y. Acad. Sci.* **1949**, *51*, 627–659. [CrossRef]

56. Dong, X.; Kimura, T.; Revol, J.; Gray, D. Effects of ionic strength on the isotropic-chiral nematic phase transition of suspensions of cellulose crystallites. *Langmuir* **1996**, *12*, 2076–2082. [CrossRef]

57. Dong, X.; Gray, D. Effect of counterions on ordered phase formation in suspensions of charged rodlike cellulose crystallites. *Langmuir* **1997**, *13*, 2404–2409. [CrossRef]

58. Yang, C.H.; Wang, M.X.; Haider, H.; Yang, J.H.; Sun, J.Y.; Chen, Y.M.; Zhou, J.; Suo, Z. Strengthening Alginate/Polyacrylamide Hydrogels Using Various Multivalent Cations. *ACS Appl. Mater. Interfaces* **2013**, *5*, 10418–10422. [CrossRef] [PubMed]

59. Flory, P.J.; Abe, A. Statistical thermodynamics of mixtures of rodlike particles. 1. Theory for polydisperse systems. *Macromolecules* **1978**, *11*, 1119–1122. [CrossRef]

60. Lekkerkerker, H.N.W.; Coulon, P.; Van Der Haegen, R.; Deblieck, R. On the isotropic-liquid crystal phase separation in a solution of rodlike particles of different lengths. *J. Chem. Phys.* **1984**, *80*, 3427–3433. [CrossRef]

61. Odijk, T.; Lekkerkerker, H.N. Theory of the isotropic-liquid crystal phase separation for a solution of bidisperse rodlike macromolecules. *J. Phys. Chem.* **1985**, *89*, 2090–2096. [CrossRef]

62. Vroege, G.J.; Lekkerkerker, H.N.W. Phase-transitions in lyotropic colloidal and polymer liquid-crystals. *Rep. Prog. Phys.* **1992**, *55*, 1241–1309. [CrossRef]

63. Bates, M.A.; Frenkel, D. Influence of polydispersity on the phase behavior of colloidal liquid crystals: A Monte Carlo simulation study. *J. Chem. Phys.* **1998**, *109*, 6193–6199. [CrossRef]

64. Wensink, H. Effect of Size Polydispersity on the Pitch of Nanorod Cholesterics. *Crystals* **2019**, *9*, 143. [CrossRef]

65. Neto, A.M.F.; Salinas, S.R.A. *The Physics of Lyotropic Liquid Crystals: Phase Transitions and Structural Properties (Monographs on the Physics and Chemistry of Materials)*; Oxford University Press: New York, NY, USA, 2005.

66. Maier, W.; Saupe, A. Eine einfache molekulare Theorie des nematischen kristallinflüssigen Zustandes. *Z. Für Naturforschung A* **1958**, *13*, 564–566. [CrossRef]

67. Zocher, H.; Jacobsohn, K. Über Taktosole. *Kolloidchem. Beihäfte* **1929**, *28*, 167–206. [CrossRef]

68. Zocher, H. Über freiwillige Strukturbildung in Solen (Eine neue Art anisotrop flüssiger Medien). *Z. Für Anorg. Und Allg. Chem.* **1925**, *147*, 91–110. [CrossRef]

69. Puech, N.; Grelet, E.; Poulin, P.; Blanc, C.; van, der Schoot, P. Nematic droplets in aqueous dispersions of carbon nanotubes. *Phys. Rev. E* **2010**, *82*, 020702. [CrossRef] [PubMed]

70. Kaznacheev, A.; Bogdanov, M.; Taraskin, S. The nature of prolate shape of tactoids in lyotropic inorganic liquid crystals. *J. Exp. Theor. Phys.* **2002**, *95*, 57–63. [CrossRef]

71. Robinson, C. Liquid-crystalline structures in solutions of a polypeptide. *Trans. Faraday Soc.* **1956**, *52*, 571. [CrossRef]

72. Nyström, G.; Arcari, M.; Mezzenga, R. Confinement-induced liquid crystalline transitions in amyloid fibril cholesteric tactoids. *Nat. Nanotechnol.* **2018**, *13*, 330–336. [CrossRef] [PubMed]

73. Mosser, G.; Anglo, A.; Helary, C.; Bouligand, Y.; Giraud-Guille, M. Dense tissue-like collagen matrices formed in cell-free conditions. *Matrix Biol.* **2006**, *25*, 3–13. [CrossRef]

74. Sonin, A. Inorganic lyotropic liquid crystals. *J. Mater. Chem.* **1998**, *8*, 2557–2574. [CrossRef]

75. Lehmann, O. *Flüssige Kristalle und die Theorien des Lebens*; Johann Ambrosius Barth: Leipzig, Germany, 1908.

76. Friedel, G. Les états mésomorphes de la matière. *Ann. Phys.* **1922**, *18*, 273–474. [CrossRef]

77. Blanc, C. Interplay between Growth Mechanisms and Elasticity in Liquid Crystalline Nuclei. *Prog. Theor. Phys. Suppl.* **2008**, *175*, 93–102. [CrossRef]

78. Langmuir, I. The Role of Attractive and Repulsive Forces in the Formation of Tactoids, Thixotropic Gels, Protein Crystals and Coacervates. *J. Chem. Phys.* **1938**, *6*, 873–896. [CrossRef]

79. Chen, W.; Gray, D.G. Interfacial tension between isotropic and anisotropic phases of a suspension of rodlike particles. *Langmuir* **2002**, *18*, 633–637. [CrossRef]

80. Lavrentovich, O. Topological defects in dispersed liquid crystals, or words and worlds around liquid crystal drops. *Liq. Cryst.* **1998**, *24*, 117–125. [CrossRef]

81. Kurik, M.V.; Lavrentovich, O. Defects in liquid crystals: Homotopy theory and experimental studies. *Physics-Uspekhi* **1988**, *31*, 196–224. [CrossRef]

82. Volovik, G.; Lavrentovich, O. Topological dynamics of defects: Boojums in nematic drops. *Zh Eksp Teor Fiz* **1983**, *85*, 1997–2010.

83. Kurik, M.; Lavrentovich, O. Negative-positive monopole transitions in cholesteric liquid-crystals. *JETP Lett.* **1982**, *35*, 444–447.

84. Bouligand, Y.; Livolant, F. The organization of cholesteric spherulites. *J. Phys.* **1984**, *45*, 1899–1923. [CrossRef]

85. Xu, F.; Crooker, P. Chiral nematic droplets with parallel surface anchoring. *Phys. Rev. E* **1997**, *56*, 6853–6860. [CrossRef]

86. Xu, F.; Kitzerow, H.S.; Crooker, P. Director configurations of nematic-liquid-crystal droplets: Negative dielectric anisotropy and parallel surface anchoring. *Phys. Rev. E* **1994**, *49*, 3061. [CrossRef]

87. Drzaic, P.S. *Liquid Crystal Dispersions*; World Scientific: Singapore, 1995.

88. Bezic, J.; Zumer, S. Structures of the cholesteric liquid crystal droplets with parallel surface anchoring. *Liq. Cryst.* **1992**, *11*, 593–619. [CrossRef]

89. Robinson, C.; Ward, J.; Beevers, R. Liquid crystalline structure in polypeptide solutions. Part 2. *Discuss. Faraday Soc.* **1958**, *25*, 29–42. [CrossRef]

90. Lopez-Leon, T.; Fernandez-Nieves, A. Drops and shells of liquid crystal. *Colloid Polym. Sci.* **2011**, *289*, 345–359. [CrossRef]

91. Urbanski, M.; Reyes, C.G.; Noh, J.; Sharma, A.; Geng, Y.; Jampani, V.S.R.; Lagerwall, J.P. Liquid crystals in micron-scale droplets, shells and fibers. *J. Phys. Condens. Matter* **2017**, *29*, 133003. [CrossRef]

92. Candau, S.; Roy, P.; Debeauvais, F. Magnetic Field Effects in Nematic and Cholesteric Droplets Suspended in a Isotropic Liquid. *Mol. Cryst. Liq. Cryst.* **1973**, *23*, 283–297. [CrossRef]

93. Jamali, V.; Behabtu, N.; Senyuk, B.; Lee, J.; Smalyukh, I.; van der Schoot, P.; Pasquali, M. Experimental realization of crossover in shape and director field of nematic tactoids. *Phys. Rev. E* **2015**, *91*, 042507. [CrossRef] [PubMed]

94. Jeong, J.; Davidson, Z.; Collings, P.; Lubensky, T.; Yodh, A. Chiral symmetry breaking and surface faceting in chromonic liquid crystal droplets with giant elastic anisotropy. *Proc. Natl. Acad. Sci. USA* **2014**, *111*, 1742–1747. [CrossRef]

95. Nayani, K.; Fu, J.; Chang, R.; Park, J.; Srinivasarao, M. Using chiral tactoids as optical probes to study the aggregation behavior of chromonics. *Proc. Natl. Acad. Sci. USA* **2017**, *114*, 3826–3831. [CrossRef]

96. Araki, J.; Kuga, S. Effect of Trace Electrolyte on Liquid Crystal Type of Cellulose Microcrystals. *Langmuir* **2001**, *17*, 4493–4496. [CrossRef]

97. Lettinga, M.P.; Kang, K.; Imhof, A.; Derks, D.; Dhont, J.K. Kinetic pathways of the nematic–isotropic phase transition as studied by confocal microscopy on rod-like viruses. *J. Phys. Condens. Matter* **2005**, *17*, S3609. [CrossRef]

98. Tran, A.; Hamad, W.; MacLachlan, M. Tactoid Annealing Improves Order in Self-Assembled Cellulose Nanocrystal Films with Chiral Nematic Structures. *Langmuir* **2018**, *34*, 646–652. [CrossRef]

99. Li, Y.; Jun-Yan Suen, J.; Prince, E.; Larin, E.; Klinkova, A.; Thérien-Aubin, H.; Zhu, S.; Yang, B.; Helmy, A.; Lavrentovich, O.; et al. Colloidal cholesteric liquid crystal in spherical confinement. *Nat. Commun.* **2016**, *7*, 12520. [CrossRef] [PubMed]

100. Lagerwall, J.P. A phenomenological introduction to liquid crystals and colloids. In *Liquid Crystals with Nano and Microparticles: (In 2 Volumes)*; Lagerwall, J.P., Scalia, G., Eds.; World Scientific: Singapore, 2016; pp. 11–93.

101. De Vries, H. Rotary power and other optical properties of certain liquid crystals. *Acta Cryst.* **1951**, *4*, 219–226. [CrossRef]

102. Wang, P.; Hamad, W.; MacLachlan, M. Structure and transformation of tactoids in cellulose nanocrystal suspensions. *Nat. Commun.* **2016**, *7*, 11515. [CrossRef] [PubMed]

103. O'Keeffe, O.; Wang, P.; Hamad, W.; MacLachlan, M. Boundary Geometry Effects on the Coalescence of Liquid Crystalline Tactoids and Formation of Topological Defects. *J. Phys. Chem. Lett.* **2019**, *10*, 278–282. [CrossRef] [PubMed]

104. Dong, X.; Revol, J.; Gray, D. Effect of microcrystallite preparation conditions on the formation of colloid crystals of cellulose. *Cellulose* **1998**, *5*, 19–32. [CrossRef]

105. Hirai, A.; Inui, O.; Horii, F.; Tsuji, M. Phase Separation Behavior in Aqueous Suspensions of Bacterial Cellulose Nanocrystals Prepared by Sulfuric Acid Treatment. *Langmuir* **2009**, *25*, 497–502. [CrossRef]

106. Honorato-Rios, C.; Lehr, C.; Schütz, C.; Sanctuary, R.; Osipov, M.A.; Baller, J.; Lagerwall, J.P.F. Fractionation of cellulose nanocrystals: Enhancing liquid crystal ordering without promoting gelation. *NPG Asia Mater.* **2018**, *10*, 455–465. [CrossRef]

107. Beck-Candanedo, S.; Roman, M.; Gray, D. Effect of reaction conditions on the properties and behavior of wood cellulose nanocrystal suspensions. *Biomacromolecules* **2005**, *6*, 1048–1054. [CrossRef]

108. Shafiei-Sabet, S.; Hamad, W.Y.; Hatzikiriakos, S.G. Rheology of Nanocrystalline Cellulose Aqueous Suspensions. *Langmuir* **2012**, *28*, 17124–17133. [CrossRef]

109. Khandelwal, M.; Windle, A.H. Self-assembly of bacterial and tunicate cellulose nanowhiskers. *Polymer* **2013**, *54*, 5199–5206. [CrossRef]

110. Finner, S.; Schilling, T.; van der Schoot, P. Connectivity, Not Density, Dictates Percolation in Nematic Liquid Crystals of Slender Nanoparticles. *Phys. Rev. Lett.* **2019**, *122*, 097801. [CrossRef] [PubMed]

111. Kang, K.; Dhont, J.K.G. Glass Transition in Suspensions of Charged Rods: Structural Arrest and Texture Dynamics. *Phys. Rev. Lett.* **2013**, *110*, 015901. [CrossRef] [PubMed]

112. Purdy, K.; Fraden, S. Isotropic-cholesteric phase transition of filamentous virus suspensions as a function of rod length and charge. *Phys. Rev. E* **2004**, *70*, 061703. [CrossRef] [PubMed]

113. Zimmermann, K.; Hagedorn, H.; Heuck, C.C.; Hinrichsen, M.; Ludwig, H. The ionic properties of the filamentous bacteriophages Pf1 and fd. *J. Biol. Chem.* **1986**, *261*, 1653–1655. [PubMed]

114. Nordenström, M.; Fall, A.; Nyström, G.; Wågberg, L. Formation of Colloidal Nanocellulose Glasses and Gels. *Langmuir* **2017**, *33*, 9772–9780. [CrossRef]

115. Lokanathan, A.; Nykänen, A.; Seitsonen, J.; Johansson, L.; Campbell, J.; Rojas, O.; Ikkala, O.; Laine, J. Cilia-mimetic hairy surfaces based on end-immobilized nanocellulose colloidal rods. *Biomacromolecules* **2013**, *14*, 2807–2813. [CrossRef]

116. Petrova, A.; Herold, C.; Petrov, E. Conformations and membrane-driven self-organization of rodlike fd virus particles on freestanding lipid membranes. *Soft Matter* **2017**, *13*, 7172–7187. [CrossRef]

117. Honorato-Rios, C.; Kuhnhold, A.; Bruckner, J.R.; Dannert, R.; Schilling, T.; Lagerwall, J.P. Equilibrium Liquid Crystal Phase Diagrams and Detection of Kinetic Arrest in Cellulose Nanocrystal Suspensions. *Front. Mater.* **2016**, *3*, 21. [CrossRef]

118. Lettinga, M.P. Kolloid-Tagung 2019. In Proceedings of the 49th Conference of the German Colloid Society, Stuttgart, Germany, 23–25 September 2019.

119. Phan-Xuan, T.; Thuresson, A.; Skepö, M.; Labrador, A.; Bordes, R.; Matic, A. Aggregation behavior of aqueous cellulose nanocrystals: The effect of inorganic salts. *Cellulose* **2016**, *23*, 3653–3663. [CrossRef]

120. Urena-Benavides, E.E.; Ao, G.; Davis, V.A.; Kitchens, C.L. Rheology and Phase Behavior of Lyotropic Cellulose Nanocrystal Suspensions. *Macromolecules* **2011**, *44*, 8990–8998. [CrossRef]

121. Shafeiei-Sabet, S.; Hamad, W.Y.; Hatzikiriakos, S.G. Influence of degree of sulfation on the rheology of cellulose nanocrystal suspensions. *Rheol. Acta* **2013**, *52*, 741–751. [CrossRef]

122. Bruckner, J.; Kuhnhold, A.; Honorato-Rios, C.; Schilling, T.; Lagerwall, J. Enhancing Self-Assembly in Cellulose Nanocrystal Suspensions Using High-Permittivity Solvents. *Langmuir* **2016**, *32*, 9854–9862. [CrossRef] [PubMed]

123. Habibi, Y. Key advances in the chemical modification of nanocelluloses. *Chem. Soc. Rev.* **2014**, *43*, 1519–1542. [CrossRef] [PubMed]

124. Eyley, S.; Thielemans, W. Surface modification of cellulose nanocrystals. *Nanoscale* **2014**, *6*, 7764–7779. [CrossRef] [PubMed]

125. Heux, L.; Chauve, G.; Bonini, C. Nonflocculating and chiral-nematic self-ordering of cellulose microcrystals suspensions in nonpolar solvents. *Langmuir* **2000**, *16*, 8210–8212. [CrossRef]

126. Elazzouzi-Hafraoui, S.; Putaux, J.L.; Heux, L. Self-assembling and Chiral Nematic Properties of Organophilic Cellulose Nanocrystals. *J. Phys. Chem. B* **2009**, *113*, 11069–11075. [CrossRef]

127. Frka-Petesic, B.; Jean, B.; Heux, L. First experimental evidence of a giant permanent electric-dipole moment in cellulose nanocrystals. *EPL* **2014**, *107*, 28006. [CrossRef]

128. Nguyen, T.D.; Hamad, W.Y.; MacLachlan, M.J. CdS Quantum Dots Encapsulated in Chiral Nematic Mesoporous Silica: New Iridescent and Luminescent Materials. *Adv. Funct. Mater.* **2014**, *24*, 777–783. [CrossRef]

129. Shopsowitz, K.E.; Kelly, J.A.; Hamad, W.Y.; MacLachlan, M.J. Biopolymer Templated Glass with a Twist: Controlling the Chirality, Porosity, and Photonic Properties of Silica with Cellulose Nanocrystals. *Adv. Funct. Mater.* **2014**, *24*, 327–338. [CrossRef]

130. Shopsowitz, Kevin, E.; Stahl, A.; Hamad, Wadood, Y.; MacLachlan, Mark, J. Hard Templating of Nanocrystalline Titanium Dioxide with Chiral Nematic Ordering. *Angew. Chem., Int. Ed.* **2012**, *51*, 6886–6890. [CrossRef]

131. Kelly, J.A.; Shopsowitz, K.E.; Ahn, J.M.; Hamad, W.Y.; MacLachlan, M.J. Chiral Nematic Stained Glass: Controlling the Optical Properties of Nanocrystalline Cellulose-Templated Materials. *Langmuir* **2012**, *28*, 17256–17262. [CrossRef] [PubMed]

132. Qi, H.; Roy, X.; Shopsowitz, K.E.; Hui, J.K.H.; MacLachlan, M.J. Liquid-Crystal Templating in Ammonia: A Facile Route to Micro- and Mesoporous Metal Nitride/Carbon Composites. *Angew. Chem. Int. Ed.* **2010**, *49*, 9740–9743. [CrossRef] [PubMed]

133. Shopsowitz, K.; Qi, H.; Hamad, W.; MacLachlan, M. Free-standing mesoporous silica films with tunable chiral nematic structures. *Nature* **2010**, *468*, 422–425. [CrossRef] [PubMed]

134. Yao, K.; Meng, Q.; Bulone, V.; Zhou, Q. Flexible and Responsive Chiral Nematic Cellulose Nanocrystal/Poly(ethylene glycol) Composite Films with Uniform and Tunable Structural Color. *Adv. Mater.* **2017**, *29*, 1701323. [CrossRef]

135. Wu, T.; Li, J.; Li, J.; Ye, S.; Wei, J.; Guo, J. A bio-inspired cellulose nanocrystal-based nanocomposite photonic film with hyper-reflection and humidity-responsive actuator properties. *J. Mater. Chem. C* **2016**, *4*, 9687–9696. [CrossRef]

136. Giese, M.; De Witt, J.; Shopsowitz, K.; Manning, A.; Dong, R.; Michal, C.; Hamad, W.; MacLachlan, M. Thermal switching of the reflection in chiral nematic mesoporous organosilica films infiltrated with liquid crystals. *ACS Appl. Mater. Interfaces* **2013**, *5*, 6854–6859. [CrossRef]

137. Kelly, J.; Shukaliak, A.; Cheung, C.; Shopsowitz, K.; Hamad, W.; MacLachlan, M. Responsive photonic hydrogels based on nanocrystalline cellulose. *Angew. Chem. Int. Ed.* **2013**, *52*, 8912–8916. [CrossRef]

138. Cheung, C.C.Y.; Giese, M.; Kelly, J.A.; Hamad, W.Y.; MacLachlan, M.J. Iridescent Chiral Nematic Cellulose Nanocrystal/Polymer Composites Assembled in Organic Solvents. *ACS Macro Lett.* **2013**, *2*, 1016–1020. [CrossRef]

139. Giese, M.; Khan, Mostofa, K.; Hamad, Wadood, Y.; MacLachlan, Mark, J. Imprinting of Photonic Patterns with Thermosetting Amino-Formaldehyde-Cellulose Composites. *ACS Macro Lett.* **2013**, *2*, 818–821. [CrossRef]

140. Khan, M.; Giese, M.; Yu, M.; Kelly, J.; Hamad, W.; MacLachlan, M. Flexible mesoporous photonic resins with tunable chiral nematic structures. *Angew. Chem. Int. Ed.* **2013**, *52*, 8921–8924. [CrossRef]

141. Shopsowitz, Kevin, E.; Hamad, Wadood, Y.; MacLachlan, Mark, J. Flexible and Iridescent Chiral Nematic Mesoporous Organosilica Films. *J. Am. Chem. Soc.* **2012**, *134*, 867–870. [CrossRef] [PubMed]

142. Terpstra, A.S.; Arnett, L.P.; Manning, A.P.; Michal, C.A.; Hamad, W.Y.; MacLachlan, M.J. Iridescent Chiral Nematic Mesoporous Organosilicas with Alkylene Spacers. *Adv. Opt. Mater.* **2018**, *6*, 1800163. [CrossRef]

143. Rao, A.; Divoux, T.; McKinley, G.; Hart, A. Shear melting and recovery of crosslinkable cellulose nanocrystal-polymer gels. *Soft Matter* **2019**, *15*, 4401–4412. [CrossRef] [PubMed]

144. Okita, Y.; Fujisawa, S.; Saito, T.; Isogai, A. TEMPO-Oxidized Cellulose Nanofibrils Dispersed in Organic Solvents. *Biomacromolecules* **2011**, *12*, 518–522. [CrossRef] [PubMed]

145. Wu, W.; Song, R.; Xu, Z.; Jing, Y.; Dai, H.; Fang, G. Fluorescent cellulose nanocrystals with responsiveness to solvent polarity and ionic strength. *Sens. Actuators B* **2018**, *275*, 490–498. [CrossRef]

146. Espinosa, S.C.; Kuhnt, T.; Foster, E.J.; Weder, C. Isolation of Thermally Stable Cellulose Nanocrystals by Phosphoric Acid Hydrolysis. *Biomacromolecules* **2013**, *14*, 1223–1230. [CrossRef] [PubMed]

147. Viet, D.; Beck-Candanedo, S.; Gray, D. Dispersion of cellulose nanocrystals in polar organic solvents. *Cellulose* **2007**, *14*, 109–113. [CrossRef]

148. Sánchez, P.; Tsubaki, S.; Pádua, A.; Wada, Y. Kinetic analysis of microwave-enhanced cellulose dissolution in ionic solvents. *Phys. Chem. Chem. Phys.* **2020**, *22*, 1003–1010. [CrossRef]

149. Lethesh, K.; Evjen, S.; Venkatraman, V.; Shah, S.; Fiksdahl, A. Highly efficient cellulose dissolution by alkaline ionic liquids. *Carbohydr. Polym.* **2020**, *229*, 115594. [CrossRef]

150. Rajeev, A.; Basavaraj, M.G. Confinement effect on spatio-temporal growth of spherulites from cellulose/ionic liquid solutions. *Polymer* **2019**, *185*, 121927. [CrossRef]

151. Okura, H.; Wada, M.; Serizawa, T. Dispersibility of HCl-treated Cellulose Nanocrystals with Water-dispersible Properties in Organic Solvents. *Chem. Lett.* **2014**, *43*, 601–603. [CrossRef]

152. Gauche, C.; Felisberti, M.I. Colloidal Behavior of Cellulose Nanocrystals Grafted with Poly(2-alkyl-2-oxazoline)s. *ACS Omega* **2019**, *4*, 11893–11905. [CrossRef] [PubMed]

153. Hu, Z.; Berry, R.M.; Pelton, R.; Cranston, E.D. One-Pot Water-Based Hydrophobic Surface Modification of Cellulose Nanocrystals Using Plant Polyphenols. *ACS Sustain. Chem. Eng.* **2017**, *5*, 5018–5026. [CrossRef]

154. Fumagalli, M.; Sanchez, F.; Boisseau, S.M.; Heux, L. Gas-phase esterification of cellulose nanocrystal aerogels for colloidal dispersion in apolar solvents. *Soft Matter* **2013**, *9*, 11309. [CrossRef]

155. Frka-Petesic, B.; Radavidson, H.; Jean, B.; Heux, L. Dynamically Controlled Iridescence of Cholesteric Cellulose Nanocrystal Suspensions Using Electric Fields. *Adv. Mater.* **2017**, *29*, 1606208. [CrossRef]

156. Kovalenko, A. Predictive Multiscale Modeling of Nanocellulose Based Materials and Systems. *IOP Conf. Ser. Mater. Sci. Eng.* **2014**, *64*, 012040. [CrossRef]

157. Gousse, C.; Chanzy, H.; Excoffier, G.; Soubeyrand, L.; Fleury, E. Stable suspensions of partially silylated cellulose whiskers dispersed in organic solvents. *Polymer* **2002**, *43*, 2645–2651. [CrossRef]

158. Yi, J.; Xu, Q.; Zhang, X.; Zhang, H. Chiral-nematic self-ordering of rodlike cellulose nanocrystals grafted with poly(styrene) in both thermotropic and lyotropic states. *Polymer* **2008**, *49*, 4406–4412. [CrossRef]

159. Xu, Q.; Yi, J.; Zhang, X.; Zhang, H. A novel amphotropic polymer based on cellulose nanocrystals grafted with azo polymers. *Eur. Polym. J.* **2008**, *44*, 2830–2837. [CrossRef]

160. Goffin, A.; Habibi, Y.; Raquez, J.; Dubois, P. Polyester-grafted cellulose nanowhiskers: A new approach for tuning the microstructure of immiscible polyester blends. *ACS Appl. Mater. Interfaces* **2012**, *4*, 3364–3371. [CrossRef]

161. Ljungberg, N.; Bonini, C.; Bortolussi, F.; Boisson, C.; Heux, L.; Cavaillé, J. New nanocomposite materials reinforced with cellulose whiskers in atactic polypropylene: Effect of surface and dispersion characteristics. *Biomacromolecules* **2005**, *6*, 2732–2739. [CrossRef] [PubMed]

162. Araki, J.; Wada, M.; Kuga, S. Steric stabilization of a cellulose microcrystal suspension by poly (ethylene glycol) grafting. *Langmuir* **2001**, *17*, 21–27. [CrossRef]

163. Siqueira, G.; Bras, J.; Dufresne, A. New process of chemical grafting of cellulose nanoparticles with a long chain isocyanate. *Langmuir* **2010**, *26*, 402–411. [CrossRef] [PubMed]

164. Tian, C.; Fu, S.; Habibi, Y.; Lucia, L. Polymerization topochemistry of cellulose nanocrystals: A function of surface dehydration control. *Langmuir* **2014**, *30*, 14670–14679. [CrossRef]

165. Blachechen, L.S.; de Mesquita, J.P.; de Paula, E.L.; Pereira, F.V.; Petri, D.F.S. Interplay of colloidal stability of cellulose nanocrystals and their dispersibility in cellulose acetate butyrate matrix. *Cellulose* **2013**, *20*, 1329–1342. [CrossRef]

166. Braun, B.; Dorgan, J. Single-step method for the isolation and surface functionalization of cellulosic nanowhiskers. *Biomacromolecules* **2009**, *10*, 334–341. [CrossRef]

167. Yuan, H.; Nishiyama, Y.; Wada, M.; Kuga, S. Surface acylation of cellulose whiskers by drying aqueous emulsion. *Biomacromolecules* **2006**, *7*, 696–700. [CrossRef]

168. Zhou, Q.; Brumer, H.; Teeri, T.T. Self-Organization of Cellulose Nanocrystals Adsorbed with Xyloglucan Oligosaccharide-Poly(ethylene glycol)-Polystyrene Triblock Copolymer. *Macromolecules* **2009**, *42*, 5430–5432. [CrossRef]

169. Salajkova, M.; Berglund, Lars, A.; Zhou, Q. Hydrophobic cellulose nanocrystals modified with quaternary ammonium salts. *J. Mater. Chem.* **2012**, *22*, 19798–19805. [CrossRef]

170. Xu, Y.; Dai, Y.; Nguyen, T.; Hamad, W.; MacLachlan, M. Aerogel materials with periodic structures imprinted with cellulose nanocrystals. *Nanoscale* **2018**, *10*, 3805–3812. [CrossRef]

171. There is a confusing discrepancy between the plotted data, from which these numbers were taken, and the text, which states a pitch change from $\sim$80 μm to $\sim$10 μm.

172. Mu, X.; Gray, D. Formation of chiral nematic films from cellulose nanocrystal suspensions is a two-stage process. *Langmuir* **2014**, *30*, 9256–9260. [CrossRef] [PubMed]

173. The percentage is given as 10% *w/v* without further clarification. While this mixing of mass and volume in a percent indication is quite commonly used in some communities, it causes problems because the meaning is not clearly defined. In addition, a percentage should be dimensionless, whereas here we have the dimension kg/m^3. We assume that 1% here means 1 g in 100 mL, but this interpretation may be incorrect.

174. Frka-Petesic, B.; Kamita, G.; Guidetti, G.; Vignolini, S. Angular optical response of cellulose nanocrystal films explained by the distortion of the arrested suspension upon drying. *Phys. Rev. Mater.* **2019**, *3*, 045601. [CrossRef]

175. Oseen, C.W. The theory of liquid crystals. *Trans. Faraday Soc.* **1933**, *29*, 883. [CrossRef]

176. Fergason, J.L. Cholesteric Structure-1 Optical Properties. *Mol. Cryst.* **1966**, *1*, 293–307. [CrossRef]

177. Dreher, R.; Meier, G. Optical properties of cholesteric liquid crystals. *Phys. Rev. A* **1973**, *8*, 1616. [CrossRef]

178. Berreman, D.; Scheffer, T. Bragg reflection of light from single-domain cholesteric liquid-crystal films. *Phys. Rev. Lett.* **1970**, *25*, 577–581. [CrossRef]

179. Berreman, D.W.; Scheffer, T.J. Reflection and Transmission by Single-Domain Cholesteric Liquid Crystal Films: Theory and Verification. *Mol. Cryst. Liq. Cryst.* **1970**, *11*, 395–405. [CrossRef]

180. Kim, S.T.; Finkelmann, H. Cholesteric liquid single-crystal elastomers (LSCE) obtained by the anisotropic deswelling method. *Macromol. Rapid Commun.* **2001**, *22*, 429–433. [CrossRef]

181. Finkelmann, H.; Kim, S.; Munoz, A.; Palffy-Muhoray, P.; Taheri, B. Tunable mirrorless lasing in cholesteric liquid crystalline elastomers. *Adv. Mater.* **2001**, *13*, 1069–1072. [CrossRef]

182. Kizhakidathazhath, R.; Geng, Y.; Jampani, V.S.R.; Charni, C.; Sharma, A.; Lagerwall, J.P.F. Facile Anisotropic Deswelling Method for Realizing Large-Area Cholesteric Liquid Crystal Elastomers with Uniform Structural Color and Broad-Range Mechanochromic Response. *Adv. Funct. Mater.* **2019**, 1909537. [CrossRef]

183. Mao, Y.; Terentjev, E.M.; Warner, M. Cholesteric elastomers: Deformable photonic solids. *Phys. Rev. E* **2001**, *64*, 041803. [CrossRef] [PubMed]

184. Revol, J.; Godbout, L.; Dong, X.; Gray, D.; Chanzy, H.; Maret, G. Chiral Nematic Suspensions of Cellulose Crystallites—Phase-separation and Magnetic-field Orientation. *Liq. Cryst.* **1994**, *16*, 127–134. [CrossRef]

185. Kimura, F.; Kimura, T.; Tamura, M.; Hirai, A.; Ikuno, M.; Horii, F. Magnetic alignment of the chiral nematic phase of a cellulose microfibril suspension. *Langmuir* **2005**, *21*, 2034–2037. [CrossRef] [PubMed]

186. Frka-Petesic, B.; Guidetti, G.; Kamita, G.; Vignolini, S. Controlling the Photonic Properties of Cholesteric Cellulose Nanocrystal Films with Magnets. *Adv. Mater.* **2017**, *29*, 1701469. [CrossRef] [PubMed]

187. Larson, R.G. Transport and deposition patterns in drying sessile droplets. *AIChE J.* **2014**, *60*, 1538–1571. [CrossRef]

188. Anyfantakis, M.; Baigl, D. Manipulating the Coffee-Ring Effect: Interactions at Work. *ChemPhysChem* **2015**, *16*, 2726–2734. [CrossRef]

189. Deegan, R.D.; Bakajin, O.; Dupont, T.F.; Huber, G.; Nagel, S.R.; Witten, T.A. Contact line deposits in an evaporating drop. *Phys. Rev. E* **2000**, *62*, 756. [CrossRef]

190. Deegan, R.D.; Bakajin, O.; Dupont, T.F.; Huber, G.; Nagel, S.R.; Witten, T.A. Capillary flow as the cause of ring stains from dried liquid drops. *Nature* **1997**, *389*, 827–829. [CrossRef]

191. Mu, X.; Gray, D.G. Droplets of cellulose nanocrystal suspensions on drying give iridescent 3-D "coffee-stain" rings. *Cellulose* **2015**, *22*, 1103–1107. [CrossRef]

192. Dupas, Lagerwall and Anyfantakis, in preparation.

193. Shimura, K.; Uchiyama, N.; Kasai, K. Prevention of evaporation of small-volume sample solutions for capillary electrophoresis using a mineral-oil overlay. *Electrophoresis* **2001**, *22*, 3471–3477. [CrossRef]

194. Gençer, A.; Schütz, C.; Thielemans, W. Influence of the Particle Concentration and Marangoni Flow on the Formation of Cellulose Nanocrystal Films. *Langmuir* **2016**, *33*, 228–234. [CrossRef] [PubMed]

195. Hu, H.; Larson, R.G. Marangoni effect reverses coffee-ring depositions. *J. Phys. Chem. B* **2006**, *110*, 7090–7094. [CrossRef] [PubMed]

196. Varanakkottu, S.; Anyfantakis, M.; Morel, M.; Rudiuk, S.; Baigl, D. Light-Directed Particle Patterning by Evaporative Optical Marangoni Assembly. *Nano. Lett.* **2016**, *16*, 644–650. [CrossRef] [PubMed]

197. Pan, J.; Hamad, W.; Straus, S. Parameters Affecting the Chiral Nematic Phase of Nanocrystalline Cellulose Films. *Macromolecules* **2010**, *43*, 3851–3858. [CrossRef]

198. Beck, S.; Bouchard, J.; Berry, R. Controlling the Reflection Wavelength of Iridescent Solid Films of Nanocrystalline Cellulose. *Biomacromolecules* **2011**, *12*, 167–172. [CrossRef]

199. Hirscher, M.; Becher, M.; Haluska, M.; Dettlaff-Weglikowska, U.; Quintel, A.; Duesberg, G.; Choi, Y.; Downes, P.; Hulman, M.; Roth, S. Hydrogen storage in sonicated carbon materials. *Appl. Phys. A Mater. Sci. Process.* **2001**, *72*, 129–132. [CrossRef]

200. Nguyen, T.; Hamad, W.; MacLachlan, M. Tuning the iridescence of chiral nematic cellulose nanocrystals and mesoporous silica films by substrate variation. *Chem. Commun.* **2013**, *49*, 11296–11298. [CrossRef]

201. Saha, P.; Davis, V.A. Photonic Properties and Applications of Cellulose Nanocrystal Films with Planar Anchoring. *ACS Appl. Nano Mater.* **2018**, *1*, 2175–2183. [CrossRef]

202. Guidetti, G.; Atifi, S.; Vignolini, S.; Hamad, W. Flexible Photonic Cellulose Nanocrystal Films. *Adv. Mater.* **2016**, *28*, 10042–10047. [CrossRef]

203. Bardet, R.; Belgacem, N.; Bras, J. Flexibility and color monitoring of cellulose nanocrystal iridescent solid films using anionic or neutral polymers. *ACS Appl. Mater. Interfaces* **2015**, *7*, 4010–4018. [CrossRef]

204. Mitov, M. Cholesteric Liquid Crystals with a Broad Light Reflection Band. *Adv. Mater.* **2012**, *24*, 6260–6276. [CrossRef] [PubMed]

205. Fernandes, S.; Almeida, P.; Monge, N.; Aguirre, L.; Reis, D.; de Oliveira, C.; Neto, A.; Pieranski, P.; Godinho, M. Mind the Microgap in Iridescent Cellulose Nanocrystal Films. *Adv. Mater.* **2017**, *29*, 1603560. [CrossRef] [PubMed]

206. Kang, K.H.; Lim, H.C.; Lee, H.W.; Lee, S.J. Evaporation-induced saline Rayleigh convection inside a colloidal droplet. *Phys. Fluids* **2013**, *25*, 042001. [CrossRef]

207. Song, W.; Lee, J.K.; Gong, M.S.; Heo, K.; Chung, W.J.; Lee, B.Y. Cellulose Nanocrystal-Based Colored Thin Films for Colorimetric Detection of Aldehyde Gases. *ACS Appl. Mater. Interf.* **2018**, *10*, 10353–10361. [CrossRef]

208. Hoeger, I.; Rojas, Orlando, J.; Efimenko, K.; Velev, Orlin, D.; Kelley, Steve, S. Ultrathin film coatings of aligned cellulose nanocrystals from a convective-shear assembly system and their surface mechanical properties. *Soft Matter* **2011**, *7*, 1957–1967. [CrossRef]

Article

Flexible and Structural Coloured Composite Films from Cellulose Nanocrystals/Hydroxypropyl Cellulose Lyotropic Suspensions

Diogo V. Saraiva, Ricardo Chagas, Beatriz M. de Abreu, Cláudia N. Gouveia, Pedro E. S. Silva, Maria Helena Godinho and Susete N. Fernandes *

i3N/CENIMAT, Department of Materials Science, NOVA School of Science and Technology, NOVA University Lisbon, Campus de Caparica, 2829-516 Caparica, Portugal; dv.saraiva@campus.fct.unl.pt (D.V.S.); r.chagas@fct.unl.pt (R.C.); b.abreu@campus.fct.unl.pt (B.M.d.A.); c.gouveia@campus.fct.unl.pt (C.N.G.); pes.silva@fct.unl.pt (P.E.S.S.); mhg@fct.unl.pt (M.H.G.)
* Correspondence: sm.fernandes@fct.unl.pt

Received: 3 January 2020; Accepted: 12 February 2020; Published: 16 February 2020

Abstract: Lyotropic colloidal aqueous suspensions of cellulose nanocrystals (CNCs) can, after solvent evaporation, retain their chiral nematic arrangement. As water is removed the pitch value of the suspension decreases and structural colour-generating films, which are mechanically brittle in nature, can be obtained. Increasing their flexibility while keeping the chiral nematic structure and biocompatible nature is a challenging task. However, if achievable, this will promote their use in new and interesting applications. In this study, we report on the addition of different amounts of hydroxypropyl cellulose (HPC) to CNCs suspension within the coexistence of the isotropic-anisotropic phases and infer the influence of this cellulosic derivative on the properties of the obtained solid films. It was possible to add 50 wt.% of HPC to a CNCs aqueous suspension (to obtain a 50/50 solids ratio) without disrupting the LC phase of CNCs and maintaining a left-handed helical structure in the obtained films. When 30 wt.% of HPC was added to the suspension of CNCs, a strong colouration in the film was still observed. This colour shifts to the near-infrared region as the HPC content in the colloidal suspension increases to 40 wt.% or 50 wt.% The all-cellulosic composite films present an increase in the maximum strain as the concentration of HPC increases, as shown by the bending experiments and an improvement in their thermal properties.

Keywords: liquid crystal; cellulose nanocrystals; hydroxypropyl cellulose; lyotropic; chiral nematic

---

1. Introduction

Recently, much interest has been given by the scientific community to cellulose nanocrystals (CNCs)-based materials, since these nanoparticles present a set of advantages that built on the ones derived from their precursor, the biopolymer cellulose. For instance, CNCs are lightweight, present a high Young's modulus, a high surface area and superficial hydrophilicity [1]. It is well accepted that cellulose fibrils, present in the cell walls of plants, show a supramolecular arrangement of crystalline domains within some amorphous regions that can be separated with an appropriate acid treatment. If CNCs are extracted by sulphuric acid hydrolysis from a cellulosic biomass, such as cotton or wood pulp, and when suspended in water above a critical concentration (C*), they can self-assemble into a lyotropic chiral nematic liquid crystalline (LC) phase [2]. In this LC phase, the local orientation of the nanoparticles is described by a common axis called director ($\vec{n}$), which modulates in a helical manner and the distance required for the director to complete a rotation of 360° is defined as the pitch value, P. The stabilisation of this colloidal suspension and its self-assembly into a lyotropic phase are attributed to the introduction of ester-sulphate groups on the surface of the CNCs during the process of acid hydrolysis and its quantity is dependent on the hydrolytic conditions [3]. The lyotropic aqueous suspensions of CNCs present a left-handed chiral nematic organisation, with pitch values in the order of a few tens of micrometres which decrease as the concentration of CNCs increases [4,5]. Starting from a CNCs colloidal suspension with a low concentration of nanoparticles, for instance ~3 wt.%, it is possible to obtain a small volume fraction of anisotropic phase (ϕ, 0.15, Figure 1a). If the solvent is slowly removed, for example using the solvent casting technique, and since the pitch values decrease with increasing chiral nematic interaction during the drying process coloured and iridescent films can be obtained [4,6,7]. These films present strong optical activity produced from the reflection of light of the helical axis and depend on the orientation and the pitch value of this helix. These films will selectively reflect left-handed circularly polarised (LCP) light and transmit right-handed circularly polarised (RCP) light. This selectively structural colouration is owed to the left-handed helical structure retained from the chiral nematic phase of the precursor suspension. The specific reflected wavelength (λ_0) can be related with the helicoidal pitch by the de Vries equation:

$$\lambda_0 = \bar{n}P\cos\theta \tag{1}$$

where $\bar{n}$ is the average refractive index of the mesophase and θ is the angle between the incident light and the helical axis [8].

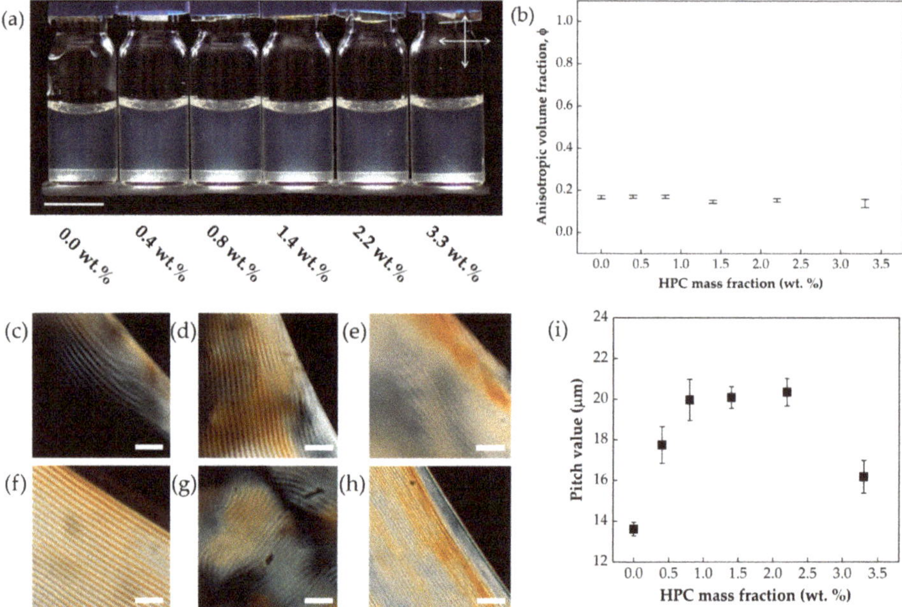

Figure 1. (**a**) Photograph obtained between crossed-polars of the phase separation observed on a lyotropic CNCs suspension with concentration 3.3 wt.% (first vial) and CNCs/HPC aqueous suspensions with different concentrations of HPC added while the concentration in CNCs is constant; (**b**) Phase diagram of CNCs/HPC/H_2O system, represented by the volume fraction, ϕ, of the anisotropic phase as a function of HPC concentration where no apparent change is observed when compared with the pristine lyotropic LC CNCs/H_2O system; (**c–h**) POM images obtained in transmission mode with crossed-polars of neat CNCs/H_2O and CNCs/HPC/H_2O suspensions; (**i**) chiral nematic pitch value plotted as a function of the HPC content in the mixtures of CNCs/HPC/H_2O. Scale bars: (**a**) 1 mm; (**c–e**) 50 µm.

The structural colouration observed in the CNCs films has been related to several parameters that influence the colloidal suspension as, for example: CNCs concentration; ionic strength [3]; energy given with ultrasonic treatment [9]; exposure to magnetic field [10] or electric field [11] and the drying rate of the solvent [12] combined with planar anchoring [13]. Although these interesting studies allow us to obtain iridescent CNCs films, due to their crystalline nature, they present low mechanical strength (highly brittle), which can restrict their range of application.

Several research studies can be found where the addition of a small molecule or polymer to an unmodified CNCs colloidal suspension allows it to retain the chiral nematic arrangement while inducing some flexibility in the final film. For instance, Yao et al. were able to retain structural colouration after the addition of polyethylene glycol (up to 50wt.% CNC/PEG) to the CNCs suspension. The final product was humidity-responsive, since exposure to humidity increases the pitch value of the films in a reversible process, and became flexible as the authors increased the PEG content in the composite system [14]. Similar phenomena were reported when glycerol was added to the CNCs suspension by He et al. [15] and by Xu et al. [16]. Wan et al. were able to produce a waterborne polyurethane elastomer CNCs and retain the cholesteric organisation after solvent evaporation and cross-linking [17]. The authors added the waterborne polyurethane elastomer, a water-insoluble polymer based on latex spheres (178 nm), to a CNCs suspension and were able to produce tuneable iridescent flexible films, that are dynamically humidity responsive. Gray and co-worker were able to show that the addition of the non-ionic molecule D-glucose to a CNCs suspension leads to an increase of the pitch value of the obtained films and a shift to the red region of the electromagnetic spectrum of light [18]. More recently, Qu et al. were able to show that the D-glucose also acts as a plasticiser and enables active tuning of the left-handed helical organisation while under mechanical deformation [19]. The use of 30 wt.% of long CNCs, extract from tunicates, in a lyotropic suspension of short CNCs, derived from wood, gave rise to a composite system that presents improved mechanical properties (Young's modulus, tensile strength and strain to failure) and when bent they do not break [20]. This example is, to the best of our knowledge, the first where the LC phase of the CNCs is retained in an all-cellulosic based composite film, so keeping in mind the high number of cellulose derivatives, and especially the water-soluble ones, an array of different all-cellulosic composites systems can be examined.

Hydroxypropyl cellulose is a cellulose derivative, where some hydroxyl groups of the anhydroglucose monomer unit of the cellulose are substituted by hydroxypropyl moieties [21]. This substitution leads to a change in solubility, when compared with its parent cellulose, since HPC is soluble in common organic solvents [22] and water, below a critical solution temperature (LCST) of approximately 42 °C [21]. It is highly used as thickener, emulsifier and coating in the pharmaceutical industry, but can also be found in the food industry [21]. Similarly, to what is observed for CNCs, HPC can present a lyotropic liquid crystalline behaviour when dissolved in water at concentrations higher than 42 wt.%. However, for this derivative, the handedness of the chiral nematic arrangement is right-handed, so it will reflect RCP light and transmit LCP light. The films derived from an LC phase of unmodified HPC present the same handedness [23]. The use of HPC as a matrix in an HPC/CNCs composite system was already discussed, although, in this case, a shear-casting technique was used to produce iridescent films. In this composite system, the small amount of CNCs in the suspension did not affect their optical properties but instead acted as a reinforcement agent by enhancing the anisotropic mechanical properties. The authors attributed the observed iridescence to the modulation of the surface into bands that appear perpendicular to the shear direction and act as a diffraction grating [24].

Herein, we report the addition of hydroxypropyl cellulose into a LC CNCs colloidal suspension and study the effect of the presence of this non-ionic water-soluble polymer in the chiral nematic organisation of the lyotropic suspension and produced films. Furthermore, the influence of the cellulosic derivative HPC in the CNCs matrix was perceived by means of chemical and physical characterisation, as, for instance, their thermal behaviour and flexibility.

2. Materials and Methods

Cellulose nanocrystals were obtained from the acid hydrolysis of microcrystalline cellulose (MCC; Aldrich Avicel®, ~50 μm particle size, derived from cotton) with sulphuric acid (Sigma-Aldrich, 95–97% purity), as described elsewhere by Fernandes et al. [25], however a reaction time of 40 min instead of 130 min was used. A lower reaction time was selected since the authors confirmed that the particles produced by these conditions were very similar to the ones synthesized using longer reaction times. For instance, the nanoparticles present similar length (153 ± 34 nm vs. 152 ± 65 nm, respectively) and similar amount of ester sulphate groups per 100 anhydroglucose units (3.92 vs. 4.39, respectively). The resulting suspension was prepared with three 10-min cycles of sonication using a UP400 S ultrasonic processor (460 W, 24 kHz, Heilscher Ultrasonics GmbH, Berlin, Germany, totalling an energy input of 12.7 kJ g^{-1} CNC). CNCs suspensions with a gravimetrically determined concentration of 3.0 ± 0.1 wt.% and 3.32 ± 0.05 wt.% were used in its acid form (pH $\cong$ 2.5) and showcases 0.76% of total sulphur content, as determined by elemental analysis (using a Thermo Finnigan-CE Instruments, Flash EA 1112 CHNS Series Analyzer, ±0.3%, San Jose, CA, USA). This value, according to the empirical formula $C_6H_{10}O_5$–(SO_3) and inputted in equation $S(\%) = 100n \times S/[6C + 10H + (5 + 3n)O + nS]$ [26], is equivalent to 3.92 ester sulphate groups (OSO_3H) per 100 anhydroglucose units. The average length and width of the CNC nanoparticles were measured to be 153 ± 34 nm and 6 ± 2 nm, respectively, amounting to an aspect ratio of 26 (see Figure S1, Supplementary Information). These values were obtained from 200 measurements using Gwyddion software (version 2.52, http://gwyddion.net/ [27]) taken from AFM images (Asylum Research MFP-3D standalone system in tapping mode, Santa Barbara, CA, USA, with commercially available silicon AFM probes, scanning frequency of 300 kHz, k = 26 N/m; for particle size distribution). The obtained nanoparticle dimensions are in accordance with the values described in literature for a similar hydrolysis time. To produce the phase diagram, an aqueous CNCs suspension with a concentration of 11 ± 1 wt.% was obtained from an osmotic bath using polyethylene glycol (PEG), as based on the method described by Frka-Petesic et al. [10]. The initial CNCs suspension was placed into a dialysis membrane (Spectrum Spectra/Por®4 membrane, molecular cut-off 12–14 kDa), which was left for 8 h in an aqueous solution of 10 wt.% poly(ethylene) glycol (PEG, Sigma-Aldrich, 35 kDa), under magnetic stirring.

The obtained CNCs suspension was diluted into samples with CNCs content ranging from 1.5 to 8.5 wt.%. The samples were allowed to set until separation of isotropic-anisotropic phases is observed and the anisotropic volume fraction changed no further (1 month), at this stage photographs were taken between crossed-polars. The anisotropic fraction value for each sample is an average of 30 measurements and error bar taken as a standard deviation of all the measurements (using ImageJ software, version 1.52a http://imagej/Nih.GOV/ij [28]); measurements were taken in the centre of each vial. POM images were observed using an Olympus BX-51 polarised optical microscope (Tokyo, Japan), connected to a SCHOTT KL2500 LCD cold light source. An equipped camera (Olympus DP73) along with Olympus Stream Basic 1.9 software were used for image capture. In order to observe the samples by polarised optical microscopy (POM), each suspension was placed into individual rectangle hollow glass tubes (VitroTubes™, 50 × 4 mm^2, 0.4 mm path length).

Hydroxypropyl cellulose (HPC, Sigma-Aldrich, $\overline{M_w}$ = 80,000) was added to a ~3 wt.% CNCs aqueous suspension in different ratios, to give rise to films with CNC/HPC ratios of 90/10, 80/20, 70/30, 60/40 and 50/50. These values correspond to suspensions with 3.3 wt.% CNC/x wt.% HPC%/y wt.% H_2O, where x is equal to 0.0, 0.4, 0.8, 1.4, 2.2, 3.3 wt. and y is equal to 96.7, 96.3, 95.9, 95.3, 94,5, 93,4, respectively. Henceforth, while referring the colloidal suspension the sample identifier will be x wt.% CNC/y wt.% HPC/(100−x−y) wt.% H_2O. In the solid films, the sample identifier is the CNC/HPC ratio. The resulting suspensions were homogenised by means of magnetic stirring until complete HPC dissolution and allowed to set until the presence of the anisotropic phase was observed. CNCs and CNCs/HPC composite films were obtained by solvent evaporation. The suspensions were placed into 35 mm polystyrene Petri dishes (1–2 mL) and solvent evaporation occurred in a chamber with controlled temperature and relative humidity at 4 °C and 60–70% of RH, respectively, for two to three weeks. The thickness of each CNC film was averaged from 10 measurements using a Mitutoyo digital micrometre. The films were kept in a desiccator (with relative humidity of 40%) until further analyses.

Photographs of CNCs films and CNCs/HPC composite films were obtained with a Canon EOS 550D coupled with an EF-S 60 mm Canon macro-lens (Tokyo, Japan), at an 8° angle perpendicularly to the substrate, under uncollimated unpolarised white light illumination. Thermogravimetric analyses (TGA/DSC) were performed using a Netzsch 449 F3 Jupiter®simultaneous thermal analyser (Selb, Germany). Each sample was heated from 25 to 550 °C, at a heating rate of 10 °C/min under nitrogen. Circular dichroism (CD) experiments were performed using an Applied Photophysics Chirascan™ CD spectrometer (Surrey, UK). The scanned wavelengths ranged from 200 and 800 nm, with a bandwidth of 1 nm and a scanning rate of 200 nm/s. Spectrophotometric analysis in UV/VIS/NIR (PerkinElmer Lambda 950, Waltham, WA, USA) was performed in reflective mode with an integrating sphere, with measured wavelength data ranging from 200 to 2500 nm. The structural assembly of CNC and CNC/HPC composite films was observed by SEM (Scanning Electron Microscope) using a Carl Zeiss AURIGA CrossBeam SEM Workstation, Oberkochen, Germany equipped with an Oxford energy-dispersive X-ray spectrometer. The in-lens mode was used with an accelerating voltage of 2 kV and a 20 μm aperture size. As preparation for SEM observation cross-sections of the films, samples were fractured, mounted onto aluminium stubs and coated with a thin carbon layer using a Q150T ES pumped coater from Quorum (Washington D.C., WA, USA). ImageJ software (version 1.52a) was used for pitch size measurements from SEM images of the cross-section of each film, and from POM images of each CNCs/HPC composite suspension. In both methods, at least 30 measurements were done per sample.

Bend-testing was conducted using cylindrical mandrels apparatus with a semi-suspended pliable platform (BRAIVE Instruments, Liège, Belgium), and a set of mandrels varying in diameter. A total of 10 mandrels were used, ranging in diameter from 25 to 2 mm. Briefly, within 1 s, each film was folded to 180° to form an inverted "U" (Figure S2) shaped bend over the mandrel, maintaining contact with it [29]. The shape was maintained for 2 s before release. Each sample was bend-tested using successively smaller mandrels, with the sample being inspected after being bent around each mandrel. The bend procedure was continued until cracks were visible or until the sample did not experience any cracking with the smallest mandrel, in which case the sample was recorded as having not cracked (NC). 3 replicas were tested for each sample.

3. Results and Discussion

From an LC colloidal aqueous suspension with a concentration of CNCs of 3.3 wt.%, different mixtures of CNC/HPC with ratios of 10, 20, 30, 40 and 50 were prepared. In the range of concentrations of HPC used in this study we did not observe the gelation as detected by Hu et al. when adding small amounts of other polysaccharides (e.g., hydroxyethyl cellulose or hydroxypropyl guar) to a diluted suspension of CNCs [30]. The CNCs/HPC mixtures were allowed to settle until phase separation was observed, as shown in the photograph obtained between crossed-polars presented in Figure 1a. From this image, it is possible to note that, macroscopically, the introduction of the macromolecule of HPC in the LC-CNCs suspension seems not to influence the liquid crystalline phase, if one compares the birefringence obtained in the CNCs/HPC/H$_2$O mixtures and the CNCs/H$_2$O suspension. Indeed, this observation is confirmed by the determination of the volume of anisotropic fraction, since all mixtures present a similar value of ϕ of ~0.15, observed in Figure 1b. A similar result was macroscopically observed by He et al. for the composite colloidal suspensions of CNCs/glycerol [15]. Still, the authors could observe an almost linear dependence in the pitch value of the suspension as the content of the polyol glycerol increases (from 0.85 μm for 0 wt.% glycerol to 1.15 μm 40 wt.% glycerol). The preparation of sealed rectangular vials allows us to observe the suspensions with polarised microscopy. In Figure 1c–h, the fingerprint texture observed in the anisotropic phase of the colloidal mixtures of CNCs/HPC/H$_2$O can be observed. If the pitch value is plotted as a function of the concentration of HPC in the colloidal suspension, depicted in Figure 1i, one can also see a direct dependence of the pitch value with the HPC content up to 0.8 wt.%, the subsequent further increase of HPC in the system does not seem to affect the pitch value of the suspension. Moreover, when 3.3 wt.% of HPC is introduced in the colloidal suspension, a reduction of 4 μm in the pitch value is observed (comparing with the system with 2.2 wt.% of HPC); however, this value is still a higher value than the one observed for the neat CNCs colloidal suspension.

To prepare iridescent CNCs and CNCs/HPC films the solvent was allowed to slowly evaporate in a solvent casting method. The infrared (IR) spectra (Figure 2a) of CNCs and the different CNCs/HPC composites show the typical IR bands related to the cellulose backbone, such as 1160, 1030 and 895 cm^{-1} from C-O-C (asymmetrical stretching) and glycosidic bonds in cellulose [1,31]. Another characteristic signal in all samples containing CNCs is the absorption band at 820 cm^{-1} corresponding to the S-O bonds, confirming the presence of sulphate groups resulting from sulphuric acid hydrolysis [32]. The band around 3400 cm^{-1} corresponds to the O-H stretching and the bands at 2900 and 2974 cm^{-1} to the C-H stretching. In the HPC sample, the characteristic bands at 2876 and 2900 cm^{-1} correspond to the C-H vibration due to the aliphatic chain within the hydroxypropyl moiety and the sharp band at 1455 cm^{-1} to the deformation vibration of the -CH$_2$ group [33]. A decrease in the transmittance, particularly in the 2876 to 2900 cm^{-1} region, was registered in all composite samples with increasing content of HPC (as can be seen in the remaining spectra in Supplementary Information Figure S5) when compared with the HPC sample.

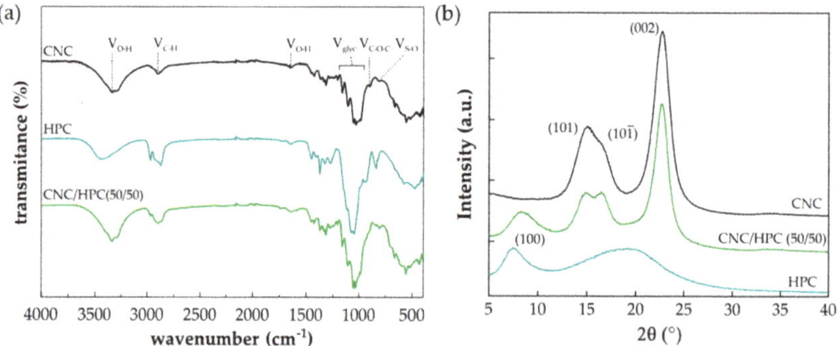

Figure 2. (**a**) FTIR spectra of CNCs, HPC and CNC/HPC (50/50) films; (**b**) XRD patterns of neat CNCs, HPC and CNC/HPC (50/50) composite films.

XRD diffractogram of CNCs (black line in Figure 2b) display the characteristic peaks of cellulose nanocrystals, with peaks at 2θ = 15.0°, 16.3° and 22.7° endorsed to the crystallographic planes (101), (10Ī), (002), where the latter is the main peak and characteristic of the crystalline region of cellulose [32]. The diffractogram of HPC film (teal line in Figure 2b) exhibits a peak at 2θ = 7.5° attributed to the HPC crystallographic plane (100), and the presence of a broad peak centred at 2θ ≈ 19.2° is ascribed to the amorphous region [34]. As expected, the presence of HPC in the composite system CNC/HPC (50/50) does not affect the supramolecular arrangement of cellulose since all the characteristic lattices of cellulose are present (olive line in Figure 2b). The presence of HPC in the composite is confirmed by the appearance of the HPC crystallographic plane (100) at a 2θ = 8.3°. The remaining samples show similar XRD spectra (Supplementary Figure S7). Nevertheless, the crystallographic plane of HPC centred at 2θ = 7.5° is more evident in the samples with a higher content of HPC.

In Figure 3a–c,j–l, photographs of the obtained films, captured under visible light and against a black surface can be seen. Up to 20 wt.% of HPC the composite films display a strong colouration, starting with the neat CNCs film that presents a colouration predominately green-yellow (Figure 3a). Adding 10 wt.% of HPC to the suspension a strong yellow-orange appearance is achieved (Figure 3b), while 20 wt.% of HPC in the composite mixture leads to a film with strong red colouration that coexists with areas without colouration (Figure 3c) as the one marked with a dashed rectangle. Composite films with 30 wt.% of HPC are predominately non-coloured, however, a large outer ring of strong red colouration can be seen (Figure 3j). Further increase in the HPC content leads to films with no iridescence and develops a matte finish (Figure 3k,l). This type of matte finish was reported recently by Chan et al. on cross-linked HPC coloured films [35] and the authors attributed this observation to a large multiplicity of pitches within the sample. From SEM images obtained from the cross-section of our films, it is easy to see that the chiral nematic arrangement on the composite's samples, up to an intake of 30 wt.% of HPC (Figure 3d–f,m), is retained after solvent evaporation, although the pitch value (defined as the value measure for each 2 consecutive layers) increases. Whilst the films derived from a suspension with 40 and 50 wt.% of HPC present no visible colouration, the SEM images from their cross-section demonstrate that the chiral nematic arrangement is also maintained (Figure 3n,o). From the SEM images and by applying Equation (1), it is possible to estimate the maximum wavelength reflected by this structural organisation at the nanoscale [36] (see Table S2 of the Supplementary Information and Figure 4b and subsequent discussion).

Figure 3g–i,p–r show the transmission POM images of the neat CNCs and CNC/HPC films. In Supplementary Figure S8 POM images of fingerprints observed at the centre of each film are presented. These were added in order to show the presence of the fingerprint texture. From these latter images, it is perceptible that this type of texture becomes more noticeable as the content of HPC in the composite films increases. If fingerprints are visible with POM this means that the helical axis of the chiral nematic arrangement was not completely aligned perpendicular to the film surface, hence a complete planar alignment was not achieved [13]. Since, the effect is even more noticeable when the content of HPC in the composite mixture increases, it can be assumed that the presence of HPC somehow lowers the concentration at which the glassy state or kinetic arrest occurs. The latter is defined as the state, that besides the presence of the solvent, no further change in the pitch value is observed, which induces the lock of the chiral nematic organisation of this point into the film structure [37]. Indeed, Mu et al. attribute the observed increase of the pitch value of the films to the fact that D–(+)–glucose added to the CNCs suspension promotes gelation and therefore hindered further decreases in the pitch values as the solvent evaporates [18]. The increase in the wavelength reflected by the structure of the films for the composite mixtures of CNCs with glycerol [15,16] or polyethylene glycol [14] was also observed. In order to support this assumption measurements of the pitch value observed by POM images on the centre of each film were determined, as can be seen from Supplementary Figure S9. From these results, one can see a slight increase in the measured pitch value as the concentration of the neutral HPC macromolecule increases in the composite system.

A UV-VIS-NIR spectrometer was used to determine the maximum wavelength reflected by each film in the dashed rectangular section denoted in the photographs of Figure 3. The set of spectra can be seen in Figure 4a. It is important to note that at the wavelength of 850 nm there is a sudden drop in the reflectance intensity that corresponds to the change in the equipment from the visible region to near-infrared. When the maximum wavelength for each spectrum is plotted as a function of HPC concentration (Figure 4b—black squares) it is noticeable that only the film samples of neat CNCs and 90/10 wt.% CNCs/HPC reflect light within the visible range of the electromagnetic spectrum. Although the 80/20 CNC/HPC sample present some red coloration the measurements were performed in the darker area observed in the photograph. Samples with 30, 40 and 50 wt.% of HPC reflect radiation in the near-infrared region. However, and as observed in the UV-VIS-NIR spectrum of the sample 50/50 wt.% CNCs/HPC, its intensity is very low, which might be related to the strong matte finish appearance of this film. As mentioned above, the determination of the average pitch value for each film can be done by measuring two consecutive chiral nematic layers from SEM images (Figure 3d–f,m–o). The average refractive index, n, of neat CNC and HPC is 1.56 and 1.33, respectively [24]. For each composite system, n was estimated considering the weight fraction of each cellulosic component in the mixture. By using these values (summarised in Table S2) in Equation (1), it is possible to estimate the maximum wavelength as a function of HPC content depicted in Figure 4b—blue dots. In general, when similar areas are used a good agreement can be observed between the maximum reflected wavelength and the one estimated by the pitch values determined by SEM and the de Vries equation.

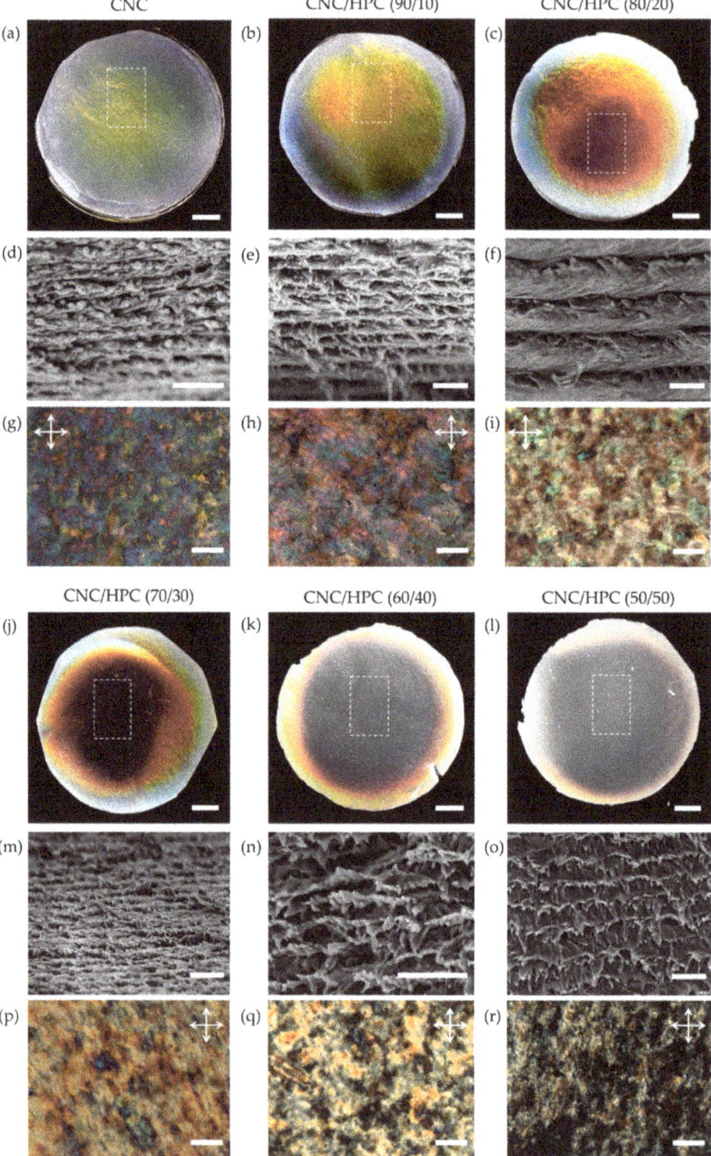

Figure 3. (**a–c,j–l**) Photographs of neat CNCs and CNC/HPC films with different content of HPC taken under uncollimated unpolarised white light; (**d–f,m–o**) SEM images of the central region of the cross-section of the CNCs and CNC/HPC films, where the chiral nematic arrangement can be observed; (**g–i,p–r**) POM images obtained with crossed-polars and in transmission mode of the central regions of neat CNCs and CNCs/HPC films. Scale bars: (**a–c** and **j–l**) 5 mm, (**d–f**) 500 nm, (**m–o**) 1 µm, (**g–i** and **p–r**) 100 µm. POM images. In the photographs, a dashed white rectangle was added that marks the area observed with the UV/VIS/IR spectroscopy.

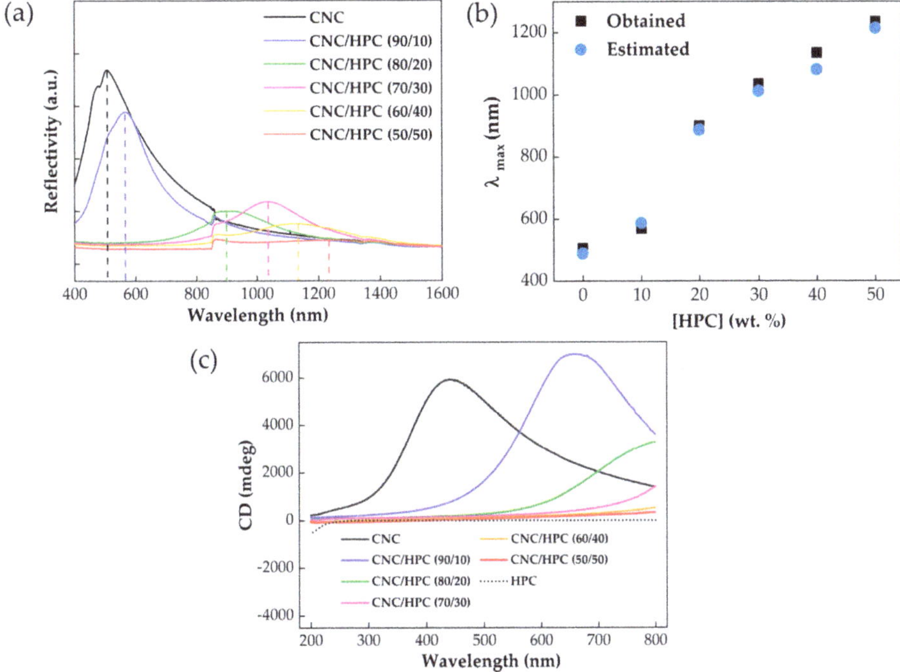

Figure 4. (a) Reflectance spectra captured from the marked region of CNCs and CNC/HPC composite films across the visible and near-infrared regions of the electromagnetic spectrum of light, where it can be seen that the increase in HPC content leads to an increase of the maximum wavelength reflected to the near-infrared region; (b) maximum reflected wavelength plotted as a function of HPC concentration in the composite system, black squares represent the obtained values from reflectance spectra, whilst blue dots represent the estimated values using Equation (1), where the average pitch value was measured from SEM images, like the ones presented in Figure 3; (c) corresponding circular dichroism spectra of CNCs, CNC/HPC, and of a neat HPC film.

Since CNCs give rise to left-handed chiral nematic organisation, circular dichroism (CD) was used in order to determine whether the handedness of the composite films was maintained after solvent evaporation. From the CD spectra (Figure 4c), it is possible to confirm that CNC/HPC composite films with a maximum of HPC of 20 wt.% are left-handed since all spectra present a positive CD value. Although the spectra of films with 30 wt.%–50 wt.% of HPC do not present a maximum wavelength within the range of 200 and 800 nm, the shape of each spectrum seems to indicate a positive CD value. In the case of the sample with the higher content of HPC, although the spectra present positive CD values they are very low. In this series of spectra, an HPC film (obtained from solvent evaporation of a solution with 3 wt.% of HPC) was also analysed, and as expected, the CD value is zero with a decrease to negative values indicative of a right-handed organisation; however, undoubtedly with a maximum wavelength further in the ultraviolet region of the electromagnetic spectrum.

With future application in mind, understanding the influence of the HPC on the thermal properties of the CNCs/HPC composite systems is significantly relevant. In Figure 5a, the thermogravimetric curves show that the CNCs thermal behaviour dictates the thermal behaviour of the composites CNC/HPC, since the curves resemble what is obtained for the neat CNCs. As presented in Supplementary Information Figure S10b, CNCs present a two-step pyrolysis process, whilst cellulose and the derivative hydroxypropyl cellulose (dashed line in Figure 5a) present a first-order pyrolysis reaction. The thermograms in Figure 5b also showed that a slight shift to higher temperatures is obtained with an increasing amount of HPC. Indeed, if the onset of the first degradation process is estimated, an increase of 10 °C, 16 °C and 21 °C on the composite system with 30 wt.% (T = 154 °C), 40 wt.% (T = 160 °C) and 50 wt.% (T = 165 °C) respectively, is observed when compared with the neat CNCs (T = 144 °C). Similar tendencies were observed in composite systems of CNCs/PEG [14]. In this work, the authors related this thermal stability to the formation of a uniform PEG layer onto the surface of the CNCs protecting them from degradation.

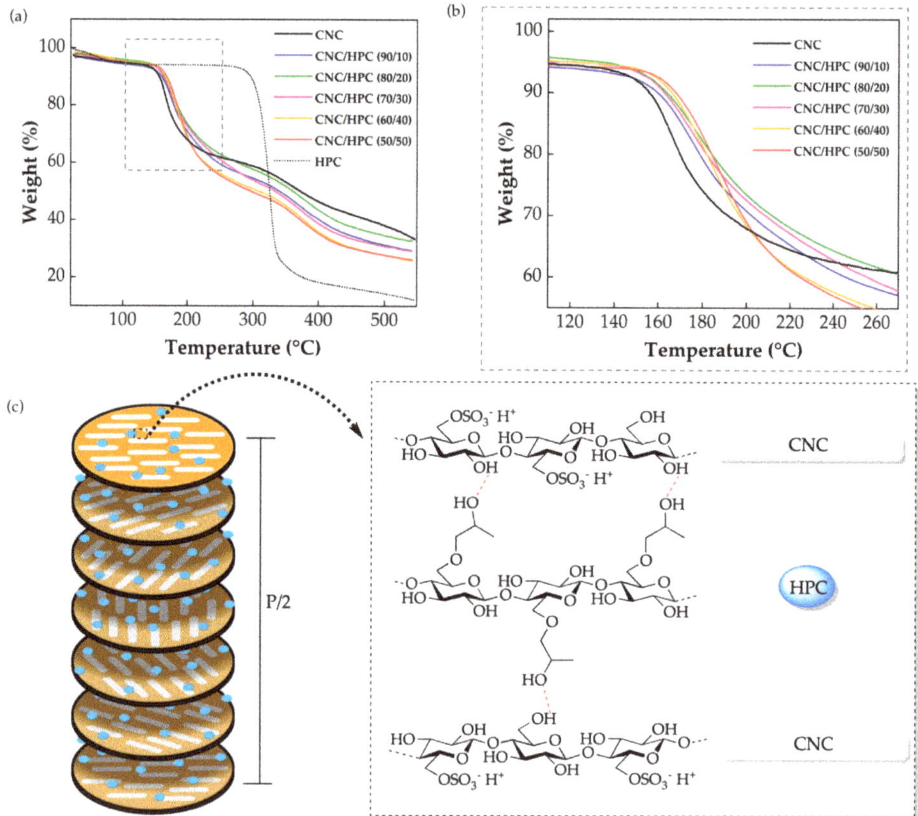

Figure 5. (**a**) Thermogravimetric analysis curves of neat CNCs, HPC and CNCs/HPC composites; (**b**) inset of the thermogravimetric curves in the range of temperatures of 110–260 °C; (**c**) proposed chiral nematic arrangement (right) where the CNCs rods (white anisotropic particles) are surrounded by HPC macromolecules (blue dots), the left scheme shows some possible hydrogen bonding (red dashed lines) between the OH groups of CNCs and HPC. Note that the representation of the number of substituents on the CNC or HPC schematics is not factual.

In our system and considering the analysis of FTIR, XRD and TGA analysis, one can hypothesise that, since CNCs and HPC have similar backbone chemical structure, hydrogen bonding might occur between the free OH groups of these two cellulosic derivatives as depicted in the scheme of Figure 5c. These interactions might be responsible not only for the improvement in the thermal stability but also for the shift of the chiral nematic structure to higher pitch values observed in the suspensions and composite films. While this manuscript was under submission another paper appeared in the journal Carbohydrate Polymers where CNCs/HPC films were also produced [38]. The authors attributed the change in the pitch value mainly to the effect that HPC has in the electrostatic interactions of CNCs layers and the volume effect of the HPC macromolecules.

Resistance to cracking of CNC films was analysed using a bend-test procedure in which a surface strain was applied to the sample without adding overall tensile load [39]. The strain necessary to induce surface cracking was determined by bending the samples over mandrels.

The strain ε, given in percentage for each sample was determined, based on the mandrel that induces the first surface cracks and the thickness of the sample, by the following equation:

$$\varepsilon = \left(\frac{t}{d+t}\right) \times 100 \tag{2}$$

where t is the thickness of the sample and d the diameter of the mandrel. This equation was derived from the bend-test configuration shown in Figure S3.

From the values of strain, determined by Equation (2), it is possible to note that as the HPC content increases in the composite system the films became less brittle since the maximum strain value increases (Table 1). Up to a concentration of HPC of 20 wt.%, no significant change in the composite behaviour while being bent is observed, when compared with the neat CNCs films. If 30 wt.% of HPC is used the strain value increases from 0.7%, from neat CNCs films, to 4.0% in the composite, showing an increase in flexibility as can be seen in Figure 6, where a sequence of pictures shows a composite film being bent. On the composite system with 40 wt.% of HPC only one sample breaks when submitted to the bend-test, whereas an HPC content of 50 wt.% leads to tests where the samples did not crack.

Table 1. Strain and strain average values, given in percentage, determined with Equation (2) for each tested sample. DNC = Did not crack; N/A = not applicable.

CNCs/HPC (wt.%)	ε			$\bar{\varepsilon}$ (%)
	Sample 1	Sample 2	Sample 3	
100/0	0.5	0.6	0.8	0.7 ± 0.1
90/10	0.9	0.7	0.8	0.8 ± 0.1
80/20	1.1	0.9	1.0	1.0 ± 0.1
70/30	3.2	4.9	3.8	4.0 ± 0.7
60/40	5.1	DNC	DNC	N/A
50/50	DNC	DNC	DNC	N/A
0/100	DNC	DNC	DNC	N/A

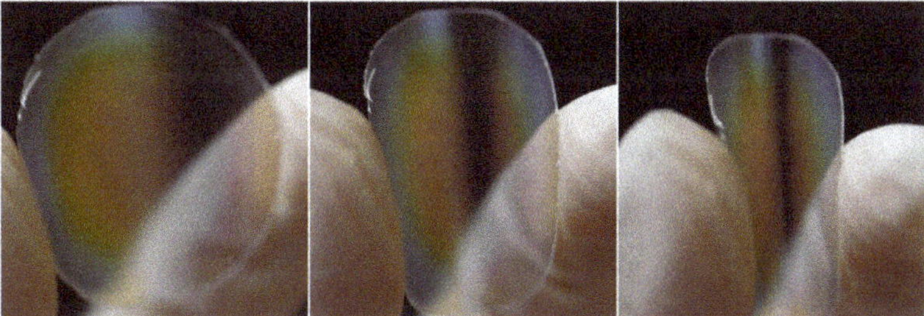

Figure 6. Sequence of images of a test where a CNCs/HPC iridescent film, with 30 wt.% of HPC, is bent (film diameter 35 mm, thickness 106 µm) before cracks appeared.

4. Summary

To summarise, in this work we were able to demonstrate that the addition of HPC to a suspension of CNCs with a small fraction of CNCs (of 3 wt.%) in the range of 10 to 40 wt.% leads to an increase in the pitch value of suspension from 13 µm to 21 µm, and 16 µm in the case of 50 wt.% of HPC in the composite, while maintaining the chiral nematic arrangement. If the solvent is slowly evaporated from these isotropic-LC suspensions, the helicoidal structure is still obtained and an increase in the reflected wavelength in the films is observed, as demonstrated by SEM observation. In contrast, a concentration of 20 wt.% of HPC in the suspension still gives rise to a structurally coloured composite CNCs/HPC film, predominantly red, a further increase in HPC content generates a shift in the reflected wavelength to the near-infrared region. The films present increased flexibility, demonstrated by the bend-testing procedure, and they do not break when a content of HPC of 40 wt.% or 50 wt.% was added. The addition of HPC also induced some improvement in the thermal degradation behaviour of the composite films by comparison with the neat CNCs. By changing the HPC content in the mixtures studied, we were able to obtain films that present iridescence or a matte finish, demonstrating the high versatility of this system. The next task is to explore the use of this CNCs suspension derived from cellulose microcrystals and the addition of HPC with higher average molecular weight; however, some modification in the pitch value of the initial suspension will have to be performed in order to be able to maintain the wavelength reflected within the visible range of the electromagnetic spectrum of light. With these systems, it is expected to obtain composite films with improved flexibility that do not crack even with lower HPC contents. We foresee that these types of low-cost all-cellulose based photonic materials can be used in commercial applications such as circularly polarised light sensors, decorative coatings, active inks and optically variable devices as security features.

Supplementary Materials: The following are available online at http://www.mdpi.com/2073-4352/10/2/122/s1, Figure S1: (a) Representative portion of an AFM image of dispersed CNC used for particle size distribution; (b), (c) CNC length and width distribution histograms, respectively, overlaid with normal distribution curves; Figure S2: Schematic representation of the bend-test configuration showing a sample bent around a mandrel; Figure S3: (a) Phase diagram of the CNCs/H2O system represented by the volume fraction ϕ of the anisotropic to isotropic phase as a function of the concentration of CNCs in the colloidal suspensions; (b–d) Crossed-polarised optical microscope images obtained at the bottom of the hollow glass tube in transmission mode of CNCs suspension with a CNC concentration of 2.6, 4.5, and 8.5 wt.%, respectively (insets of the red square regions showing the banded fingerprint in a small area). Scale bars: (b–d) 50 μm; inset (b–d) 25 μm; Figure S4: FTIR spectra of microcrystalline cellulose (MCC-Avicel®PH-101) and cellulose nanocrystals (CNC). The vertical dashed lines mark various cellulose-characteristic absorption bands. Figure S5: FTIR spectra of CNC, HPC and CNC/HPC composite films in different ratios; Figure S6: X-ray diffraction spectra for MCC and CNC samples, with labelled crystalline peaks; Figure S7: X-ray diffraction spectra for CNC, HPC and CNC/HPC composites, with labelled crystalline peaks; Figure S8: POM images in greyscale obtained with crossed-polars and in transmission mode of the central regions of neat CNCs and CNCs/HPC films, where fingerprint textures are most visible. Scale bars: 25 μm; Figure S9: Estimated pitch values measured at the surface of the films CNC and CNC/HPC as a function of HPC composition in the composite system; Figure S10: TGA and DSC curves for samples a) MCC and b) CNC; Figure S11: TGA and respective DTG curves for CNC, HPC and CNC/HPC composites. The bottom-right graph displays two separate CNC batches produced under conditions identical to the batch of CNCs. Table S1:. Crystalline peak locations for the MCC and CNC samples' XRD spectra, along with their respective calculated crystallinity index; Table S2: Various measured values from CNC and CNC/HPC composite films; Table S3. Estimates of the amount of moisture in each CNC, HPC and CNC/HPC composite film analysed by TGA/DTG. These values are measured from the TGA curves in Figure S11.

Author Contributions: Conceptualisation and project supervision, S.N.F.; Methodology, D.V.S., B.M.d.A., C.N.G., R.C. and P.E.S.S. contributed equally to this work; D.V.S., B.M.d.A. and C.N.G. produced and characterised the cellulose nanocrystals by acid hydrolysis methodology, the lyotropic suspensions, mixtures and films. D.V.S., R.C. and P.E.S.S. did the optical and mechanical bending characterisation of the lyotropic suspensions and films; Writing, reviewing and editing S.N.F. and M.H.G. All authors have read and agreed to the published version of the manuscript.

Funding: This work is funded by FEDER funds through the COMPETE 2020 Program, National Funds through FCT - Portuguese Foundation for Science and Technology and POR Lisboa2020, under the projects with references POCI- 01-0145-FEDER-007688 (Reference UID/CTM/50025), PTDC/CTM-BIO/6178/2014, M-ERA- NET2/0007/2016 (CellColor) and PTDC/CTM-REF/30529/2017 (NanoCell2SEC).

Acknowledgments: We are thankful to Elizabete Ferreira for the CD experiments and Nuno Basílico for patience and the supply of the cell quartz used in the CD experiments.

Conflicts of Interest: The authors declare no conflict of interest.

References

1. Klemm, D.; Kramer, F.; Moritz, S.; Lindstrom, T.; Ankerfors, M.; Gray, D.; Dorris, A. Nanocelluloses: A new family of nature-based materials. *Angew. Chem. Int. Ed. Engl.* **2011**, *50*, 5438–5466. [CrossRef]
2. Revol, J.F.; Bradford, H.; Giasson, J.; Marchessault, R.H.; Gray, D.G. Helicoidal self-ordering of cellulose microfibrils in aqueous suspension. *Int. J. Biol. Macromol.* **1992**, *14*, 170–172. [CrossRef]
3. Abitbol, T.; Kam, D.; Levi-Kalisman, Y.; Gray, D.G.; Shoseyov, O. Surface charge influence on the phase separation and viscosity of cellulose nanocrystals. *Langmuir* **2018**, *34*, 3925–3933. [CrossRef] [PubMed]
4. Parker, R.M.; Guidetti, G.; Williams, C.A.; Zhao, T.; Narkevicius, A.; Vignolini, S.; Frka-Petesic, B. The self-assembly of cellulose nanocrystals: Hierarchical design of visual appearance. *Adv. Mater.* **2018**, *30*, 1704477. [CrossRef] [PubMed]
5. Wang, P.X.; Hamad, W.Y.; MacLachlan, M.J. Structure and transformation of tactoids in cellulose nanocrystal suspensions. *Nat. Commun.* **2016**, *7*, 11515. [CrossRef]
6. Almeida, A.P.C.; Canejo, J.P.; Fernandes, S.N.; Echeverria, C.; Almeida, P.L.; Godinho, M.H. Cellulose-based biomimetics and their applications. *Adv. Mater.* **2018**, *30*, 1703655. [CrossRef]
7. Fernandes, S.N.; Lopes, L.F.; Godinho, M.H. Recent advances in the manipulation of circularly polarised light with cellulose nanocrystal films. *Curr. Opin. Solid State Mater. Sci.* **2019**, *23*, 63–73. [CrossRef]
8. De Vries, H. Rotatory power and other optical properties of certain liquid crystals. *Acta Cryst.* **1951**, *4*, 219–226. [CrossRef]
9. Beck, S.; Bouchard, J.; Berry, R. Controlling the reflection wavelength of iridescent solid films of nanocrystalline cellulose. *Biomacromolecules* **2011**, *12*, 167–172. [CrossRef]
10. Frka-Petesic, B.; Guidetti, G.; Kamita, G.; Vignolini, S. Controlling the photonic properties of cholesteric cellulose nanocrystal films with magnets. *Adv. Mater.* **2017**, *29*, 1701469. [CrossRef]
11. Frka-Petesic, B.; Radavidson, H.; Jean, B.; Heux, L. Dynamically controlled iridescence of cholesteric cellulose nanocrystal suspensions using electric fields. *Adv. Mater.* **2017**, *29*, 1606208. [CrossRef] [PubMed]
12. Zhao, T.H.; Parker, R.M.; Williams, C.A.; Lim, K.T.P.; Frka-Petesic, B.; Vignolini, S. Printing of responsive photonic cellulose nanocrystal microfilm arrays. *Adv. Funct. Mater.* **2019**, *29*, 1804531. [CrossRef]
13. Saha, P.; Davis, V.A. Photonic properties and applications of cellulose nanocrystal films with planar anchoring. *Acs Appl. Nano Mater.* **2018**, *1*, 2175–2183. [CrossRef]
14. Yao, K.; Meng, Q.; Bulone, V.; Zhou, Q. Flexible and responsive chiral nematic cellulose nanocrystal/poly(ethylene glycol) composite films with uniform and tunable structural color. *Adv. Mater.* **2017**, *28*, 1701323. [CrossRef]
15. He, Y.D.; Zhang, Z.L.; Xue, J.; Wang, X.H.; Song, F.; Wang, X.L.; Zhu, L.L.; Wang, Y.Z. Biomimetic optical cellulose nanocrystal films with controllable iridescent color and environmental stimuli-responsive chromism. *Acs Appl. Mater. Interfaces* **2018**, *10*, 5805–5811. [CrossRef]
16. Xu, M.; Li, W.; Ma, C.; Yu, H.; Wu, Y.; Wang, Y.; Chen, Z.; Li, J.; Liu, S. Multifunctional chiral nematic cellulose nanocrystals/glycerol structural colored nanocomposites for intelligent responsive films, photonic inks and iridescent coatings. *J. Mater. Chem. C* **2018**, *6*, 5391–5400. [CrossRef]
17. Wan, H.; Li, X.; Zhang, L.; Li, X.; Liu, P.; Jiang, Z.; Yu, Z.-Z. Rapidly responsive and flexible chiral nematic cellulose nanocrystal composites as multifunctional rewritable photonic papers with eco-friendly inks. *Acs Appl. Mater. Interfaces* **2018**, *10*, 5918–5925. [CrossRef]
18. Mu, X.; Gray, D.G. Formation of chiral nematic films from cellulose nanocrystal suspensions is a two-stage process. *Langmuir* **2014**, *30*, 9256–9260. [CrossRef]
19. Qu, D.; Zheng, H.; Jiang, H.; Xu, Y.; Tang, Z. Chiral photonic cellulose films enabling mechano/chemo responsive selective reflection of circularly polarized light. *Adv. Opt. Mater.* **2019**, *7*, 1801395. [CrossRef]
20. Natarajan, B.; Krishnamurthy, A.; Qin, X.; Emiroglu, C.D.; Forster, A.; Foster, E.J.; Weder, C.; Fox, D.M.; Keten, S.; Obrzut, J.; et al. Binary cellulose nanocrystal blends for bioinspired damage tolerant photonic films. *Adv. Funct. Mater.* **2018**, *28*, 1800032. [CrossRef]
21. Godinho, M.H.; Gray, D.G.; Pieranski, P. Revisiting (hydroxypropyl) cellulose (hpc)/water liquid crystalline system. *Liq. Cryst.* **2017**, 1–13. [CrossRef]
22. Werbowyj, R.S.; Gray, D.G. Liquid crystalline structure in aqueous hydroxypropyl cellulose solutions. *Mol. Cryst. Liq. Cryst.* **1976**, *34*, 97–103. [CrossRef]

23. Charlet, G.; Gray, D.G. Solid cholesteric films cast from aqueous (hydroxypropyl)cellulose. *Macromolecules* **1987**, *20*, 33–38. [CrossRef]

24. Fernandes, S.N.; Geng, Y.; Vignolini, S.; Glover, B.J.; Trindade, A.C.; Canejo, J.P.; Almeida, P.L.; Brogueira, P.; Godinho, M.H. Structural color and iridescence in transparent sheared cellulosic films. *Macromol. Chem. Phys.* **2013**, *214*, 25–32. [CrossRef]

25. Fernandes, S.N.; Almeida, P.L.; Monge, N.; Aguirre, L.E.; Reis, D.; de Oliveira, C.L.P.; Neto, A.M.F.; Pieranski, P.; Godinho, M.H. Mind the microgap in iridescent cellulose nanocrystal films. *Adv. Mater.* **2017**, *29*, 1603560. [CrossRef]

26. Hamad, W.Y.; Hu, T.Q. Structure–process–yield interrelations in nanocrystalline cellulose extraction. *Can. J. Chem. Eng.* **2010**, *88*, 392–402. [CrossRef]

27. Nečas, D.; Klapetek, P. Gwyddion: An open-source software for spm data analysis. *Cent. Eur. J. Phys.* **2012**, *10*, 181–188. [CrossRef]

28. Schneider, C.A.; Rasband, W.S.; Eliceiri, K.W. Nih image to imagej: 25 years of image analysis. *Nat. Methods* **2012**, *9*, 671–675. [CrossRef]

29. ASTM. *Standard Test Method for Flexibility Determination of Hot-Melt Adhesives by Mandrel Bend Test Method*; American Society For Testing And Materials 100 Barr Harbor Dr.; American Society For Testing And Materials: West Conshohocken, PA, USA, 1999; Volume D 3111.

30. Hu, Z.; Cranston, E.D.; Ng, R.; Pelton, R. Tuning cellulose nanocrystal gelation with polysaccharides and surfactants. *Langmuir* **2014**, *30*, 2684–2692. [CrossRef]

31. Xu, F.; Yu, J.; Tesso, T.; Dowell, F.; Wang, D. Qualitative and quantitative analysis of lignocellulosic biomass using infrared techniques: A mini-review. *Appl. Energy* **2013**, *104*, 801–809. [CrossRef]

32. Gaspar, D.; Fernandes, S.N.; Oliveira, A.G.d.; Fernandes, J.G.; Grey, P.; Pontes, R.V.; Pereira, L.; Martins, R.; Godinho, M.H.; Fortunato, E. Nanocrystalline cellulose applied simultaneously as the gate dielectric and the substrate in flexible field effect transistors. *Nanotechnology* **2014**, *25*, 094008. [CrossRef] [PubMed]

33. Zhang, Y.; Luo, C.; Wang, H.; Han, L.; Wang, C.; Jie, X.; Chen, Y. Modified adsorbent hydroxypropyl cellulose xanthate for removal of Cu^{2+} and Ni^{2+} from aqueous solution. *Desalin. Water Treat.* **2016**, *57*, 27419–27431. [CrossRef]

34. Samuels, R.J. Solid-state characterization of the structure and deformation behavior of water-soluble hydroxypropylcellulose. *J. Polym. Sci. Part A-2* **1969**, *7*, 1197–1258. [CrossRef]

35. Chan, C.L.C.; Bay, M.M.; Jacucci, G.; Vadrucci, R.; Williams, C.A.; van de Kerkhof, G.T.; Parker, R.M.; Vynck, K.; Frka-Petesic, B.; Vignolini, S. Visual appearance of chiral nematic cellulose-based photonic films: Angular and polarization independent color response with a twist. *Adv. Mater.* **2019**, *31*, 1905151. [CrossRef]

36. Dumanli, A.G.; van der Kooij, H.M.; Kamita, G.; Reisner, E.; Baumberg, J.J.; Steiner, U.; Vignolini, S. Digital color in cellulose nanocrystal films. *Acs Appl. Mater. Interfaces* **2014**, *6*, 12302–12306. [CrossRef]

37. Honorato-Rios, C.; Kuhnhold, A.; Bruckner, J.R.; Dannert, R.; Schilling, T.; Lagerwall, J.P.F. Equilibrium liquid crystal phase diagrams and detection of kinetic arrest in cellulose nanocrystal suspensions. *Front. Mater.* **2016**, *3*, art. 21. [CrossRef]

38. Su, F.; Liu, D.; Li, M.; Li, Q.; Liu, C.; Liu, L.; He, J.; Qiao, H. Mesophase transition of cellulose nanocrystals aroused by the incorporation of two cellulose derivatives. *Carbohydr. Polym.* **2020**, *233*, 115843. [CrossRef]

39. Guo, A.; Yi, G.T.; Ashmead, C.C.; Mitchell, G.G.; de Groh, K.K.; Banks, B.A. *Embrittlement of Misse 5 Polymers after 13 Months of Space Exposure*; Springer Berlin Heidelberg: Berlin/Heidelberg, Gemany, 2013; pp. 389–398.

Article

Effect of Crowding Agent Polyethylene Glycol on Lyotropic Chromonic Liquid Crystal Phases of Disodium Cromoglycate

Runa Koizumi [1], Bing-Xiang Li [1] and Oleg D. Lavrentovich [1,2,*]

[1] Advanced Materials and Liquid Crystal Institute, Chemical Physics Interdisciplinary Program, Kent State University, Kent, OH 44242, USA; rkoizumi@kent.edu (R.K.); bli15@kent.edu (B.-X.L.)

[2] Department of Physics, Kent State University, Kent, OH 44242, USA

* Correspondence: olavrent@kent.edu

Received: 20 February 2019; Accepted: 15 March 2019; Published: 19 March 2019

Abstract: Adding crowding agents such as polyethylene glycol (PEG) to lyotropic chromonic liquid crystals (LCLCs) formed by water dispersions of materials such as disodium cromoglicate (DSCG) leads to a phase separation of the isotropic phase and the ordered phase. This behavior resembles nanoscale condensation of DNAs but occurs at the microscale. The structure of condensed chromonic regions in crowded dispersions is not yet fully understood, in particular, it is not clear whether the condensed domains are in the nematic (N) or the columnar (C) state. In this study, we report on small angle X-ray scattering (SAXS) and wide-angle X-ray scattering (WAXS) measurements of mixtures of aqueous solutions of DSCG with PEG and compare results to measurements of aqueous solutions of pure DSCG. X-ray measurements demonstrate that addition of PEG to DSCG in the N phase triggers appearance of the C phase that coexists with the isotropic (I) phase. Within the coexisting region, the lateral distance between the columns of the chromonic aggregates decreases as the temperature is increased.

Keywords: chromonics; structure; physico-chemical properties

---

1. Introduction

Lyotropic Chromonic Liquid Crystals (LCLCs) are a class of liquid crystals in which the phase transitions are caused by both temperature changes as well as changes in concentration of a solute in a solvent [1–4]. They are formed by water-soluble rigid plank-like molecules with polyaromatic cores and polar peripheries [1,5]. Due to the hydrophilic nature of the peripheries and the hydrophobic nature of the core, LCLC molecules dispersed in water form elongated aggregates by stacking on top of each other, to avoid contact of water with the core. As the concentration of chromonic molecules is raised, the number and length of aggregates increase and one observes a transition from an isotropic dispersion to a uniaxial nematic phase with parallel alignment of the aggregates, followed by a formation of the columnar phase, in which the parallel aggregates pack into a two-dimensional hexagonal lattice [1–5]. Molecules capable of forming LCLCs often show interesting functionalities, such as pharmaceutical activity or light absorption in a certain spectral range [1–3,6]; chromonic type of aggregation is also met in aqueous dispersions of nucleic acids [7,8]. As compared to surfactant-based lyotropic liquid crystals, LCLCs are not toxic and can be successfully interfaced with biological cells [9,10].

LCLC aggregates show a striking similarity to the double-strand B-DNA molecules. For example, the chromonic molecules of disodium cromoglycate (DSCG) stack on top of each other with a typical separation of about 0.33–34 nm. When the polar groups of DSCG molecules are fully ionized, the line density of electric charge along the aggregate is $6e/nm$, where e is the electron's charge. Although the LCLC aggregates are not stabilized by any covalent bonds, these molecular scale similarities extend also to macroscopic behavior. It is known that neutral additives such as poly-ethylene glycol (PEG) can, in the presence of salt, condense and align macromolecular B-DNAs through the excluded volume effect [11–14]. Similar effects have been demonstrated for noncovalent assembly in chromonics, which can be controlled by adding neutral and weakly charged additives that crowd the solution [15–17]. In particular, crowding agents such as PEG were reported [15] to cause condensation of DSCG aggregates and phase separation into isotropic and ordered phases with nontrivial geometries of coexistence, such as toroids. Qualitatively, these results correlate with the theoretical models by Madden and Herzfeld [18,19] who predicted that the mixing of an aggregating chromonic with neutral nonaggregating spheres will result in the phase separation to a liquid crystalline state with a high concentration of chromonics and an isotropic (I) phase with a low concentration of chromonics and a high concentration of neutral-aggregating spheres. However, the experimental data on the structure of condensed LCLC regions in crowded dispersions remain scarce. In particular, it is not clear whether the crowding-induced phase separation produces a coexistence of the isotropic phase with a nematic (N), columnar hexagonal (C) phase, or both these phases; the result depends strongly on the concentration of ingredients and the temperature [15].

The concentration of DSCG, PEG and temperature define the shape of the phase-separated mesomorphic regions coexisting with the isotropic phase [15]. When a low (≤ 0.005 mol/kg) concentration of PEG is added to DSCG, the condensed regions appear in the form of tactoids, the interior of which is a nematic [20]. The interior order is of prime importance in defining the shape of phase-separated regions. In a nematic, the director $\hat{n}$ specifying the local orientation of chromonic aggregates, can experience splay, twist, bend, and saddle-splay. As a result, the N tactoids show a spindle-like shape that results from a balance of anisotropic surface tension with all modes of the director distortions [20]. However, if the interior is a C phase with two-dimensional positional order, only the bend of $\hat{n}$ is allowed by the requirement of columns' equidistance, and the phase-separated inclusions adopt a very different toroidal shape [15]. The size of LCLC toroids is 10 μm, nearly three orders of magnitude larger than the length scale of toroids found in DNA condensates. Because of their large size, C toroids can be studied in detail by optical microscopy [15,20], and are thus ideally suited to explore the equilibrium shape of the columnar nuclei from an isotropic environment, which is our long-term goal. However, optical studies cannot provide a direct insight into the interior ordering of the phase-separated regions, whether it is indeed a C or an N phase, perhaps close to the N-to-C phase transition. X-ray diffraction provides an excellent insight of this type for LCLCs [15–17,21,22]. The pioneering X-ray studies [15] reported a clear columnar hexagonal ordering when the concentrations of PEG added to DSCG was in the range $c_{PEG} = (0.03 - 0.07)$ mol/kg. At smaller c_{PEG}, the optical and X-ray studies could not distinguish well between the N + I coexistence and C + N + I coexistence [15]. The goal of the present work is to explore by X-ray diffraction what kind of molecular ordering is triggered by weak concentrations of PEG, $c_{PEG} = 0.011$ mol/kg and $c_{PEG} = 0.022$ mol/kg, and how this ordering depends on temperature. Furthermore, we compare the behavior of PEG-condensed phases of DSCG with the highly condensed phases of an additive-free DSCG and present the structural characteristics such as inter-and intra-aggregate separations of molecules for concentrations of DSCG that exceeds the range previously studied by Agra-Koojiman et al. [21].

2. Materials and Methods

We performed the X-ray studies of mixtures of electrically neutral polymer PEG (purchased from Sigma Aldrich, St. Louis, MO, United States, molecular weight 3.35 kg/mol) and DSCG of purity 98% (purchased from Alfa Aesar, Tewksbury, MA, United States) for two different concentrations of PEG (0.011 and 0.022 mol/kg, or 3.5 and 7 wt%, respectively). PEG of a molecular weight of 3350 g/mol was chosen because it was previously reported to cause condensation of DSCG aggregates [15,20], whereas PEG of a lower (0.60–1.50 kg/mol) molecular weight is known to produce no condensing effects on LCLCs [23]. In the experiments with PEG, the concentration of DSCG in de-ionized water (with resistivity of $\geq$18.0 MΩ cm) was fixed at $c_{DSCG} = 0.34$ mol/kg or 15 wt %; at this concentration, DSCG forms a nematic phase. PEG is known to be a crowding agent that causes aqueous DSCG solutions to phase separate into an isotropic phase and an ordered phase, such as a nematic or columnar. Since PEG partitions into the isotropic regions, it causes condensation of the ordered regions, which contain higher concentrations of DSCG as compared to the original solution [15,18,19]. In order to perform a comparative analysis of the PEG-induced condensation and the effect of a simple increase of DSCG concentration, we also explored aqueous solutions of pure DSCG, at three different concentrations ($c_{DSCG} = 0.49$, 0.65, 0.80 mol/kg or 20, 25, 29 wt %, respectively). This range extends the prior studies of pure DSCG dispersions in water reported by Agra-Kooijman et al. [21] for $c_{DSCG} = 15$, 20, 25 wt%.

About thirty-six hours prior to the X-ray measurements, DSCG mixtures were filled in 2 mm inner-diameter quartz capillaries in the isotropic phase, sealed with epoxy, and mounted into a custom-built aluminum cassette which was fit into a hot stage (Instec model HCS402) with an accuracy better than 0.01 °C. Experiments were performed at Brookhaven National Laboratory (11-BM CMS). The beamline was configured for a collimated X-ray beam of size 0.2 mm by 0.2 mm and a divergence of 0.1 mrad by 0.1 mrad with energy of 17 keV. The background scattering was collected from an empty capillary and the sample-to-detector distance was calibrated using a silver behenate calibration standard.

The optical textures of the samples were obtained by sandwiching the material between two glass slides separated by 20 μm spacers, sealing all four sides of the cell using epoxy to prevent evaporation, placing them inside a hot stage (Linkam model PE94) and making observations under an optical microscope with two crossed linear polarizers, upon cooling the sample from the isotropic phase at 0.1 °C/min.

3. Results and Discussions

In order to perform preliminary characterization of the explored mixtures, we used polarizing microscopy. The phase diagrams of aqueous solutions of the DSCG (0.34 mol/kg) + PEG mixtures and pure DSCG mixtures established by polarizing microscopy upon cooling, are shown in Figure 1a,b, respectively. Figure 1c–h show the optical textures of all mixtures, taken at different temperatures. For 0.49 mol/kg, 0.65 mol/kg, and 0.80 mol/kg DSCG solutions, cooling from the I phase results first in the formation of the biphasic I + C region, see, for example, Figure 1i, $T = 46$ °C. Further cooling produces either an I + N coexistence, Figure 1j, 0.49 mol/kg, $T = 41$ °C, or C + N coexistence, Figure 1l, $T = 36$ °C.

The I phase appears under the crossed polarizers as a dark region, while the coexisting N and C phases show birefringent textures with different arrangements of the optic axis which is also the director. The C phase that coexists with the I phase forms elongated inclusions, with the optic axis along the direction of elongation, see Figure 1k, $T = 58$ °C and Figure 1m, $T = 63$ °C, as the interfacial tension is not strong enough to overcome bulk elasticity and to make the C inclusions more compact. When the volume fraction of the C phase is large, as in Figure 1l, $T = 36$ °C, one observes characteristic textures of the so-called developable domains, in which the director experiences bend deformations but no splay nor twist. The reason is in the two-dimensional periodic order of the columnar structure; the requirement of inter-columnar equidistance prohibits splay and twist but allows bend [24]. In 0.49 mol/kg solution, cooling from the I + C state results in a formation of the I + N coexistence, Figure 1j, $T = 41$ °C, with compact inclusions of the N phase, called tactoids. Tactoids exhibit characteristic cusps at the two poles [15,25,26]. As discussed by van der Schoot [27–29] and others [25,30,31], the shape of N tactoids is a result of fine balance of the bulk elasticity that permits splay, bend, and twist deformations of the director, and an anisotropic interfacial tension that tends to align the chromonic aggregates parallel to the I–N interface.

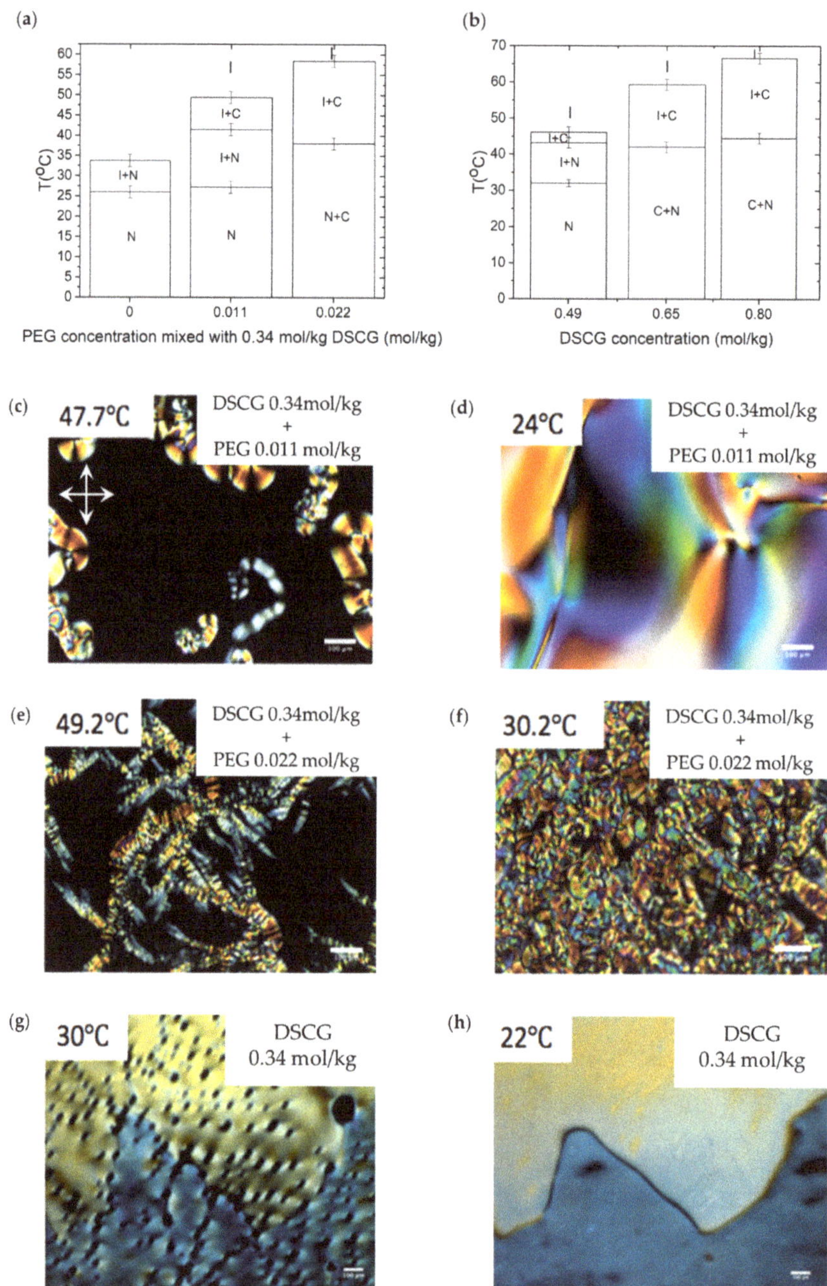

Figure 1. *Cont.*

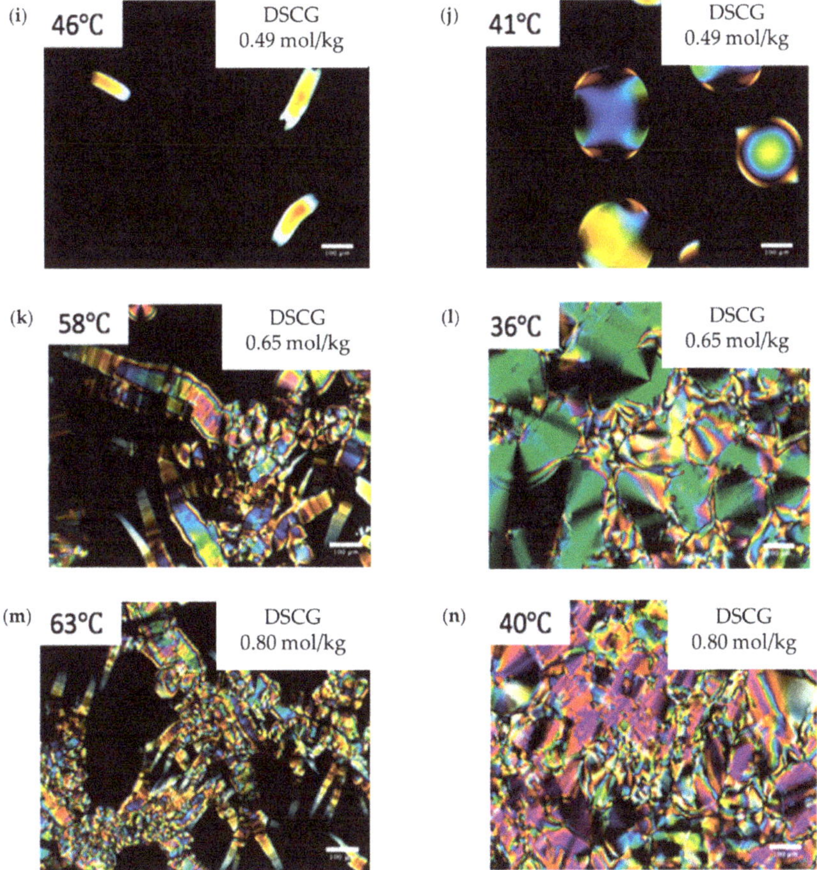

Figure 1. Phase diagram of (**a**) Mixture of 0.34 mol/kg disodium cromoglicate (DSCG) with polyethylene glycol (PEG); (**b**) Aqueous solution of pure DSCG; Optical textures of DSCG mixtures observed with crossed polarizers (**c–n**). (**c**) DSCG 0.34 mol/kg + PEG 0.011 mol/kg mixture at (**c**) $T = 47.7$ °C; (**d**) $T = 24$ °C; DSCG 0.34 mol/kg + PEG 0.022 mol/kg mixture at (**e**) $T = 49.2$ °C; (**f**) $T = 30.2$ °C; DSCG 0.34 mol/kg mixture at (**g**) $T = 30$ °C; (**h**) $T = 22$ °C; DSCG 0.49 mol/kg mixture at (**i**) $T = 46$ °C; (**j**) $T = 41$ °C; DSCG 0.65 mol/kg mixture at (**k**) $T = 58$ °C; (**l**) $T = 36$ °C; DSCG 0.80 mol/kg mixture at (**m**) $T = 63$ °C; (**n**) $T = 40$ °C. Scale bar 100 μm.

The I + C coexistence region is narrow, less than 5 °C, for 0.49 mol/kg DSCG solution, but expands as the concentration of DSCG increases, Figure 1b. In 0.34 mol/kg DSCG without any PEG, the C phase does not form; as the solution is cooled down, we observe nucleation of the N phase from the I phase. The coexistence I + N region transforms into the homogeneous N phase around room temperature. Polarizing microscopy textures in Figure 1c suggest that addition of PEG to 0.34 mol/kg DSCG causes the appearance of the C phase in coexistence with the I phase. To prove the appearance of the C phase and to characterize its structure in terms of symmetry and distance between the chromonic aggregates, we turn to small-angle-X-ray scattering (SAXS, using SAXS detector Pilatus 2M from Dectris at Brookhaven National Laboratory, Upton, NY, United States) and wide-angle-X-ray scattering (WAXS, using WAXS detector ImageStar 135 mm CCD from Photonic Science at Brookhaven National Laboratory, Upton, NY, United States) studies.

The structure of the phase can be determined by the position of the peaks obtained from SAXS and WAXS. For a columnar phase with hexagonal packing, the position of the peaks should obey the ratio $1 : \sqrt{3} : 2 : \sqrt{7}$ [32,33]. The lateral distance d, the distance between the centers of neighboring columnar stacks s, and the stacking distance w are defined as shown in Figure 2. Parameter w can be obtained from WAXS, and parameters d and s can be obtained from the position q of the first peak from SAXS:

$$d = \frac{2\pi}{q} \tag{1}$$

$$s = \frac{4\pi}{q\sqrt{3}} \tag{2}$$

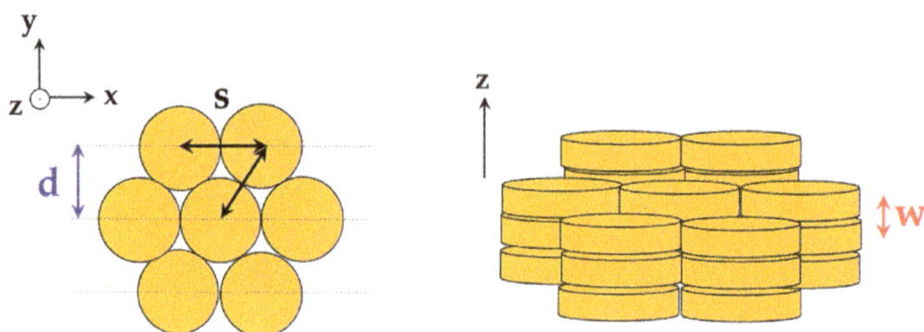

Figure 2. Definitions of the parameters d (lateral distance), s (neighboring columnar stacks), and w (stacking distance).

X-ray diffraction patterns for DSCG (0.34 mol/kg) + PEG (0.011 mol/kg) mixture confirm the phase diagram assignments shown in Figure 1a. In particular, at high temperature, $T = 44.5\ °C$, the material exhibits three strong reflexes, located at $P_1 \sim 0.192\ A^{-1}$ and $P_2 \sim 0.329\ A^{-1}$, and $P_3 \sim 0.385\ A^{-1}$. Peaks P_1 and P_2 correspond to the lateral spacings $d_1 = 2\pi/P_1 \simeq 3.27$ nm and $d_2 = 2\pi/P_2 \simeq 1.91$ nm respectively. The peak P_3 corresponds to Kapton reflexes and does not characterize the DSCG structure. The ratio of the position of peaks P_1 and P_2 is $P_2/P_1 = 1.71 \approx \sqrt{3}$, thereby indicating the existence of the C phase. There is a broad peak to the left shoulder of peak P_1, located around $0.155\ A^{-1}$. This indicates the coexistence of the I phase with the C phase, confirming phase separation that is caused as a result of the addition of PEG to DSCG.

In WAXS diffraction patterns at $T = 44.5\ °C$, Figure 3c,d, there is a peak Q_1 positioned at $Q_1 \sim 1.854\ A^{-1}$, from which we can get the stacking distance $w = 2\pi/Q_1 \simeq 0.34$ nm.

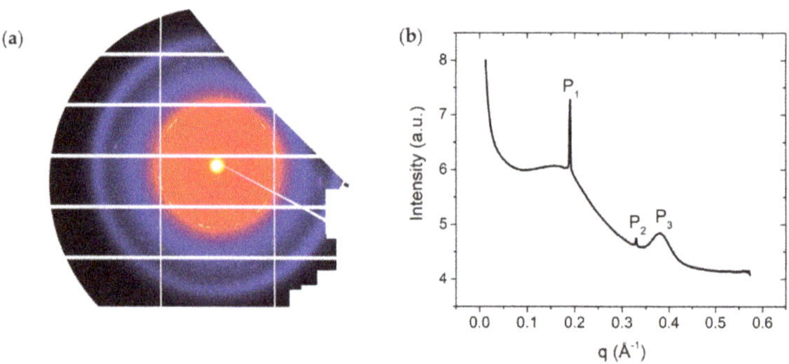

Figure 3. *Cont.*

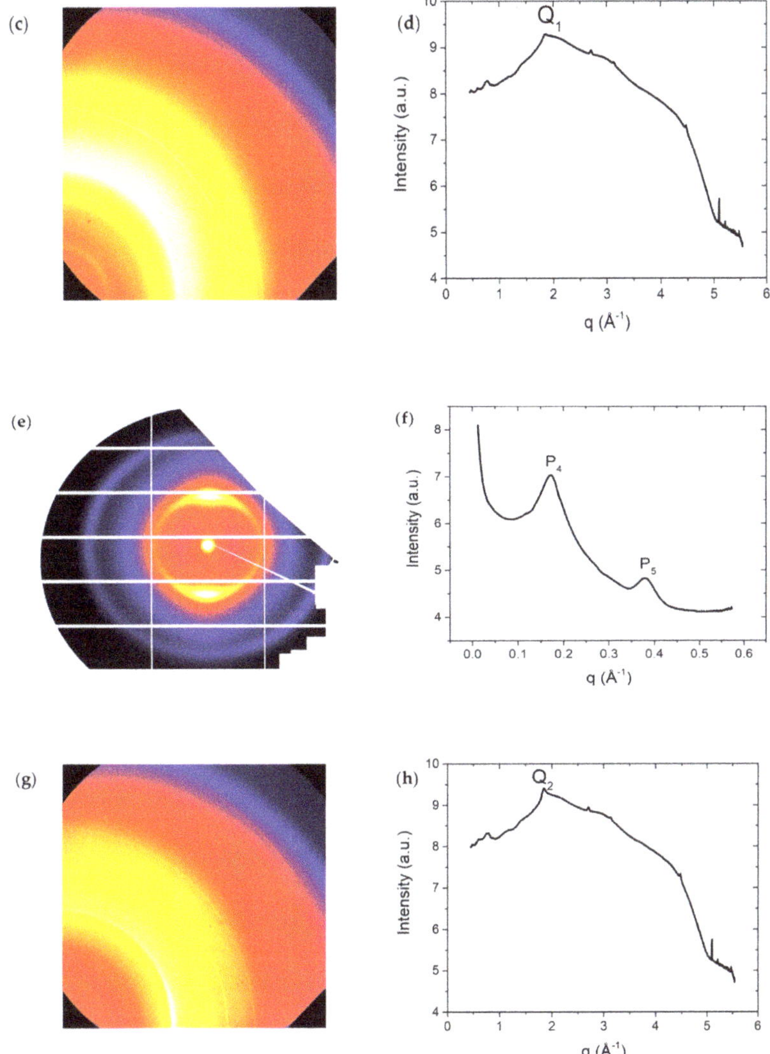

Figure 3. X-ray data for DSCG 0.34 mol/kg + PEG 0.011 mol/kg at $T = 44.5\,^{\circ}$C. (**a**) Diffraction pattern from small angle X-ray scattering (SAXS); (**b**) Intensity profile from SAXS; (**c**) Diffraction pattern from wide-angle X-ray scattering (WAXS); (**d**) Intensity profile from WAXS; (**e**) Diffraction pattern from SAXS at $T = 40.7\,^{\circ}$C; (**f**) Intensity profile from SAXS at $T = 40.7\,^{\circ}$C; (**g**) Diffraction pattern from WAXS at $T = 40.7\,^{\circ}$C; (**h**) Intensity profile from WAXS at $T = 40.7\,^{\circ}$C. Peak P_3 in (**b**), P_5 in (**f**), and the multiple peaks seen below $q = 1\,A^{-1}$ in (**d**,**h**) correspond to background scattering from Kapton windows and do not characterize the DSCG structure.

Upon further cooling down the mixture to $T = 40.7\,^{\circ}$C, the diffraction pattern changes, Figure 5e,f, and there are two strong reflexes, located at $P_4 \sim 0.173\,A^{-1}$ and $P_5 \sim 0.380\,A^{-1}$ (due to Kapton). The peak P_4 corresponds to the lateral spacing $d = 2\pi/P_4 \simeq 3.63$ nm. Peak P_4 is broad compared to peak P_1 and indicates the existence of the N phase. The change in temperature does not affect the stacking distance w. Based on WAXS diffraction patterns at $T = 40.7\,°C$, Figure 3g,h, there is a peak located at $Q_2 \sim 1.850\,A^{-1}$, corresponding to the stacking distance $w = 2\pi/Q_2 \simeq 0.34$ nm.

Based on SAXS measurements, we obtain the temperature dependences of the lateral distance d (nm, closed symbols) as well as the local concentration c (mol/kg, open symbols) for different mixtures, as shown in Figure 4. c was estimated from the following relation [34,35]

$$c = \frac{D\pi^2\rho_{DSCG}}{W_{DSCG}\left[100\left\{2\sqrt{3}s^2\rho_w + D\pi^2\rho_{DSCG}(\rho_{DSCG} - \rho_w)\right\} - D\pi^2\rho_{DSCG}\right]} \tag{3}$$

where $D \approx 1.6$ nm is the diameter of the columns [35], $W_{DSCG} = 0.512$ kg/mol is the molecular weight of DSCG, $\rho_{DSCG} = 1.55 \times 10^3$ kg/m^3 is the density of pure DSCG [35], and $\rho_w = 9.98 \times 10^2$ kg/m^3 is the density of water at room temperature.

Adding PEG into 15 wt% DSCG results in the appearance of the C phase, Figure 4a,b, which would not occur without the presence of PEG. This agrees with the phase diagram obtained from polarizing microscopy, shown in Figure 1a. Furthermore, there is a sudden drop in the lateral distance d when the mixtures transition from the I phase to the C phase, because the columns pack closely together via hexagonal packing in the C phase. Note that d grows as the temperature is decreased. This counterintuitive behavior might be related to restructuring of the molecular stacking in the aggregates, or counterionic clouds surrounding them, or both. Although the mechanism is not clear, we note that Agra-Kooijman et al. found a similar trend for DSCG 15 wt%, 20 wt%, and 25 wt% dispersions [21].

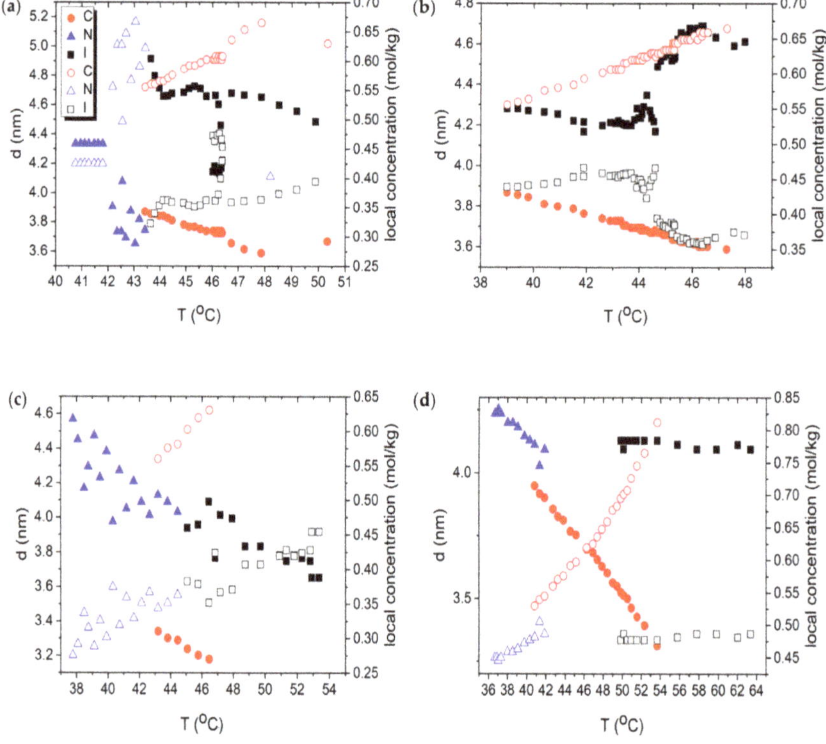

Figure 4. *Cont.*

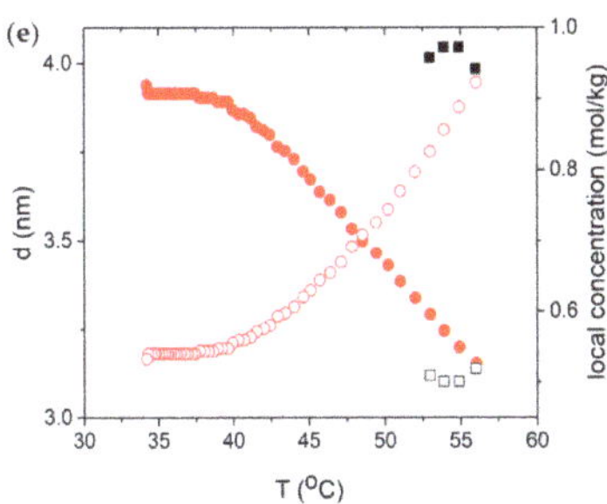

Figure 4. Temperature dependence of lateral distance d (left axis, closed symbol) and local concentration c (right axis, open symbol). Red circles correspond to C phase, blue triangles to N phase, and black squares to I phase. (**a**) DSCG 0.34 mol/kg + PEG 0.011 mol/kg; (**b**) DSCG 0.34 mol/kg + PEG 0.022 mol/kg; (**c**) DSCG 0.49 mol/kg; (**d**) DSCG 0.65 mol/kg; (**e**) DSCG 0.80 mol/kg.

The difference between the calculated local concentration c mol/kg and the actual concentration of DSCG comes from the inhomogeneity of the liquid crystalline materials in the aqueous solution of DSCG. Adding PEG to 0.34 mol/kg DSCG also raises c from 0.34 mol/kg, the initial concentration of DSCG contained in the mixture, to values in the range of 0.55 mol/kg to 0.76 mol/kg, Figure 4a,b.

The comparison of the stacking distance w and lateral distance d in the C phase at $T = 46.0\,^{\circ}\mathrm{C}$ for all mixtures is shown in Table 1. The increase in the concentration of PEG from 0.011 mol/kg to 0.022 mol/kg results in closer packing of columns, which is accompanied by a slight decrease in d. w, on the other hand, remains more or less constant for all mixtures. The values of w and d for 0.34 mol/kg DSCG and 0.80 mol/kg DSCG have been reported previously [22], and the values we found show close resemblance to previously reported values.

Table 1. Comparison of stacking distance w nm, lateral distance d nm, and distance between the centers of neighboring columnar stacks s nm for each mixture in the C phase at $T = 46.0\,^{\circ}\mathrm{C}$.

Parameters	DSCG 0.34 mol/kg, PEG 0.011 mol/kg	DSCG 0.34 mol/kg, PEG 0.022 mol/kg	DSCG 0.49 mol/kg	DSCG 0.65 mol/kg	DSCG 0.80 mol/kg
w (nm)	0.34	0.34	0.34	0.34	0.32
d (nm)	3.74	3.62	3.20	3.70	3.63
s (nm)	4.32	4.18	3.70	4.27	4.19

Small angle diffraction, or SAXS, is used to measure the average distance between the columnar aggregates. Hence, the full width at half maximum (FWHM) of the SAXS peak is used to determine the correlation length associated with interaggregate distances, $\xi_{\perp}$. The FWHM were determined by fitting the corresponding diffraction peak to a sum of Lorentzian peaks, as shown in Figure 5 for the DSCG 0.34 mol/kg + PEG 0.011 mol/kg mixture at $T = 44.5\,^{\circ}\mathrm{C}$. The entire peak can be fit well by the linear sum of two Lorentzian peaks corresponding to the I phase and C phase, thereby confirming the coexistence of these two phases at this temperature.

Figure 6 shows the temperature dependences of $\xi_{\perp}$ for all mixtures. The value of $\xi_{\perp}$ in the C phase is significantly higher than that in the N phase, by more than an order of magnitude, Figure 6, because of positional order in hexagonal packing of the columns. For all mixtures, $\xi_{\perp}$ increases as the temperature is decreased, which might be related with the decreased distance between the DSCG aggregates, and with an increased stiffness of the aggregates at lower temperatures. For DSCG 0.34 mol/kg + PEG 0.011 mol/kg and DSCG 0.34 mol/kg + PEG 0.022 mol/kg mixtures, $\xi_{\perp}$ increases rather quickly soon after the appearance of the C + I condensed phase until it saturates and becomes more or less temperature independent. The saturated value $\xi_{\perp S}$ of $\xi_{\perp}$ in the condensed I + C regions is higher for the mixture containing a higher concentration of PEG; for DSCG 0.34 mol/kg + PEG

0.022 mol/kg mixture, $\xi_{\perp S} \approx 330$ nm, Figure 6b, whereas for DSCG 0.34 mol/kg + PEG 0.011 mol/kg mixture, $\xi_{\perp S} \approx 270$ nm, Figure 6a.

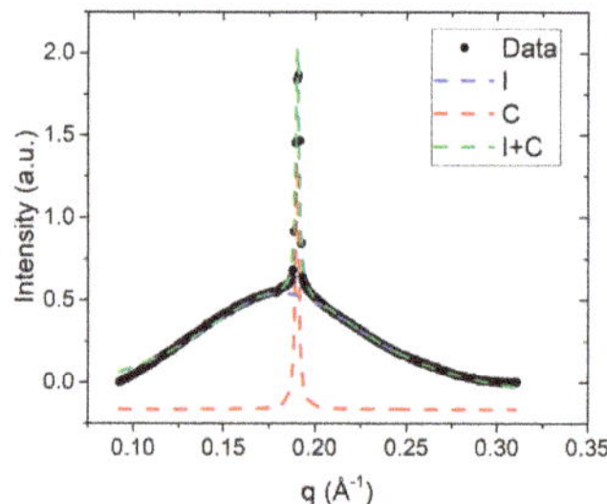

Figure 5. SAXS peak for DSCG 0.34 mol/kg + PEG 0.011 mol/kg mixture, fit to a sum of two Lorentzian peaks; I peak (blue dotted line), C peak (red dotted line), and I + C peak (green dotted line).

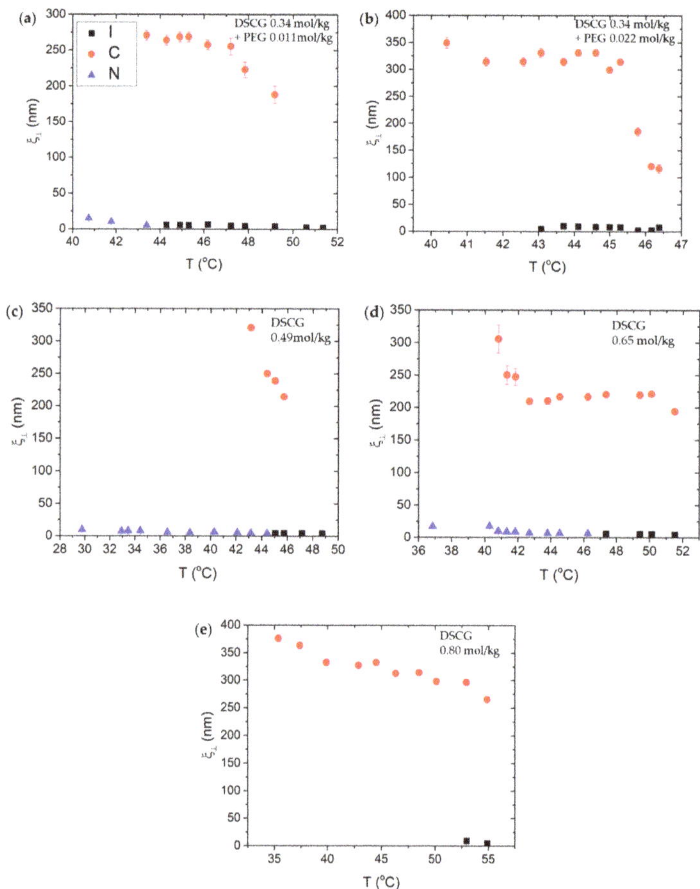

Figure 6. Temperature dependence of the positional correlation length $\xi_{\perp}$ for (**a**) DSCG 0.34 mol/kg + PEG 0.011 mol/kg; (**b**) DSCG 0.34 mol/kg + PEG 0.022 mol/kg; (**c**) DSCG 0.49 mol/kg; (**d**) DSCG 0.65 mol/kg; (**e**) DSCG 0.80 mol/kg.

Note that although $\xi_\perp$ in the N phase is more than an order of magnitude smaller than in the C phase, it still takes nonzero values that are slightly greater than in the I phase. For example, for DSCG 0.34 mol/kg + PEG 0.011 mol/kg mixture, $\xi_\perp \approx 15$ nm, and for DSCG 0.49 mol/kg, $\xi_\perp \approx 8$ nm. These values might be indicative of pre-transitional fluctuations.

Similarly, the FWHM of large angle diffraction, or WAXS, is used to determine the correlation length $\xi_\parallel$ of stacking along the aggregate axis. Figure 7 shows the temperature dependences of $\xi_\parallel$ for all mixtures.

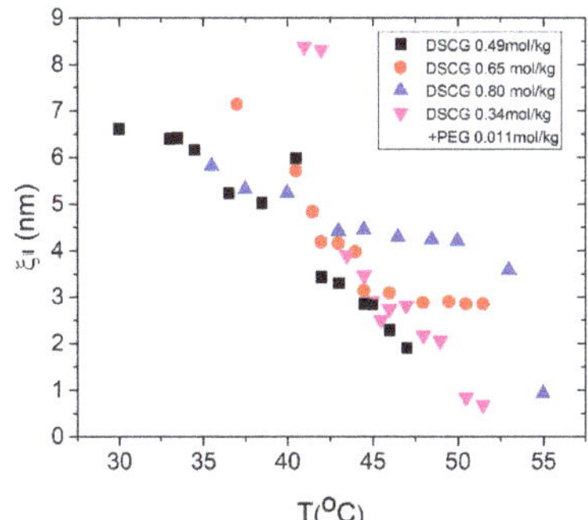

Figure 7. Temperature dependence of the positional correlation length $\xi_\parallel$ for DSCG 0.34 mol/kg + PEG 0.011 mol/kg (magenta triangle base-up), DSCG 0.49 mol/kg (black square), DSCG 0.65 mol/kg (red circle), and DSCG 0.80 mol/kg (blue triangle base-down).

Here, $\xi_\parallel$ is associated with the short scale correlated stacking of molecules along the aggregates, and does not give the actual contour length L of the entire aggregate [17]. This implies that $\xi_\parallel$ obtained from X-ray measurements might be smaller than the actual length of stacking, due to stacking defects such as molecular shift junctions, as suggested by Park et al. [17]. Indeed, if we assume that $\xi_\parallel \approx L$ and take the diameter of stacks as $D \approx 1.6$ nm to calculate the ratio of the length and width L/D for, say, 0.49 mol/kg DSCG mixture at $T = 34.5\,°\text{C}$, we find $L/D \approx 6.16/1.6 = 3.85$. If we multiply this value of L/D by the volume fraction ϕ where [17]:

$$\phi = \frac{c_{DSCG} W_{DSCG} \rho_{water}}{\rho_{DSCG} + (c_{DSCG} W_{DSCG} \rho_{water})} \tag{4}$$

We obtain $\phi L/D \approx 0.71$. Here, c_{DSCG} mol/kg is the concentration of DSCG solution in molality units and $W_{DSCG} = 0.512$ kg/mol is the molecular weight of DSCG. This value of $\phi L/D \approx 0.71$ is much smaller than the critical value $\phi L/D \approx 4$ estimated by the Onsager theory [36] for a monodisperse system of rigid rods to obtain orientational order and form the N phase. Thus, at this value of ϕ, the system should not have any orientational order unless L is of a much greater value. However, our experimental data clearly show the existence of the N phase at this temperature for this concentration, Figure 1b,c. This discrepancy confirms that $\xi_\parallel$ is smaller than L.

The most plausible reason for $\xi_\parallel$ to be smaller than the actual aggregate length L is the existence of structural defects—stacking faults that interrupt positional correlations along the aggregates' long axes. These stacking faults were proposed by Park et al. [17], to explain a discrepancy between the length of chromonic aggregates expected within the framework of the Onsager theory and the length inferred from X-ray diffraction. The defects can be in the shape of shift junctions (in a shape of a letter "C") and 3-fold Y junctions, and have been shown to contribute to the dynamic light scattering at DSCG-based LCLCs by Zhou et al. [37]. They are also evident in numerical simulations by Sidky and Whitmer [38]. Finally, NMR data and numerical simulations by Xiao et al. [39] demonstrated that similar shift junctions can be observed in another LCLC Sunset Yellow; the authors concluded that the shift junctions lead to the underestimated length of the chromonic aggregates by X-ray scattering; Y-junctions were found to be of a transient nature, devolving in pairs of disconnected stacks.

In the I + C coexistence region, $\xi_\parallel$ increases as the concentration of pure DSCG is increased, Figure 7. Furthermore, in this coexistence region, DSCG 0.34 mol/kg + PEG 0.011 mol/kg shows $\xi_\parallel$ that is higher than that in DSCG 0.49 mol/kg, but similar to $\xi_\parallel$ in DSCG 0.65 mol/kg. In other words, addition of PEG mimics the effect of increasing the concentration of DSCG.

Understanding the structure of condensed chromonic regions in crowded dispersions is crucial for the understanding of nanoscale condensation of DNAs. Previous studies provided indirect insight into the phase state and structure of the condensed domains by comparing the birefringence of the condensed and isotropic regions [15]. Here, we provide a direct characterization of the nature of the phase separated regions through the X-ray measurements.

The X-ray measurements show that the addition of PEG mimics the effect of an increased concentration of DSCG. This is illustrated by an increase in interaggregate correlation length $\xi_\perp$ and short-range stacking correlation length $\xi_\parallel$ with the addition of PEG. The phase separation and co-existence of the C and I phases in mixtures with PEG is confirmed by X-ray diffraction patterns showing hexagonal ordering and a dramatic increase of $\xi_\perp$ in the C phase, by more than an order of magnitude as compared to the I phase and the N phase. Both $\xi_\perp$ and $\xi_\parallel$ increase as the temperature is decreased for all mixtures, indicating a better ordering at lower temperatures. Furthermore, $\xi_\parallel$ also increases with the concentration of DSCG. The values of $\xi_\parallel$ in DSCG 0.34 mol/kg + PEG 0.011 mol/kg and DSCG 0.65 mol/kg are close, implying that the addition of PEG helps to enhance the short-range molecular stacking of aggregates. However, the experimental values of $\xi_\parallel$ are much smaller than what is expected by the Onsager theory [36] for the actual contour length L of the aggregates that is necessary to form an ordered phase. Although the Onsager theory, developed for monodisperse systems, cannot be used directly for the description of the polydisperse chromonic solutions, the discrepancy between $\xi_\parallel$ and the expected L is too big (by a factor of 5–6) to be explained solely by polydispersity. As discussed above, defects in molecular packing within the chromonic aggregates can contribute significantly to the inequality $\xi_\parallel < L$.

4. Summary

To summarize, in this work, we present the analysis of phase formations of mixtures of DSCG and crowding agent PEG using polarizing microscopy and X-ray diffraction measurements. Adding PEG to 0.34 mol/kg DSCG results in phase separation of the isotropic phase and the liquid crystalline phase; polarizing microscopy suggest that this liquid crystalline phase is columnar, which is confirmed by detailed X-ray measurements. The effect of adding the crowding agent PEG mimics an increase of the concentration of DSCG. We also find that in all chromonic samples explored, with and without PEG, the interaggregate separation $s = 2d/\sqrt{3}$ in the nematic and columnar hexagonal phases decreases as the temperature is increased. Increasing the content of PEG leads to a smaller $s = 2d/\sqrt{3}$ in the columnar phase, i.e., PEG-generated osmotic pressure condenses the chromonic aggregates closer to each other. The next task is to explore the shape of the columnar nuclei embedded into the isotropic environment; the shape is expected to be controlled by the compressional and bending stiffness of the material and by the anisotropic surface tension at the columnar–isotropic interface.

Author Contributions: R.K. and B.-X.L. contributed equally to this work. R.K. and B.-X.L. did the experiments. R.K., B.-X.L. and O.D.L. wrote the paper. O.D.L. conceived and directed the project.

Funding: This research was funded by NSF grant DMREF DMS-1729509 and supported by the Brookhaven National Laboratory, NSLS II Project ID 302111.

Acknowledgments: We are thankful to Masafumi Fukuto, Brookhaven National Laboratory, for the help with the X-ray experiments, and to D. Golovaty, M. C. Calderer and N. J. Walkington for discussions that stimulated this work. This research used the 11-BM CMS beamline of National Synchrotron Light Source-II (NSLS-II), Brookhaven National Laboratory (BNL), a U.S. Department of Energy User Facility operated for the Office of Science by BNL under Contract DE-SC0012704.

Conflicts of Interest: The authors declare no conflict of interest.

References

1. Lydon, J.E. Chromonic liquid crystal phases. *Curr. Opin. Colloid Interface* **1998**, *3*, 458–466. [CrossRef]
2. Tam-Chang, S.W.; Huang, L. Chromonic liquid crystals: Properties and applications as functional materials. *Chem. Commun.* **2008**, 1957–1967. [CrossRef] [PubMed]
3. Dickinson, A.J.; LaRacuente, N.D.; McKitterick, C.B.; Collings, P.J. Aggregate structure and free energy changes in chromonic liquid crystals. *Mol. Cryst. Liq. Cryst.* **2009**, *509*, 751–762. [CrossRef]
4. Park, H.S.; Lavrentovich, O.D. Lyotropic Liquid Crystals: Emerging Applications. In *Liquid Crystals beyond Displays: Chemistry, Physics, and Applications*; Li, Q., Ed.; John Wiley & Sons: Hoboken, NJ, USA, 2012.
5. Lydon, J.E. Chromonics. In *Handbook of Liquid Crystals*; Demus, D., Goodby, J.W., Gray, G.W., Speiss, H.-W., Vill, V., Eds.; Wiley-VCH: New York, NY, USA, 1998; pp. 981–1007.
6. Edwards, D.J.; Jones, J.W.; Lozman, O.; Ormerod, A.P.; Sintyureva, M.; Tiddy, G.J.T. Chromonic liquid crystal formation by Edicol Sunset Yellow. *J. Phys. Chem. B* **2008**, *112*, 14628–14636. [CrossRef] [PubMed]
7. Nakata, M.; Zanchetta, G.; Chapman, B.D.; Jones, C.D.; Cross, J.O.; Pindak, R.; Bellini, T.; Clark, N.A. End-to-end stacking and liquid crystal condensation of 6 to 20 base pair DNA duplexes. *Science* **2007**, *318*, 1276–1279. [CrossRef] [PubMed]
8. Mariani, P.; Saturni, L. Measurement of intercolumnar forces between parallel guanosine four-stranded helices. *Biophys. J.* **1996**, *70*, 2867–2874. [CrossRef]
9. Woolverton, C.J.; Gustely, E.; Li, L.; Lavrentovich, O.D. Liquid crystal effects on bacterial viability. *Liq. Cryst.* **2005**, *32*, 417–423. [CrossRef]
10. Zhou, S.; Sokolov, A.; Lavrentovich, O.D.; Aranson, I.S. Living liquid crystals. *Proc. Natl. Acad. Sci. USA* **2014**, *111*, 1265–1270. [CrossRef] [PubMed]
11. Raspaud, E.; De La Cruz, M.O.; Sikorav, J.L.; Livolant, F. Precipitation of DNA by polyamines: A polyelectrolyte behavior. *Biophys. J.* **1998**, *74*, 381–393. [CrossRef]
12. Hud, N.V.; Vilfan, I.D. Toroidal DNA Condensates: Unraveling the Fine Structure and the Role of Nucleation in Determining Size. *Annu. Rev. Biophys. Biomol. Struct.* **2005**, *34*, 295–318. [CrossRef]
13. Zhou, H.-X.; Rivas, G.; Minton, A.P. Macromolecular crowding and confinement: Biochemical, biophysical, and potential physiological consequences. *Annu. Rev. Biophys.* **2008**, *37*, 375–397. [CrossRef] [PubMed]
14. Zanchetta, G.; Nakata, M.; Buscaglia, M.; Bellini, T.; Clark, N.A. Phase separation and liquid crystallization of complementary sequences in mixtures of nanoDNA oligomers. *Proc. Natl. Acad. Sci. USA* **2008**, *105*, 1111–1117. [CrossRef] [PubMed]
15. Tortora, L.; Park, H.-S.; Kang, S.-W.; Savaryn, V.; Hong, S.-H.; Kaznatcheev, K.; Finotello, D.; Sprunt, S.; Kumar, S.; Lavrentovich, O.D. Self-assembly, condensation, and order in aqueous lyotropic chromonic liquid crystals crowded with additives. *Soft Matter* **2010**, *6*, 4157–4167. [CrossRef]
16. Park, H.S.; Kang, S.W.; Tortora, L.; Kumar, S.; Lavrentovich, O.D. Condensation of self-assembled lyotropic chromonic liquid crystal sunset yellow in aqueous solutions crowded with polyethylene glycol and doped with salt. *Langmuir* **2011**, *27*, 4164–4175. [CrossRef] [PubMed]
17. Park, H.S.; Kang, S.W.; Tortora, L.; Nastishin, Y.; Finotello, D.; Kumar, S.; Lavrentovich, O.D. Self-assembly of lyotropic chromonic liquid crystal Sunset Yellow and effects of ionic additives. *J. Phys. Chem. B* **2008**, *112*, 16307–16319. [CrossRef] [PubMed]
18. Madden, T.L.; Herzfeld, J. Exclusion of Spherical Particles from the Nematic Phase of Reversibly Assembled Rod-Like Particles. *MRS Proc.* **1991**, *248*. [CrossRef]
19. Madden, T.L.; Herzfeld, J. Liquid Crystal Phases of Self-Assembled Amphiphilic Aggregates. *Philos. Trans. Phys. Sci. Eng.* **1993**, *344*, 357–375.
20. Tortora, L.; Lavrentovich, O.D. Chiral symmetry breaking by spatial confinement in tactoidal droplets of lyotropic chromonic liquid crystal. *Proc. Natl. Acad. Sci. USA* **2011**, *108*, 5163–5168. [CrossRef]
21. Agra-Kooijman, D.M.; Singh, G.; Lorenz, A.; Collings, P.J.; Kitzerow, H.S.; Kumar, S. Columnar molecular aggregation in the aqueous solutions of disodium cromoglycate. *Phys. Rev. E* **2014**, *89*, 062504. [CrossRef]
22. Yamaguchi, A.; Smith, G.P.; Yi, Y.; Xu, C.; Biffi, S.; Serra, F.; Bellini, T.; Zhu, C.; Clark, N.A. Phases and structures of sunset yellow and disodium cromoglycate mixtures in water. *Phys. Rev. E* **2016**, *93*, 012704. [CrossRef]

23. Simon, K.A.; Sejwal, P.; Gerecht, R.B.; Luk, Y.-Y. Water-in-water emulsions stabilized by non-amphiphilic interactions: Polymer-dispersed lyotropic liquid crystals. *Langmuir* **2007**, *23*, 1453–1458. [CrossRef] [PubMed]
24. Kleman, M.; Lavrentovich, O.D. *Soft Matter Physics: An Introduction*; Springer: New York, NY, USA, 2003.
25. Kim, Y.K.; Shiyanovskii, S.V.; Lavrentovich, O.D. Morphogenesis of defects and tactoids during isotropic-nematic phase transition in self-assembled lyotropic chromonic liquid crystals. *J. Phys. Condens. Matter* **2013**, *25*, 404202. [CrossRef] [PubMed]
26. Nastishin, Y.A.; Liu, H.; Schneider, T.; Nazarenko, V.; Vasyuta, R.; Shiyanovskii, S.V.; Lavrentovich, O.D. Optical characterization of the nematic lyotropic chromonic liquid crystals: Light absorption, birefringence, and scalar order parameter. *Phys. Rev. E* **2005**, *72*, 041711. [CrossRef] [PubMed]
27. Prinsen, P.; van der Schoot, P. Shape and director-field transformation of tactoids. *Phys. Rev. E* **2003**, *68*, 021701. [CrossRef]
28. Prinsen, P.; van der Schoot, P. Continuous director-field transformation of nematic tactoids. *Eur. Phys. J. E* **2004**, *13*, 35–41. [CrossRef]
29. Prinsen, P.; van der Schoot, P. Parity breaking in nematic tactoids. *J. Phys. Condens. Matter* **2004**, *16*, 8835–8850. [CrossRef]
30. Kaznacheev, A.V.; Bogdanov, M.M.; Taraskin, S.A. The nature of prolate shape of tactoids in lyotropic inorganic liquid crystals. *J. Exp. Theor. Phys.* **2002**, *95*, 57–63. [CrossRef]
31. Zhang, C.; Acharya, A.; Walkington, N.J.; Lavrentovich, O.D. Computational modelling of tactoid dynamics in chromonic liquid crystals. *Liq. Cryst.* **2018**, *45*, 1084–1100. [CrossRef]
32. Van der Beek, D.; Petukhov, A.V.; Oversteegen, S.M.; Vroege, G.J.; Lekkerkerker, H.N. Evidence of the hexagonal columnar liquid-crystal phase of hard colloidal platelets by high-resolution SAXS. *Eur. Phys. J. E* **2005**, *16*, 253–258. [CrossRef]
33. Alexandridis, P.; Olsson, U.; Lindman, B. A Record Nine Different Phases (Four Cubic, Two Hexagonal, and One Lamellar Lyotropic Liquid Crystalline and Two Micellar Solutions) in Ternary Isothermal System of an Amphiphilic Block Copolymer and Selective Solvents (Water and Oil). *Langmuir* **1998**, *14*, 2627–2638. [CrossRef]
34. Luzzati, V. *Biological Membranes*; Chapman, D., Ed.; Academic Press: New York, NY, USA, 1968; p. 78.
35. Park, H.S. Self-Assembly of Lyotropic Chromonic Liquid Crystals: Effects of Additives and Applications. Ph.D. Dissertation, Kent State University, Kent, OH, USA, 2010.
36. Onsager, L. The effects of shape on the interaction of colloidal particles. *Ann. N. Y. Acad. Sci.* **1949**, *51*, 627–659. [CrossRef]
37. Zhou, S.; Neupane, K.; Nastishin, Y.A.; Baldwin, A.R.; Shiyanovskii, S.V.; Lavrentovich, O.D.; Sprunt, S. Elasticity, viscosity, and orientational fluctuations of a lyotropic chromonic nematic liquid crystal disodium cromoglycate. *Soft Matter* **2014**, *10*, 6571–6581. [CrossRef]
38. Sidky, H.; Whitmer, J.K. The Emergent Nematic Phase in Ionic Chromonic Liquid Crystals. *J. Phys. Chem. B* **2017**, *121*, 6691–6698. [CrossRef]
39. Xiao, W.; Hu, C.; Carter, D.J.; Nichols, S.; Ward, M.D.; Raiteri, P.; Rohl, A.L.; Kahr, B. Structural Correspondence of Solution, Liquid Crystal, and Crystalline Phases of the Chromonic Mesogen Sunset Yellow. *Cryst. Growth Des.* **2014**, *14*, 4166–4176. [CrossRef]

Review

Lyotropic Liquid Crystals from Colloidal Suspensions of Graphene Oxide

Adam P. Draude and Ingo Dierking *

Department of Physics and Astronomy, University of Manchester, Manchester M13 9PL, UK
* Correspondence: Ingo.Dierking@manchester.ac.uk

Received: 20 August 2019; Accepted: 30 August 2019; Published: 31 August 2019

Abstract: Lyotropic liquid crystals from colloidal particles have been known for more than a century, but have attracted a revived interest over the last few years. This is due to the developments in nanoscience and nanotechnology, where the liquid crystal order can be exploited to orient and reorient the anisotropic colloids, thus enabling, increasing and switching the preferential properties of the nanoparticles. In particular, carbon-based colloids like carbon nanotubes and graphene/graphene–oxide have increasingly been studied with respect to their lyotropic liquid crystalline properties over the recent years. We critically review aspects of lyotropic graphene oxide liquid crystal with respect to properties and behavior which seem to be generally established, but also discuss those effects that are largely unfamiliar so far, or as of yet of controversial experimental or theoretical outcome.

Keywords: liquid crystal; graphene oxide; lyotropic; colloid; nematic

1. Introduction

Many of the liquid crystal (LC) phases known from applications in displays are based on rod-shaped molecules, so called calamitic mesogens, which exhibit the liquid crystal behaviour as a function of changing temperature and are called thermotropic. Nevertheless, for the formation of liquid crystals only general shape anisotropy of the molecules is necessary, and in 1977, Chandrasekhar et al. [1] reported liquid crystallinity from disk-shaped molecules, called discotic liquid crystals, which are also part of the family of thermotropic LC phases. But liquid crystallinity can also be observed as a function of changing concentration of either amphiphilic molecules, or anisotropic nanoparticles and colloids, in an isotropic host fluid, often water. This leads to a wholly new class of LCs, the lyotropic phases, which are the topic of this paper and indeed of the whole journal volume of this special issue of *Crystals*. Besides those formed from amphiphilic molecules, several types of lyotropic LCs may be distinguished. Chromonics [2,3] are liquid crystals of dissolved rigid dyes, thus flat molecules, in an isotropic solvent. These stack to form super-molecular structures which then exhibit nematic and columnar phases. Inorganic liquid crystals [4] are lyotropics that can be formed by rod-like colloids, such as vanadium pentoxide, V_2O_5, and other minerals, but also by disc-shaped colloids, especially clays [5–7]. Such systems have already been observed by Langmuir in 1938 on bentonite clays [8] and form the basis of the Onsager theory, which will be discussed herein.

Somewhere in between the molecular size of chromonic dyes and the macroscopic platelets of clays, lies graphene oxide. Since the discovery of graphene in 2004 [9], a stable, two-dimensional hexagonal lattice of carbon with quite surprising and unprecedented physical properties [10], this material has attracted enormous interest (Figure 1a). This is on the one side founded by questions of fundamental research, and on the other side by the promise of a large range of possible high volume technological applications. When working with graphene in a liquid crystal environment, one will unfortunately quickly arrive at the conclusion that these composites and mixtures are by no means easy to handle, because graphene is largely insoluble in nearly all liquid solvents. This is very different for graphene–oxide (GO) which exhibits different functional groups at the edges and the plane of the graphene oxide sheets, and is readily soluble in many solvents, including water, which is a significant advantage for environmentally friendly applications (Figure 1b). At certain concentrations, roughly > 0.1–1 wt%, GO often exhibits very stable lyotropic nematic phases [11–13]. One of the interesting aspects of GO liquid crystals is the possibility of self-organized and self-ordered systems of graphene oxide. This order can then easily be manipulated by application of external stimuli, such as boundary conditions, mechanical shear, but possibly also electric and magnetic fields. After obtaining a certain directional order one could wash away or evaporate the solvent, leaving an ordered GO structure, which may even be reduced chemically or through heat application to produce reduced graphene oxide (rGO), which displays some of the originally desirable graphene properties. Lyotropic graphene oxide liquid crystals are thus of immense interest for nanotechnology and its applications. There have been several review articles on the properties of GO liquid crystals [14–16]. In this paper we want to give a short critical account of what is known, what is not yet known, and what are the controversial questions with regards to graphene oxide liquid crystals. Effects of flake size and solvent will be discussed, alignment, addition of salts and polymers, as well as electro-optic properties, and electric and magnetic field effects.

2. Graphene and Graphene Oxide

Pristine graphene is a two-dimensional hexagonal lattice of sp^2 hybridized carbon with a basis of two, giving rise to a honeycomb structure [10]. Graphene oxide is a very different material to graphene. Graphene is often obtained by mechanical exfoliation of graphite (the so-called Scotch Tape® method) or grown by chemical vapor deposition. On the other hand, graphite that has undergone treatment to become graphite oxide readily exfoliates into monolayers when dispersed in water (Figure 1) [17]. Thus, graphene oxide is often easier to produce in large quantities than graphene and is readily dispersible in solution for storage or further processing [18]. There are two main processes described in the literature for producing graphite oxide, the Hummers [19] and Brodie [20] methods.

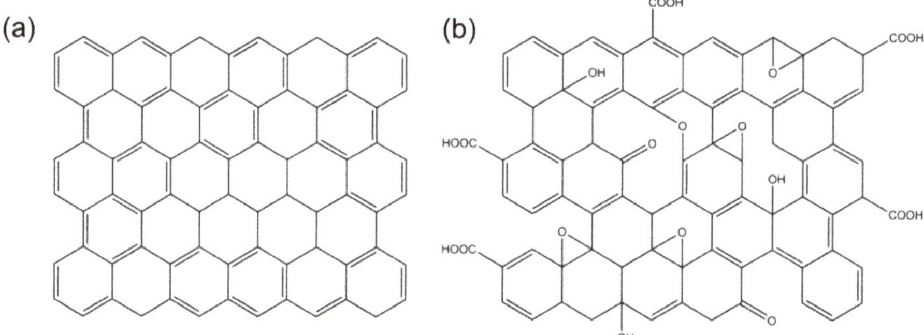

Figure 1. The structures of pristine graphene (**a**) and a model of graphene oxide (**b**). The functional groups of graphene oxide can vary depending on its method of production and its thermal and/or storage history.

The structure of graphene oxide is reminiscent of graphene but the lattice is now distorted and the covalent bonding of oxygen functional groups such as epoxy, hydroxyl and carboxyl groups [21] means much more of the carbon is now sp3 hybridized. The disruption to the sp^2 bonding network means graphene oxide is not electrically conductive. Typically, graphene oxide is produced by a modified Hummers method [22] and the degree of oxidation is characterized by X-ray photoelectron spectroscopy (XPS) measurements. From XPS measurements one can determine the ratio of oxygen to carbon atoms in the material, which is often taken as a measure of how oxidized the graphene sheet is. Graphene oxide can be reduced by thermal or chemical means, and the product is known as rGO. When GO is reduced, the first change to occur is that the C–O peak in XPS decreases in intensity relative to the C–C and the C–O=O/C=O bonds (Figure 2) [23,24]. This can be interpreted as a relative decrease in the abundance of epoxy groups (which only include C–O bonds) on the GO surface.

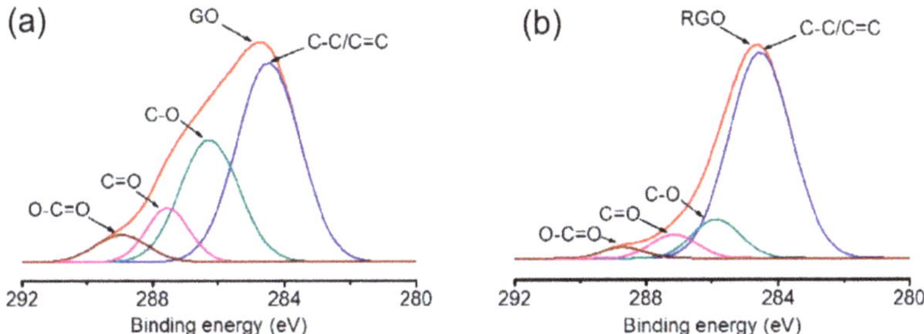

Figure 2. De-convolved X-ray photoelectron spectroscopy (XPS) spectra in the region of the C1s core electron binding energy. In graphene–oxide (GO) (**a**), three peaks associated with the bonding of oxygen to carbon are identified. In rGO (**b**), the overall intensity of these peaks is reduced relative to that originating from C-C bonds. Figure adapted from Chen et al. [23] with permission of Springer Nature.

Stable dispersions of graphene oxide in appropriate solvents can be described as a colloid—a suspension of one phase (in this case a solid) inside another. Colloids of anisodiametric (shape-anisotropic) particles are known to make lyotropic liquid crystal phases, and the same is true of graphene oxide.

3. Colloidal Liquid Crystals of Graphene Oxide

Liquid crystals are soft condensed matter systems which exhibit orientational ordering of their basic units (and in some cases, a degree of positional order) as a solid phase does, but which can also flow. As mentioned above, there are two main classes of liquid crystals—thermotropic and lyotropic (Figure 3) [25]. Thermotropic liquid crystals are systems in which phases exhibiting liquid crystal properties can be observed to occur between the solid and liquid states. Thus, temperature and pressure are the main variables of state which alters their phase. The best known example of thermotropic liquid crystals are the rod-like (calamitic) molecules found in everyday display screens [26]. Lyotropic liquid crystals are systems of at least two components, and it is their relative concentration that brings about orientational ordering of self-assembled structures [27]. Examples include the columnar and lamellar phases of amphiphilic molecules in water [28]. Amphiphilic lyotropic liquid crystals are also the most common type of all liquid crystals, since their lamellae form the basic structure of biological cell membranes [27,29]. Colloids of anisodiametric particles in liquid solvents can also possess orientational ordering at sufficiently high concentrations [30]. Though the structures that are aligning are quite different to the case of amphiphiles, they are nevertheless referred to as lyotropic liquid crystals.

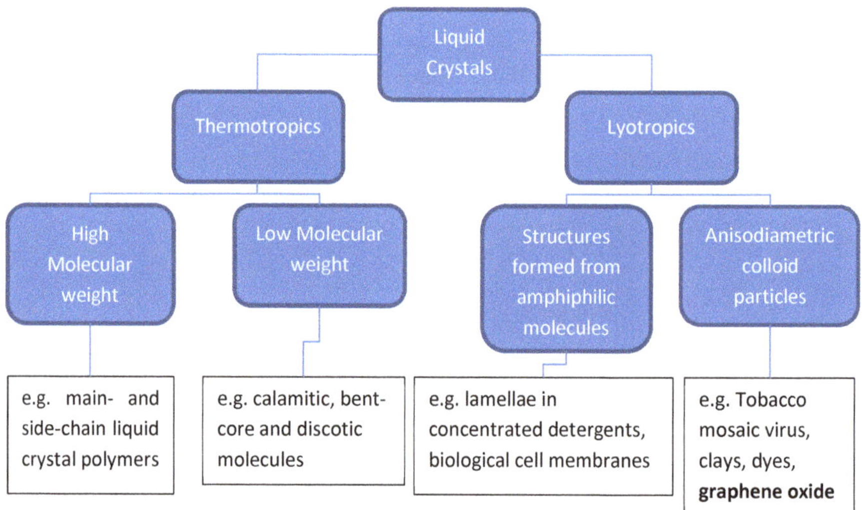

Figure 3. Flow diagram showing the nomenclature used in the field of liquid crystals. Examples are given of each type of liquid crystal (LC).

In 2010, Behabtu et al. [31] described the observation in polarized optical microscopy (POM) of birefringence in suspensions of graphene in chlorosulphonic acid. The authors interpreted these observations as being due to orientational alignment and hence lyotropic liquid crystallinity of the graphene flakes [31]. Just one year later, Kim et al. [11], Dan et al. [13], and Xu and Gao [12,32] reported that graphene oxide dispersed in water behaved in the same way.

Typical POM images of graphene oxide dispersion in the nematic phase from Xu and Gao [12] are shown in Figure 4. These images show the evolution in the microstructure of graphene oxide dispersions as the concentration is increased from the isotropic phase (image 1) through the biphase region (image 2) and into the nematic liquid crystal phase (image 3).

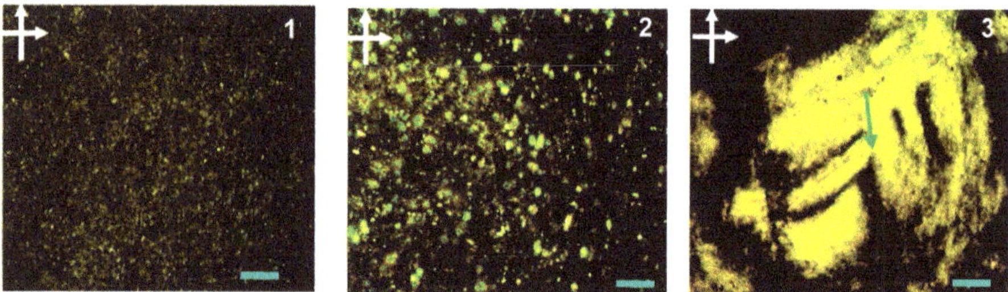

Figure 4. Polarized optical microscopy (POM) images of graphene oxide dispersions increasing in concentration from the isotropic phase ($1-5 \times 10^{-4} f_m$, where f_m is the mass fraction of GO in the dispersion) through the biphasic region in which small clusters of birefringent domains can be seen ($2-1 \times 10^{-3} f_m$) and then to the nematic phase ($3-3 \times 10^{-3} f_m$). The scale bar represents 200 µm. The arrow in the third image indicates the boundary between two nematic domains. This appears as a dark brush in polarized microscopy, analogous to the Schlieren textures of thermotropic nematic liquid crystals. Reprinted with permission from Xu and Gao [12]. Copyright (2011) American Chemical Society.

Theoretically, the lyotropic phase behaviour of anisodiametric nano- and micro-particles is described by the Onsager theory [33]. In his original work of 1949 [33], Onsager used a virial expansion method to describe the orientational ordering that had been observed in suspensions of tobacco mosaic viruses. The underlying principle of the theory is that some volume a particle would otherwise be able to move into is excluded by the presence of a further particle i.e., the particles cannot interpenetrate one another. As the volume concentration of particles increases, this means there is less and less space available to a particular particle and for anisodiametric particles orientational ordering spontaneously emerges at a critical concentration. The transition is known as an entropy-driven transition [34]. Thus far, all known theories about entropy driven phase transitions in colloidal lyotropics have used the approximation that the particles are inflexible.

Forsyth et al. [35] extended the Onsager model to the hard-disk fluid in 1977. Their findings showed that the phase behaviour of disks was distinctly different from rods. When the aspect ratio (diameter/thickness) of the discs increases, transitions to the biphase and the nematic phase occur at lower volume concentrations. The reverse happens for rods; as their aspect ratio (length/diameter) increases so does the minimum volume concentration required for the onset of orientational ordering.

Bates and Frenkel [36], almost two decades later, have argued that the extension of Onsager's theory from rods to discs is invalid. Furthermore, they argued that since real systems are polydisperse, model systems should be as well. Their approach was to use Monte–Carlo simulations of infinitely thin disks with varying polydispersity ($\sigma = \langle D \rangle / \sqrt{(\langle D^2 \rangle - \langle D \rangle^2)}$, where $\langle D \rangle$ is the average diameter of the disk). Their results are summarized in Figure 5. For ease of interpretation, these results have been converted into units typically used in experimental graphene oxide research (i.e., micrometers, vol %) using the method of Shen et al. [37]. It can be seen that the width of the biphasic region increases with increasing polydispersity, but the range of polydispersity investigated in the simulations does not stretch to that typically encountered for graphene oxide dispersions [37].

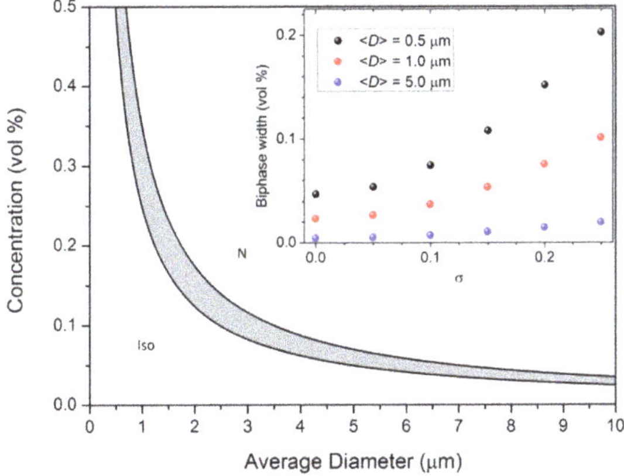

Figure 5. The phase diagram for a hard-disk fluid with 25% polydispersity from simulations by Bates and Frenkel. The isotropic (Iso) and nematic (N) phases are labelled and the shaded part corresponds to the biphasic region. Inset: the effect of polydispersity on the width of the biphase.

When the concentration of graphene oxide is further increased, other phases can be observed. The first phase to appear with increasing concentration is the lamellar phase. In the lamellar phase, the graphene oxide sheets form layers which lead to Bragg reflection in small-angle X-ray scattering (SAXS) [12]. On the basis of ellipsometry measurements and freeze-fracture cryo-SEM (scanning electron microscopy), Xu and Gao have also proposed that this lamellar phase is in fact chiral. They propose the structure of this phase to be an analogue of the twist grain boundary phases of thermotropic liquid crystals, with lamellar blocks of GO twisted relative to one another forming a helical superstructure. This observation is remarkable, since neither the graphene oxide nor the solvent is chiral, so it is not clear from where the chirality of the system arises. When the graphene oxide is covalently functionalized with polyacrylonitrile, there is also evidence of a chiral nematic phase, in which the GO sheets form a helical structure but are not yet in lamellae. These results were established using ellipsometry and SAXS [38]. When the sheet-to-sheet interaction is modified with salt, the phase diagram becomes more complex, and this will be discussed later in this review. Meanwhile in the next section, the nematic phase is discussed in more detail.

4. Effects of Flake Size and Solvent

Recently, Ahmad et al. [39] have investigated the dependence of the electro-optical properties of GO dispersions on the graphene oxide flake size. We will return to electro-optic effects later but we note here that the authors' experimentally determined phase diagram is quite different to that produced from the simulations of Bates and Frenkel [36] (Figure 6). There are various explanations as to why this could be the case. Firstly, GO sheets are not rigid but in fact are highly flexible [40]. Secondly, they are not infinitely thin but have a finite thickness of around 1 nm as determined by atomic force microscopy (AFM) [24,41]. Furthermore, the polydispersity of GO systems cannot usually be approximated by a normal distribution (as the simulations do), and is usually asymmetric with a tail to higher particle sizes (skewed distribution, Pearson distribution). Ahmad et al. [39] determined the phase boundaries through polarized photography of graphene oxide dispersions contained within a Pasteur pipette. Under such conditions, the isotropic phase appears black. However, it is possible to observe transient birefringence caused by flow driven alignment in the isotropic phase [37] (see the next section for more detail), which could be misinterpreted as the onset of the biphasic region. Furthermore, the transition from the two-phase region to the nematic phase is effectively continuous, making the boundary difficult to determine by either photography of macroscopic samples or microscopy.

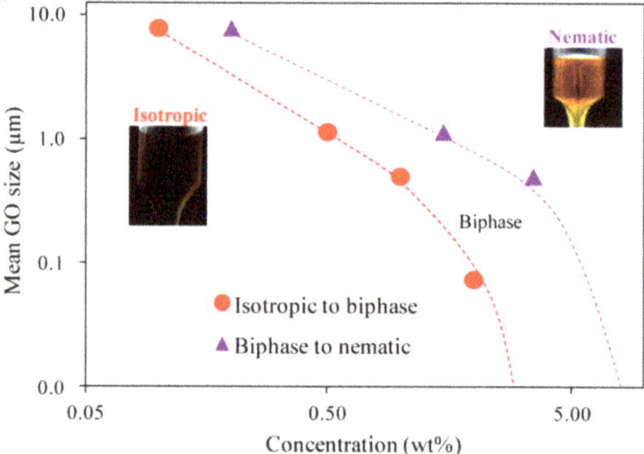

Figure 6. Experimentally determined lyotropic phase diagram for graphene oxide dispersed in water. The inset figures show photographs of sample held in a Pasteur pipette between crossed polarizers. It is by such photographs the authors interpreted the location of the phase boundaries. Adapted with permission from ref [39], The Optical Society.

The overall trend of the phase diagram is at odds with the theoretical model of Bates and Frenkel [36] with the phase boundaries shifting to higher concentration with decreasing mean particle size – the inverse of the theoretical prediction. The reasons for this are not currently well understood. What is known is that current theoretical models are too restrictive and do not accurately represent an experimental system. GO sheets are not, for example, infinitely thin, ridged and of approximately equal size. Although it still appears that the primary driver for the onset of liquid crystallinity is excluded volume effects, there are other ways flakes can interact with each other, as will be discussed herein.

Since, in principle, the phase behaviour of colloidal systems of graphene oxide is only dependent on concentration, the isotropic liquid which acts as the solvent should not play any part, and all solvents that can disperse graphene oxide without aggregation, flocculation or sedimentation should be equal. It has been widely reported [21,41,42] that water, dimethylformamide (DMF) and N-methyl-2-pyrrolidone (NMP) are the best of the common laboratory solvents for dispersing graphene oxide. Accordingly, lyotropic liquid crystalline behavior was found in all three of these in another paper by Ahmad et al. [42] (Figure 7).

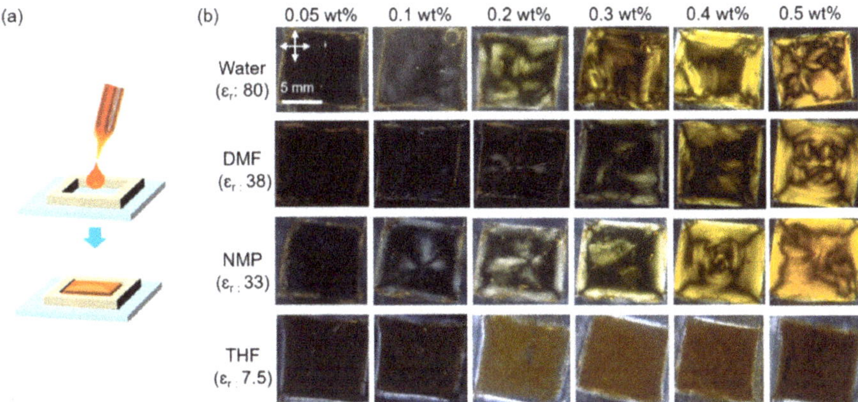

Figure 7. Polarizing optical photographs of dispersions of graphene oxide in different solvents and at different concentrations are shown (**b**). In tetrahydrofuran (THF), GO tends to flocculate, meaning spontaneous orientational order does not occur. The apparatus used to contain the dispersions for observation is shown in part (**a**). Reprinted from Ahmad et al. [42], with the permission of AIP Publishing.

Ionization of the polar oxygen functional groups on the GO surface in solvents leads to GO being negatively charged, and thus electrostatic interactions play a role in the dispersion of graphene oxide and the pair-wise interaction. By measuring the conductivity of GO dispersions in the three different solvents Ahmad et al. [42] showed that the surface ionization of GO varies depending on the solvent, with water having the highest conductivity and hence aqueous GO having the highest surface ionization.

5. Alignment of GO Liquid Crystals

The alignment of conventional thermotropic liquid crystals can be achieved using surface treatment of the glass used to confine them for study. Examples include rubbed polyimide for planar alignment and the use of surfactants such as C-TAB or DMOAP for homeotropic alignment. Graphene oxide liquid crystals are a very different system however, with flakes being tens of thousands of times larger than thermotropic liquid crystal molecules. In 2011, Dan et al. [13] reported that nematic GO confined in a rectangular capillary aligned spontaneously with the average GO sheet normal (the optical axis in this case) being parallel to the capillary axis. However, they suspected this was a metastable state [13]. Recently Al-Zangana et al. [43] have shown that graphene oxide sheets in solution tend to prefer to lie parallel to glass. Thus, the director aligns perpendicularly to a glass surface. This has been demonstrated by the construction of both covered and uncovered glass channels constructed from microscope slides and coverslips (Figure 8). In polarized optical microscopy, the channel appears bright when its axis is placed at 45° to each of the polarizers, indicating uniform alignment of the director field perpendicular to the axis of the channel, along its length. When a glass plate is placed on top of this channel, the middle section becomes dark, irrespective of the position of the channel relative to the polarizers. This indicates local homeotropic alignment of the director field i.e. the graphene oxide is parallel to the glass plate.

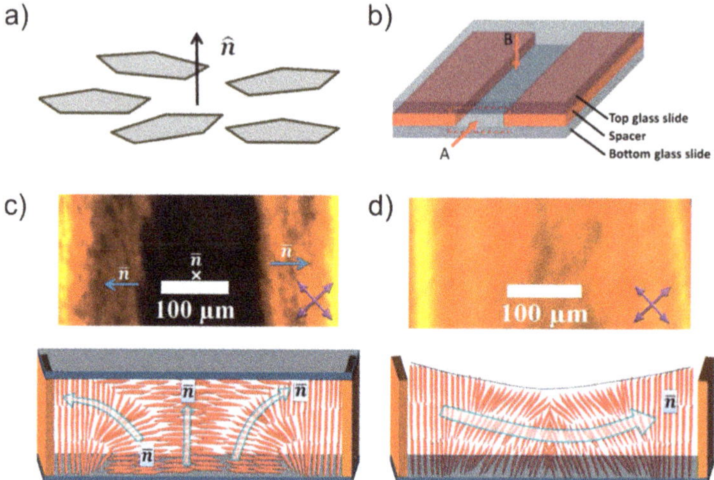

Figure 8. (**a**) Schematic diagram showing the orientation of GO flakes with respect to the director, n. (**b**) The confining geometry used to photograph GO dispersions (POM) in parts (**c**,**d**). In (**c**) the top glass plate is present and the LC becomes homeotropic in the centre of the channel. In part (**d**) the top plate is removed and the boundary conditions are set by the vertical walls. The director configurations in each case are also shown below the POM pictures. (**b**–**d**) reproduced and adapted from Al-Zangana et al. [43] under the terms of the Creative Commons Attribution 3.0 licence.

Shear flows can also cause GO liquid crystals in the isotropic phase to be birefringent. This has been demonstrated by Shen et al. [37] by using magnetic stirring and manual shaking of samples between crossed polarizers (Figure 9). The birefringence observed can be explained by the graphene oxide flakes feeling a torque caused by the velocity field of the solvent. This causes local alignment of the graphene oxide flakes, and hence birefringence, where the change in velocity of the solvent is greatest.

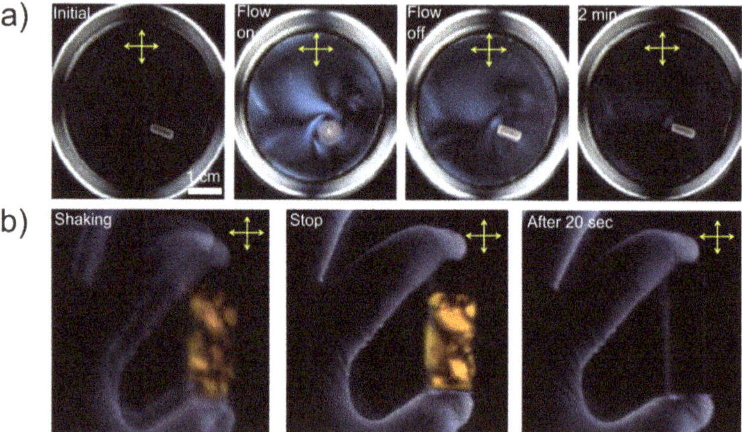

Figure 9. Time series of polarized photographs showing the temporary flow-induced alignment of graphene oxide flakes in the isotropic phase by (**a**) magnetic stirring and (**b**) manual shaking. Reprinted by permission from Springer Nature: Nature Materials, Shen et al. [37] © 2014.

Tkacz et al. [44] have also shown that dispersions of reduced graphene oxide can be used to produce highly aligned films on glass substrates. The authors proposed that diffusion of sheets in the fluid, assisted by capillary forces, causes the self-assembly of the sheets on the substrate.

6. Ions and Polymers in GO Liquid Crystals

Xu and Gao [12] have demonstrated that electrostatic interactions also play a role in the lyotropic behaviour of graphene oxide to some degree. To reveal this, they measured the phase behaviour by observing POM images as a function of salt (NaCl) concentration. If the NaCl concentration was too high, the repulsive force between the sheets was reduced such that flocculation occurred. Small amounts of NaCl did however influence the width of the two-phase region significantly (Figure 10). The authors also measured the zeta-potential of the dispersion for each concentration studied and found the absolute value decreased with increasing NaCl content. These results indicate that both electrostatic and excluded volume effects play a role in the lyotropic behaviour of graphene oxide, and that the overall stability of the nematic phase is reduced in an electrolyte environment. However, Zhao et al. [45] have shown that by coating the GO flakes with polyelectrolytes, the nematic phase can remain stable even in NaCl concentrations of up to 6.2 M, depending on the polyelectrolyte employed.

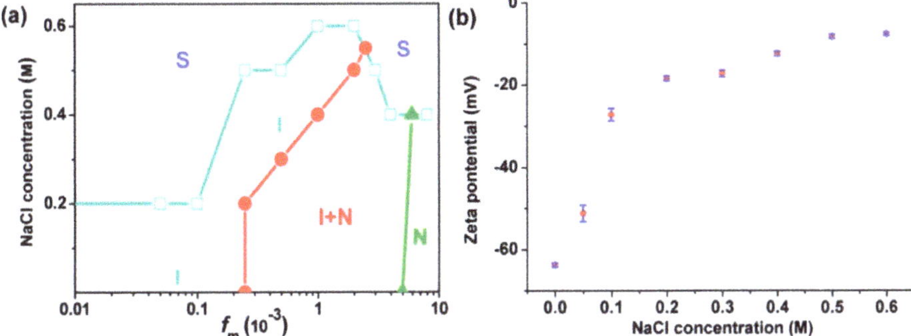

Figure 10. The dependence of (**a**) the phase diagram of a GO-LC (where f_m is the mass fraction of GO) and (**b**) the zeta potential of the GO dispersion on the concentration of NaCl salt in the solution. Reprinted with permission from Xu and Gao. [12] Copyright (2011) American Chemical Society.

Konkena and Vasudevan [46] have also demonstrated that by varying the GO and NaCl salt concentrations, a number of different phases or arrested states can be formed (Figure 11). By arrested states, the authors refer to structures which are not at global minima of the thermodynamic potential, thus metastable [46]. At high GO concentrations, but low salt concentrations where electrostatic repulsion dominates the interactions, a glass is observed. This is attributed as a Wigner glass, which is analogous to the Wigner crystals of electrons in solids [47,48]. Wigner glasses have also been observed in dispersions of other plate-like materials such as laptonite [49]. GO glasses and gels do not flow, unlike nematic liquid crystals, and so can be easily identified by inverting the container of the GO dispersion. To further demonstrate a glass experimentally, the authors performed dilution experiments. The glass, in which GO interactions are predominantly repulsive, melts upon the addition of further solvent whereas the states observed at higher salt concentrations do not [46] Konkena and Vasudevan further suggest that at higher salt concentrations, the dominant GO interaction is an attractive one which leads to a phase which appears dark between crossed polarizers but turns bright when sheared. These observations point towards homeotropic alignment of the graphene oxide sheets i.e. collectively the sheets are in the same plane as the substrate and perpendicular to the light path. This behaviour is attributed to the onset of a columnar phase by the authors, based on the appearance of the POM textures [46] though recent X-ray scattering experiments by Rubim et al. suggest that this observation could be a lamellar phase [50].

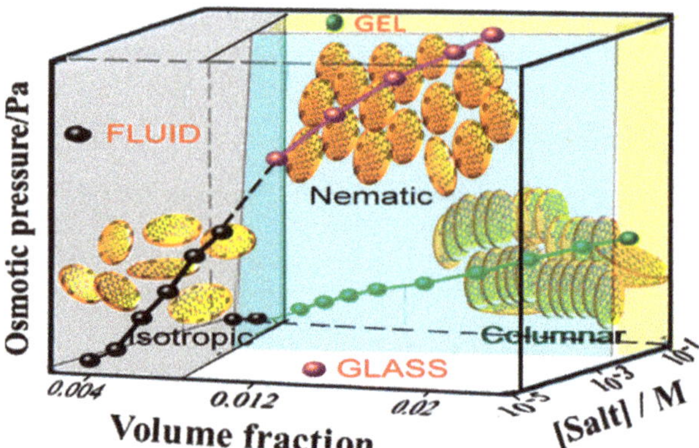

Figure 11. The phase diagram of graphene oxide dispersions as determined by Konkena and Vasudevan [46] displaying isotropic, nematic, columnar and glass phases. Reprinted with permission from ref. [46], Copyright (2014) American Chemical Society.

Graphene oxide liquid crystals are often observed with water as the solvent. Typically, the water is deionized and at pH ~7 for a fresh dispersion. However, the LC behaviour can also be observed at pH 2, 6 and 9 as reported by Tkacz et al. [51] As the pH increases, the fraction of dissociated carboxylic groups changes (the pK_a value is ~4 for these groups [52]) and so too does the width of the biphasic region between the isotropic and nematic phases. This work further suggests that electrostatic interactions between GO sheets are important for the onset of lyotropic behavior. Furthermore, at biphasic concentrations, highly ordered phase-separated nematic droplets were observed [51].

The addition of polymers to a colloidal liquid crystal system is often performed to better stabilize the dispersion via steric hindrance or to cause gelation. Recently, Shim et al. [53] have reported that the addition of polyethylene glycol (PEG) to GO LCs causes the onset of the nematic phase to occur at lower concentrations of GO in the total volume. Furthermore, they report that the Wigner glass transition of the GO dispersion is pushed up to higher GO concentrations. [53] Therefore the overall effect of the addition of PEG widens the nematic phase. The concentrations of PEG required to observe these effects are rather high (Figure 12).

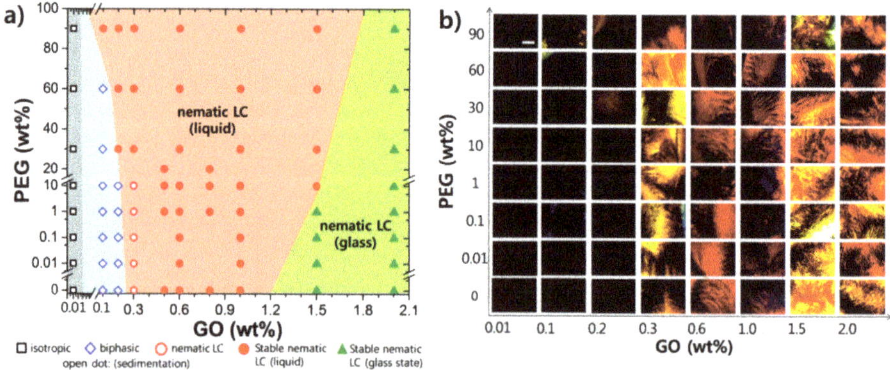

Figure 12. (**a**) The phase diagram of a graphene oxide liquid crystal with added polyethylene glycol (PEG) as determined by Shim et al. [53]. (**b**) Typical POM images for different GO/waterPEG mixtures. Weak birefringence can be seen as the PEG concentration is increased in an initially isotropic GO dispersion. Republished with permission of the Royal Society of Chemistry, from Shim et al. [53]; permission conveyed through Copyright Clearance Center, Inc.

The addition of PEG to a concentration of 20 wt% in the mixture causes the macroscopic viscosity of the whole dispersion to decrease by a factor of 20–30 relative to that of GO in pure water, for the same concentration of GO. This, the authors rightly suggest, points to a need for a greater understanding of the micro-structure of the polymer-solvent-GO system at the GO surface. In a follow-up work by the same group [54], a selection of molecular weights of PEG were used. It was shown that PEG with a larger average molecular weight (10,000 gmol^{-1}) reduced electrostatic repulsion between the sheets, which is the origin of the glass transition in GO. Reduction of this repulsive energy was offered as explanation for a decrease in the macro-viscosity of the LC. If PEG with a lower molecular weight (400 gmol^{-1}) was added, the viscosity did not decrease by the same amount.

7. Electro-Optic Effects

Birefringence induced by electric fields can be observed in a variety of otherwise optically isotropic materials. These effects are commonly grouped into one of two categories, the Pockels effect or the Kerr effect, depending on whether the induced birefringence is proportional to the magnitude of the applied electric field or its square, respectively [55]. The Pockels effect only occurs in crystals which lack inversion symmetry. Dispersions of graphene oxide in the isotropic, biphasic and nematic phases have displayed electric field-induced birefringence [37,39,42,56]. In the case of the nematic phase, this corresponds most likely to a reorientation of the director field as a whole, as is the case with thermotropic nematics. In the less concentrated phases, a Kerr effect is observed where each flake, from whichever orientation it was in, aligns itself to the electric field. This behavior of a GO flake in an electric field can be described by the Maxwell–Wagner–O'Konski model [57,58]. The key parameters of the GO in this framework are the aspect ratio and the ionic conductivity at the surface. It is known that in aqueous dispersions GO is negatively charged with an ionic double layer at its surface [12].

Shen et al. [37] measured the Kerr coefficient of aqueous dispersions of graphene oxide and found values of the order 10^{-5} mV^{-2}. This value is very high in comparison to other materials known to have high Kerr coefficients such as thermotropic liquid crystal blue phases ($\sim 10^{-9}$ mV^{-2}) [59–61] or nitrobenzene (10^{-12} mV^{-2}) [62]. In a qualitatively similar system consisting of suspended platelets of clay the Kerr coefficient was four orders of magnitude lower than that of graphene oxide [62]. Shen et al. [37] also observed that the electric field-induced birefringence decreased as the concentration was increased through the biphasic region and that in the concentrated nematic phase there was barely any effect of the electric field at up to 20 Vmm^{-1}. By analogy, thermotropic nematic liquid crystals require much higher field strengths to switch the director (~ 0.7 Vμm^{-1}). In contrast, Ahmad et al. [42] measured a significant field-induced birefringence in the nematic phase with field strengths of 10 Vmm^{-1} or lower (Figure 13). Interestingly, they also found that the maximum field-induced birefringence of a given flake size distribution depended on what solvent it was dispersed in [42]. Water was found to outperform both DMF and NMP at high frequency (10 kHz), but electrophoretic drift of the GO particles was observed at lower frequencies (100 Hz). Both of these results were attributed to the higher ionic content of the aqueous dispersion [42] which occurs because the surface functional groups of graphene oxide are easily ionized in aqueous dispersions. The maximum field-induced birefringence has been shown to be affected by the flake size and distribution [63] (Figure 13) but interestingly also the number of cleaning steps taken in the production of the graphene oxide [64].

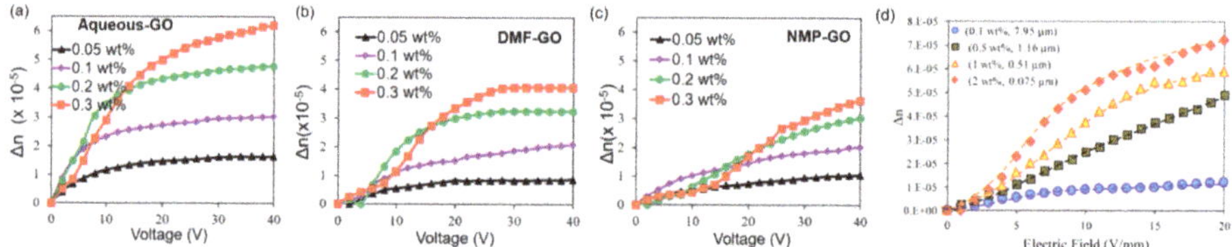

Figure 13. The induced birefringence of GO dispersions as a function of voltage for three different solvents: water (**a**), DMF (**b**) and NMP (**c**). Reprinted from Ahmad et al. [42], with the permission of AIP Publishing. (**d**) The field induced birefringence of different dispersions with different average flake sizes and concentrations as a function of electric field. Reprinted with permission from ref. [39], The Optical Society.

The time taken for the graphene oxide dispersions to become fully aligned to the electric field is known as the response time. The relaxation time is defined as the time taken for the system to relax once the electric field is removed. Both the response and relaxation times for graphene oxide dispersions have been found to be just under half a second, depending on the flake size distribution and concentration (Figure 14) [39,63]. This has been shown to be affected by flake size and in particular the flake size distribution [39,63]. The higher the fraction of large flakes (>1 μm) in a mixture which otherwise contains mainly small flakes (<1 μm) the slower the response to electric fields and the slower the relaxation time when the field is removed. This has been attributed to increased inter-particle friction when larger flakes are present [63].

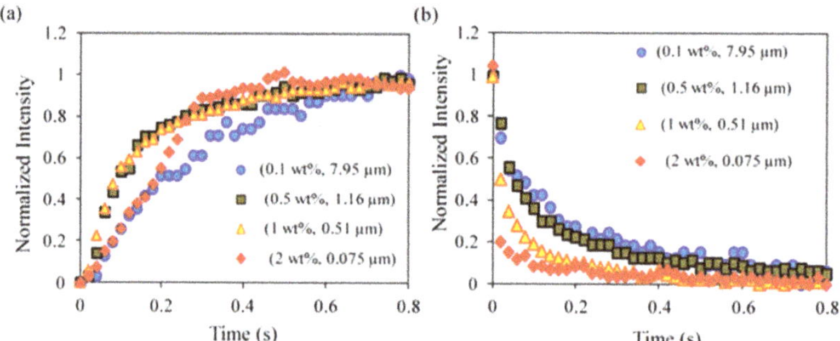

Figure 14. Response times of the aqueous dispersions shown in Fig.11d. (**a**) on-time (**b**) off-time. Adapted with permission from ref. [39], The Optical Society.

Graphene oxide in aqueous dispersions ages with time. This is due to C–C bond cleavage and the formation of vinylogous carboxylic acids [65]. This reaction also releases protons and decreases the pH of the dispersion. Consequently, the electro-optic performance has been shown to decrease with time [66]. Over the course of one week of storage at room temperature, Shen et al. [66] found that the field-induced birefringence fell by approximately 25% as compared to fresh GO. However, they did find that re-cleaning the material with fresh water in a centrifuge restores the performance completely [66].

8. Effect of Magnetic Fields

Early in the story of graphene oxide lyotropic liquid crystals, Kim et al. [11] reported that over a long time (5 h) a magnetic field (0.25T) could cause a nematic graphene oxide dispersion to align (Figure 15).

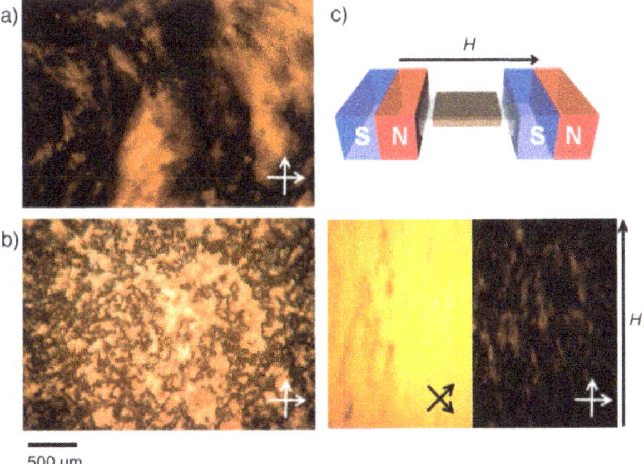

Figure 15. Alignment of aqueous graphene oxide sheets using a magnetic field as reported by Kim et al. [11]. (**a**) POM texture of the nematic GO LC immediately after preparation. (**b**) The texture after 3h with no external field present. (**c**) Schematic diagram showing the magnetic field set-up and POM textures of the aligned GO LC. Reproduced from ref. [11] with permission from John Wiley and Sons.

Since that time, it appears that this effect has not been independently demonstrated. Recently, we have performed small angle X-ray scattering measurements of the nematic phase of graphene oxide with an in situ magnetic field of approximately 0.9 T, applied perpendicularly to the X-ray beam. The experiments used a SAXSLAB Ganesha 300XL machine. This uses a copper Kα X-ray source and a 2D Pilatus 300K detector which is movable along the beam axis so the sample-detector distance can be adjusted. The experimental set-up is well described by Figure 1 in Sims et al. [67]. The diffraction pattern was initially anisotropic, presumably due to an average preferred orientation in the diffraction capillary used to contain the dispersion, or from shear alignment as it entered. After approximately 19 hours in the magnetic field, there was no change in the alignment of the GO sheets and they remained out of alignment with the direction of the field (Figure 16c). The graphene oxide used had the XPS spectrum shown in Figure 16a and the size distribution shown in Figure 16b. The stark difference in behaviour we observe compared to that of Kim et al. [11] could be due to the degree of oxidation of the GO. Recently Diamantopoulou et al. [68] have demonstrated that the magnetic response of GO and rGO is very different. This suggests that the precise conditions during the fabrication and storage of the graphene oxide, which affect the degree of reduction, could have an effect on its magneto-optic response.

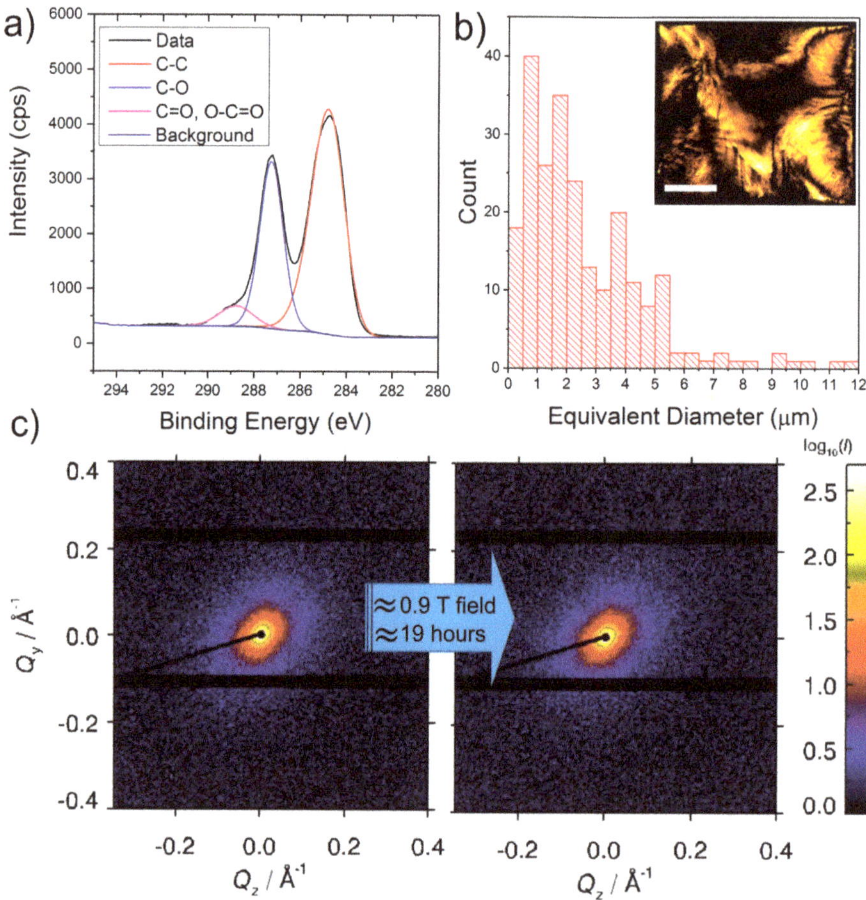

Figure 16. XPS and size distribution of the graphene oxide used in our magnetic field experiment (**a**,**b**). The inset shows a typical POM texture of the GO (scale of 500 µm). (**c**) Small-angle X-ray scattering (SAXS) patterns produced by this GO at a 0.5 wt% dispersion in DMF. The magnetic field is in the horizontal direction. After leaving the sample overnight in the field, no alignment effect was observed.

9. Applications of Graphene Oxide Colloids Lyotropic Behaviour

Graphene oxide dispersions are an ideal example of stable dispersions of high aspect ratio plate-like particles. The stability of such dispersions opens up a variety of solution-based processing routes. The spontaneous self-assembly into nematic and lamellar phases with increasing concentration has been shown to assist in the manufacture of novel three-dimensional architectures. An example of this is in the production of graphene oxide membranes [69]. Such membranes have received increased interest over recent years due to their potential use in the desalination of water [69,70]. The technology has arisen from the observation that water permeates the membranes rapidly, faster even than helium [71]. Membranes have been fabricated by vacuum filtration of a GO dispersion [71] or by evaporation of the solvent [72]. However, in 2016, Akbari et al. [69] showed that the shear alignment of a concentrated dispersion in the nematic state on a substrate can produce membranes with very even topography over a large area (Figure 17b). A further example of nematic behavior leading to ordered three-dimensional architectures is the observation of photonic behavior in GO dispersions. Li et al. [73] found that for a dispersion of high aspect-ratio sheets with low polydispersity, Bragg reflections could be observed in natural light at nematic concentrations (Figure 17c). The exact mechanism for this is not fully understood, but the authors noted that a sensitive balance of forces and flake sizes was key to the appearance of reflections, since the addition of NaCl or sonication (to reduce the flake size) both removed Bragg reflections [73]. Nevertheless, appropriately designed, such photonic crystals could be used in novel inks, or since their colour is based on concentration, humidity sensors. Li et al. also propose that if such structures could be controlled by external fields, GO-based photonic crystals could also be used for energy-efficient colour display applications [73].

Since the liquid crystallinity of graphene oxide dispersions was first discovered [11], the anisotropic ordering and self-assembly properties have been exploited in the production of fibrous materials. Kim et al. have shown that fibres drawn from concentrated GO dispersions with added poly(acrylic acid) have highly ordered microstructures and display birefringence (Figure 17d) [11]. The rheological properties of nematic graphene oxide dispersions make them favourable for wet-spinning of fibres [14] which was first demonstrated by Xu and Gao in 2011 [32]. A topic at the forefront of graphene applications is that of smart clothing and wearable electronics. GO can be reduced to restore some of graphene's conductivity, and conductive fibres using liquid crystal-assisted assembly, followed by reduction, are already at a mature stage of research [14,16,74].

As well as the potential for display technology based on the Kerr effect of GO dispersions in the isotropic phase (as discussed earlier in this review) He et al. have reported on a re-writable reflective display using nematic-phase GO dispersions [75]. When the concentrated dispersion is sheared in the plane of the surface, the surface becomes more reflective due to the larger amount of GO-surface area interacting with the light. Then, using a stick-like implement, the GO director field can be disturbed, and one is able to create writing on the surface (Figure 17a).

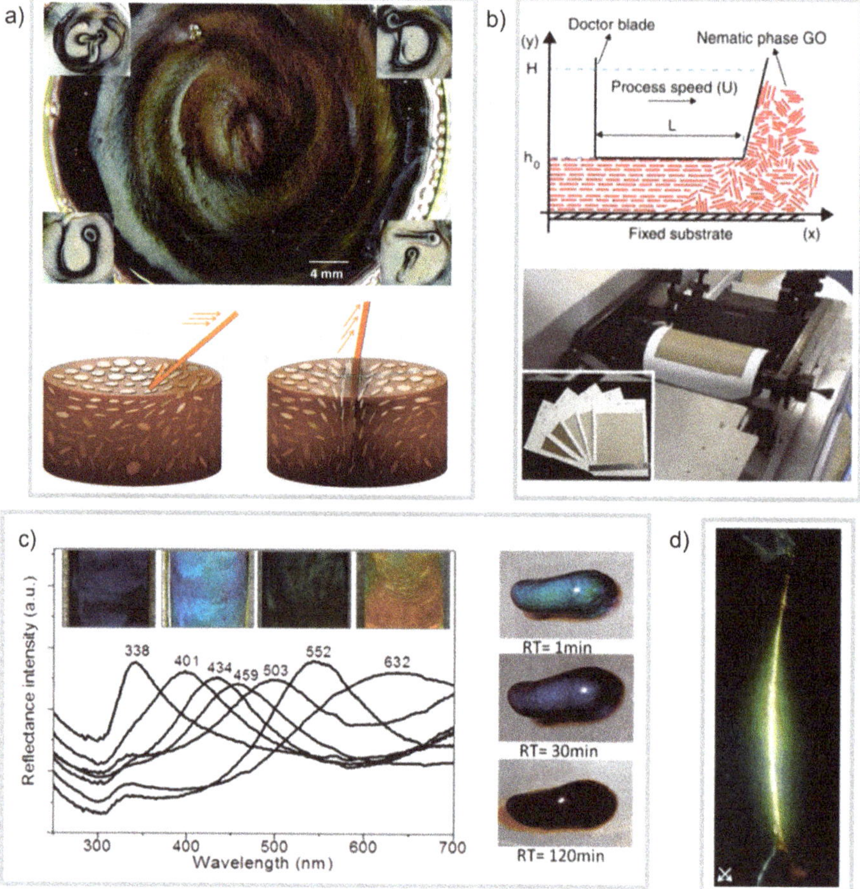

Figure 17. (**a**) A rewritable reflective display based on a graphene oxide nematic liquid crystals. Through application of a rod, writing can be achieved by reorientation of the director without the need for polarizers. Republished with permission of the Royal Society of Chemistry from He et al. [75]; permission conveyed through Copyright Clearance Center, Inc. (**b**) Shear-alignment of nematic GO can be exploited in the fabrication of large-area, highly uniform membranes for desalination applications. Reproduced under a Creative Commons Attribution 4.0 International License from Akbari et al. [69] (**c**) For a sensitive range of flake size and concentrations, Li et al. [73] observed selective reflection of visible light in a gel-like GO LC, with the wavelength of reflective light increasing with time as water evaporated and concentration of GO increased. Reprinted with permission from ref. [73]. Copyright (2014) American Chemical Society. (**d**) This polarized photograph represents the first demonstration of a hand-drawn firer composed of a graphene oxide liquid crystal and poly(acrylic acid). Orientation of the flakes along the length of the fibre leads to a high level of birefringence. Reproduced from ref. [11] with permission from John Wiley and Sons.

10. Conclusions

In the present paper we have reviewed the general properties of lyotropic graphene oxide liquid crystals. We especially discussed some characterization methods of the employed graphene oxide sheets or flakes, the occurrence of liquid crystallinity with respect to flake size and solvent, including a comparison to a simple theoretical description, which was found to only poorly describe the experimental observations. Alignment properties of GO liquid crystals were discussed, as were the effects of added salt and polymers. Only few investigations have been carried out so far concerning the effect of applied electric and magnetic fields. Some of the reported results indicate a controversial and even opposing behaviour, whose underlying possible reasons were considered. Graphene oxide liquid crystals are certainly a very interesting, timely and worthwhile field of investigation, with a potential for the development of fundamental liquid crystal and soft matter understanding, as well as for future applications in the areas of nanotechnology, templating, electro-optic devices, all the way to possibilities in fibre composites or the desalination of sea water.

Author Contributions: A.P.D. and I.D. together conceived of the subject matter and structure for this review. A.P.D. wrote the main text created the figures and sought copyright permissions where reproductions were necessary. I.D. supervised the project and contributed to the writing of the manuscript. A.P.D. prepared and characterized the graphene oxide samples and analysed the data from the experiments.

Funding: This research was funded by The Engineering and Physical Sciences Research Council (EPSRC) of the United Kingdom, grant number EP/L01548X/1.

Acknowledgments: The authors would like to acknowledge Maria Iliut and Aravind Vijayaraghavan for kind provision of the GO used in the original experiments described in Section 7. Furthermore we would like to thank Robert M. Richardson and Andrew Thomas for assistance with the SAXS and XPS experiments, respectively.

Conflicts of Interest: The authors declare no conflict of interest.

References

1. Chandrasekhar, S.; Sadashiva, B.K.; Suresh, K.A. Liquid crystals of disc-like molecules. *Pramana J. Phys.* **1977**, *9*, 471–480. [CrossRef]
2. Lydon, J. Chromonic liquid crystalline phases. *Liq. Cryst.* **2011**, *38*, 1663–1681. [CrossRef]
3. Park, H.S.; Lavrentovich, O.D. Lyotropic Chromonic Liquid Crystals. In *Emerging Applications. In Liquid Crystals Beyond Displays, Li, Q., Ends.*; Wiley: Hoboken, NJ, USA, 2012.
4. Sonin, A.S. Inorganic lyotropic liquid crystals. *J. Mater. Chem.* **1998**, *8*, 2557–2574. [CrossRef]
5. Gabriel, J.C.P.; Sanchez, C.; Davidson, P. Observation of nematic liquid-crystal textures in aqueous gels of smectite clays. *J. Phys. Chem.* **1996**, *100*, 11139–11143. [CrossRef]
6. Davidson, P.; Penisson, C.; Constantin, D.; Gabriel, J.C.P. Isotropic, nematic, and lamellar phases in colloidal suspensions of nanosheets. *Proc. Natl. Acad. Sci. USA* **2018**, *115*, 6662–6667. [CrossRef] [PubMed]
7. Connolly, J.; Van Duijneveldt, J.S.; Klein, S.; Pizzey, C.; Richardson, R.M. Manipulation of modified clay particles in a nematic solvent by a magnetic field. *J. Phys. Condens. Matter* **2007**, *19*, 156103. [CrossRef]
8. Langmuir, I. The Role of Attractive and Repulsive Forces in the Formation of Tactoids, Thixotropic Gels, Protein Crystals and Coacervates. *J. Chem. Phys.* **1938**, *6*, 873–896. [CrossRef]
9. Novoselov, K.S.; Geim, A.K.; Morozov, S.V.; Jiang, D.; Zhang, Y.; Dubonos, S.V.; Grigorieva, I.V.; Firsov, A.A. Electric Field Effect in Atomically Thin Carbon Films. *Science* **2004**, *306*, 666–669. [CrossRef] [PubMed]
10. Proctor, J.E.; Armada, D.M.; Vijayaraghavan, A. *An Introduction to Graphene and Carbon Nanotubes*; CRC Press: Boca Raton, FL, USA, 2017.
11. Kim, J.E.; Han, T.H.; Lee, S.H.; Kim, J.Y.; Ahn, C.W.; Yun, J.M.; Kim, S.O. Graphene oxide liquid crystals. *Angew. Chem. Int. Ed.* **2011**, *50*, 3043–3047. [CrossRef]
12. Xu, Z.; Gao, C. Aqueous liquid crystals of graphene oxide. *ACS Nano* **2011**, *5*, 2908–2915. [CrossRef]
13. Dan, B.; Behabtu, N.; Martinez, A.; Evans, J.S.; Kosynkin, D.V.; Tour, J.M.; Pasquali, M.; Smalyukh, I.I. Liquid crystals of aqueous, giant graphene oxide flakes. *Soft Matter* **2011**, *7*, 11154–11159. [CrossRef]
14. Padmajan Sasikala, S.; Lim, J.; Kim, I.H.; Jung, H.J.; Yun, T.; Han, T.H.; Kim, S.O. Graphene oxide liquid crystals: A frontier 2D soft material for graphene-based functional materials. *Chem. Soc. Rev.* **2018**, *47*, 6013–6045. [CrossRef] [PubMed]
15. Lin, F.; Tong, X.; Wang, Y.; Bao, J.; Wang, Z.M. Graphene oxide liquid crystals: synthesis, phase transition, rheological property, and applications in optoelectronics and display. *Nanoscale Res. Lett.* **2015**, *10*, 1–16. [CrossRef] [PubMed]
16. Liu, Y.; Xu, Z.; Gao, W.; Cheng, Z.; Gao, C. Graphene and Other 2D Colloids: Liquid Crystals and Macroscopic Fibers. *Adv. Mater.* **2017**, *29*, 1606794. [CrossRef] [PubMed]

17. Li, D.; Müller, M.B.; Gilje, S.; Kaner, R.B.; Wallace, G.G. Processable aqueous dispersions of graphene nanosheets. *Nat. Nanotechnol.* **2008**, *3*, 101–105. [CrossRef]
18. Novoselov, K.S.; Fal'Ko, V.I.; Colombo, L.; Gellert, P.R.; Schwab, M.G.; Kim, K. A roadmap for graphene. *Nature* **2012**, *490*, 192–200. [CrossRef] [PubMed]
19. Hummers, W.S.; Offeman, R.E. Preparation of Graphitic Oxide. *J. Am. Chem. Soc.* **1958**, *80*, 1339. [CrossRef]
20. Brodie, B.C., XIII. On the atomic weight of graphite. *Philos. Trans. R. Soc. Lond.* **1859**, *149*, 249–259.
21. Dreyer, D.R.; Park, S.; Bielawski, C.W.; Ruoff, R.S. The chemistry of graphene oxide. *Chem. Soc. Rev.* **2010**, *39*, 228–240. [CrossRef]
22. Marcano, D.C.D.; Kosynkin, D.D.V.; Berlin, J.M.J.; Sinitskii, A.; Sun, Z.; Slesarev, A.; Alemany, L.B.; Lu, W.; Tour, J.M. Improved synthesis of graphene oxide. *ACS Nano* **2010**, *4*, 4806–4814. [CrossRef]
23. Chen, C.; Long, M.; Xia, M.; Zhang, C.; Cai, W. Reduction of graphene oxide by an in-situ photoelectrochemical method in a dye-sensitized solar cell assembly. *Nanoscale Res. Lett.* **2012**, *7*, 1–5. [CrossRef] [PubMed]
24. Rodrigues, A.F.; Newman, L.; Lozano, N.; Mukherjee, S.P.; Fadeel, B.; Bussy, C.; Kostarelos, K. A blueprint for the synthesis and characterisation of thin graphene oxide with controlled lateral dimensions for biomedicine. *2D Mater.* **2018**, *5*, 035020. [CrossRef]
25. Chandrasekhar, S. *Liquid Crystals*; Cambridge University Press: Cambridge, UK, 1992.
26. Castellano, J.A. *Liquid Gold: The Story of Liquid Crystal Displays and the Creation of an Industry*; World Scientific Pub Co Inc.: Singapore, 2005.
27. Petrov, A.G. *The Lyotropic State of Matter: Molecular Physics and Living Matter Physics*; CRC Press: Boca Raton, FL, USA, 1999.
28. Neto, A.M.F.; Salinas, S.R.A.; Press, O.U. *The Physics of Lyotropic Liquid Crystals: Phase Transitions and Structural Properties*; Oxford University Press: Oxford, UK, 2005.
29. Garti, N.; Somasundaran, P.; Mezzenga, R. *Self-Assembled Supramolecular Architectures: Lyotropic Liquid Crystals*; John Wiley & Sons, Inc: Hoboken, NJ, USA, 2012.
30. Dierking, I.; Al-Zangana, S. Lyotropic Liquid Crystal Phases from Anisotropic Nanomaterials. *Nanomaterials* **2017**, *7*, 305. [CrossRef] [PubMed]
31. Behabtu, N.; Lomeda, J.R.; Green, M.J.; Higginbotham, A.L.; Sinitskii, A.; Kosynkin, D.V.; Tsentalovich, D.; Parra-Vasquez, A.N.G.; Schmidt, J.; Kesselman, E.; et al. Spontaneous high-concentration dispersions and liquid crystals of graphene. *Nat. Nanotechnol.* **2010**, *5*, 406–411. [CrossRef] [PubMed]
32. Xu, Z.; Gao, C. Graphene chiral liquid crystals and macroscopic assembled fibres. *Nat. Commun.* **2011**, *2*, 571–579. [CrossRef] [PubMed]
33. Onsager, L. The Effects of Shape on the Interaction of Colloidal Particles. *Ann. N. Y. Acad. Sci.* **1949**, *51*, 627–659. [CrossRef]
34. Frenkel, D. Entropy-driven phase transitions. *Phys. A Stat. Mech. its Appl.* **1999**, *263*, 26–38. [CrossRef]
35. Forsyth, P.A.; Marčelja, S.; Mitchell, D.J.; Ninham, B.W. Onsager transition in hard plate fluid. *J. Chem. Soc. Faraday Trans. 2 Mol. Chem. Phys.* **1977**, *73*, 84–88. [CrossRef]
36. Bates, M.A.; Frenkel, D. Nematic-isotropic transition in polydisperse systems of infinitely thin hard platelets. *J. Chem. Phys.* **1999**, *110*, 6553–6559. [CrossRef]
37. Shen, T.Z.; Hong, S.H.; Song, J.K. Electro-optical switching of graphene oxide liquid crystals with an extremely large Kerr coefficient. *Nat. Mater.* **2014**, *13*, 394–399. [CrossRef]
38. Liu, Z.; Xu, Z.; Hu, X.; Gao, C. Lyotropic liquid crystal of polyacrylonitrile-grafted graphene oxide and its assembled continuous strong nacre-mimetic fibers. *Macromolecules* **2013**, *46*, 6931–6941. [CrossRef]
39. Ahmad, R.T.M.; Hong, S.H.; Shen, T.Z.; Song, J.K. Optimization of particle size for high birefringence and fast switching time in electro-optical switching of graphene oxide dispersions. *Opt. Express* **2015**, *23*, 4435. [CrossRef] [PubMed]
40. Poulin, P.; Jalili, R.; Neri, W.; Nallet, F.; Divoux, T.; Colin, A.; Aboutalebi, S.H.; Wallace, G.; Zakri, C. Superflexibility of graphene oxide. *Proc. Natl. Acad. Sci. USA* **2016**, *113*, 11088–11093. [CrossRef] [PubMed]
41. Paredes, J.I.; Villar-Rodil, S.; Martínez-Alonso, A.; Tascón, J.M.D. Graphene Oxide Dispersions in Organic Solvents. *Langmuir* **2008**, *24*, 10560–10564. [CrossRef] [PubMed]
42. Ahmad, R.T.M.; Hong, S.H.; Shen, T.Z.; Masud, A.R.; Song, J.K. Effect of solvents on the electro-optical switching of graphene oxide dispersions. *Appl. Phys. Lett.* **2016**, *108*, 251903. [CrossRef]
43. Al-Zangana, S.; Iliut, M.; Turner, M.; Vijayaraghavan, A.; Dierking, I. Confinement effects on lyotropic nematic liquid crystal phases of graphene oxide dispersions. *2D Mater.* **2017**, *4*, 041004. [CrossRef]

44. Tkacz, R.; Oldenbourg, R.; Fulcher, A.; Miansari, M.; Majumder, M. Capillary-force-assisted self-assembly (CAS) of highly ordered and anisotropic graphene-based thin films. *J. Phys. Chem. C* **2014**, *118*, 259–267. [CrossRef]

45. Zhao, X.; Xu, Z.; Xie, Y.; Zheng, B.; Kou, L.; Gao, C. Polyelectrolyte-stabilized graphene oxide liquid crystals against salt, pH, and serum. *Langmuir* **2014**, *30*, 3715–3721. [CrossRef]

46. Konkena, B.; Vasudevan, S. Glass, gel, and liquid crystals: Arrested states of graphene oxide aqueous dispersions. *J. Phys. Chem. C* **2014**, *118*, 21706–21713. [CrossRef]

47. Wigner, E. On the Interaction of Electrons in Metals. *Phys. Rev.* **1934**, *46*, 1002–1011. [CrossRef]

48. Sciortino, F.; Tartaglia, P. Glassy colloidal systems. *Adv. Phys.* **2005**, *54*, 471–524. [CrossRef]

49. Bonn, D.; Tanaka, H.; Wegdam, G.; Kellay, H.; Meunier, J. Aging of a colloidal "Wigner" glass. *Europhys. Lett.* **1999**, *45*, 52–57. [CrossRef]

50. Leite Rubim, R.; Abrantes Barros, M.; Missègue, T.; Bougis, K.; Navailles, L.; Nallet, F. Highly confined stacks of graphene oxide sheets in water. *Eur. Phys. J. E* **2018**, *41*, 30. [CrossRef] [PubMed]

51. Tkacz, R.; Oldenbourg, R.; Mehta, S.B.; Miansari, M.; Verma, A.; Majumder, M. PH dependent isotropic to nematic phase transitions in graphene oxide dispersions reveal droplet liquid crystalline phases. *Chem. Commun.* **2014**, *50*, 6668–6671. [CrossRef] [PubMed]

52. Konkena, B.; Vasudevan, S. Understanding aqueous dispersibility of graphene oxide and reduced graphene oxide through p K a measurements. *J. Phys. Chem. Lett.* **2012**, *3*, 867–872. [CrossRef] [PubMed]

53. Shim, Y.H.; Lee, K.E.; Shin, T.J.; Kim, S.O.; Kim, S.Y. Wide concentration liquid crystallinity of graphene oxide aqueous suspensions with interacting polymers. *Mater. Horiz.* **2017**, *4*, 1157–1164. [CrossRef]

54. Shim, Y.H.; Lee, K.E.; Shin, T.J.; Kim, S.O.; Kim, S.Y. Tailored Colloidal Stability and Rheological Properties of Graphene Oxide Liquid Crystals with Polymer-Induced Depletion Attractions. *ACS Nano* **2018**, *12*, 11399–11406. [CrossRef]

55. Band, Y.B. *Light and Matter: Electromagnetism, Optics, Spectroscopy and Lasers*; Wiley: Hoboken, NJ, USA, 2006.

56. Hong, S.H.; Shen, T.Z.; Song, J.K. Electro-optical Characteristics of Aqueous Graphene Oxide Dispersion Depending on Ion Concentration. *J. Phys. Chem. C* **2014**, *118*, 26304–26312. [CrossRef]

57. O'Konski, C.T. Electric properties of macromolecules. v. theory of ionic polarization in polyelectrolytes. *J. Phys. Chem.* **1960**, *64*, 605–619.

58. Dozov, I.; Paineau, E.; Davidson, P.; Antonova, K.; Baravian, C.; Bihannic, I.; Michot, L.J. Electric-Field-Induced Perfect Anti-Nematic Order in Isotropic Aqueous Suspensions of a Natural Beidellite Clay. *J. Phys. Chem. B* **2011**, *115*, 7751–7765. [CrossRef]

59. Chen, Y.; Xu, D.; Wu, S.T.; Yamamoto, S.I.; Haseba, Y. A low voltage and submillisecond-response polymer-stabilized blue phase liquid crystal. *Appl. Phys. Lett.* **2013**, *102*, 141116. [CrossRef]

60. Hisakado, Y.; Kikuchi, H.; Nagamura, T.; Kajiyama, T. Large electro-optic Kerr effect in polymer-stabilized liquid-crystalline blue phases. *Adv. Mater.* **2005**, *17*, 96–98. [CrossRef]

61. Haseba, Y.; Kikuchi, H.; Nagamura, T.; Kajiyama, T. Large electro-optic kerr effect in nanostructured chiral liquid-crystal composites over a wide temperature range. *Adv. Mater.* **2005**, *17*, 2311–2315. [CrossRef]

62. Jiménez, M.L.; Fornasari, L.; Mantegazza, F.; Mourad, M.C.D.; Bellini, T. Electric Birefringence of Dispersions of Platelets. *Langmuir* **2012**, *28*, 251–258. [CrossRef] [PubMed]

63. Ahmad, R.T.M.; Shen, T.Z.; Masud, A.R.; Ekanayaka, T.K.; Lee, B.; Song, J.K. Guided Electro-Optical Switching of Small Graphene Oxide Particles by Larger Ones in Aqueous Dispersion. *Langmuir* **2016**, *32*, 13458–13463. [CrossRef] [PubMed]

64. Shen, T.Z.; Hong, S.H.; Song, J.K. Effect of centrifugal cleaning on the electro-optic response in the preparation of aqueous graphene-oxide dispersions. *Carbon N. Y.* **2014**, *80*, 560–564. [CrossRef]

65. Dimiev, A.M.; Alemany, L.B.; Tour, J.M. Graphene Oxide. Origin of Acidity, Its Instability in Water, and a New Dynamic Structural Model. *ACS Nano* **2013**, *7*, 576–588. [CrossRef] [PubMed]

66. Shen, T.Z.; Hong, S.H.; Guo, J.K.; Song, J.K. Deterioration and recovery of electro-optical performance of aqueous graphene-oxide liquid-crystal cells after prolonged storage. *Carbon N. Y.* **2016**, *105*, 8–13. [CrossRef]

67. Sims, M.T.; Abbott, L.C.; Richardson, R.M.; Goodby, J.W.; Moore, J.N. Considerations in the determination of orientational order parameters from X-ray scattering experiments. *Liq. Cryst.* **2018**, *00*, 1–14. [CrossRef]

68. Diamantopoulou, A.; Glenis, S.; Zolnierkiwicz, G.; Guskos, N.; Likodimos, V. Magnetism in pristine and chemically reduced graphene oxide. *J. Appl. Phys.* **2017**, *121*, 043906. [CrossRef]

69. Akbari, A.; Sheath, P.; Martin, S.T.; Shinde, D.B.; Shaibani, M.; Banerjee, P.C.; Tkacz, R.; Bhattacharyya, D.; Majumder, M. Large-area graphene-based nanofiltration membranes by shear alignment of discotic nematic liquid crystals of graphene oxide. *Nat. Commun.* **2016**, *7*, 1–12. [CrossRef] [PubMed]
70. Mohan, V.B.; tak Lau, K.; Hui, D.; Bhattacharyya, D. Graphene-based materials and their composites: A review on production, applications and product limitations. *Compos. Part B Eng.* **2018**, *142*, 200–220. [CrossRef]
71. Nair, R.R.; Wu, H.A.; Jayaram, P.N.; Grigorieva, I.V.; Geim, A.K. Unimpeded Permeation of Water Through Helium-Leak–Tight Graphene-Based Membranes. *Science* **2012**, *335*, 442–444. [CrossRef] [PubMed]
72. Chen, C.; Yang, Q.H.; Yang, Y.; Lv, W.; Wen, Y.; Hou, P.X.; Wang, M.; Cheng, H.M. Self-assembled free-standing graphite oxide membrane. *Adv. Mater.* **2009**, *21*, 3007–3011. [CrossRef]
73. Li, P.; Wong, M.; Zhang, X.; Yao, H.; Ishige, R.; Takahara, A.; Miyamoto, M.; Nishimura, R.; Sue, H.J. Tunable Lyotropic Photonic Liquid Crystal Based on Graphene Oxide. *ACS Photonics* **2014**, *1*, 79–86. [CrossRef]
74. Xin, G.; Yao, T.; Sun, H.; Scott, S.M.; Shao, D.; Wang, G.; Lian, J. Highly thermally conductive and mechanically strong graphene fibers. *Science* **2015**, *349*, 1083–1087. [CrossRef] [PubMed]
75. He, L.; Ye, J.; Shuai, M.; Zhu, Z.; Zhou, X.; Wang, Y.; Li, Y.; Su, Z.; Zhang, H.; Chen, Y.; et al. Graphene oxide liquid crystals for reflective displays without polarizing optics. *Nanoscale* **2015**, *7*, 1616–1622. [CrossRef]

Review

The Effects of Size and Shape Dispersity on the Phase Behavior of Nanomesogen Lyotropic Liquid Crystals

Fatima Hamade, Sadat Kamal Amit, Mackenzie B. Woods and Virginia A. Davis *

Department of Chemical Engineering, 212 Ross Hall, Auburn University, Auburn, AL 36849, USA;
fzh0014@auburn.edu (F.H.); sza0111@auburn.edu (S.K.A.); mlb0061@auburn.edu (M.B.W.)
* Correspondence: davisva@auburn.edu

Received: 7 July 2020; Accepted: 10 August 2020; Published: 18 August 2020

Abstract: Self-assembly of anisotropic nanomaterials into fluids is a key step in producing bulk, solid materials with controlled architecture and properties. In particular, the ordering of anisotropic nanomaterials in lyotropic liquid crystalline phases facilitates the production of films, fibers, and devices with anisotropic mechanical, thermal, electrical, and photonic properties. While often considered a new area of research, experimental and theoretical studies of nanoscale mesogens date back to the 1920s. Through modern computational, synthesis, and characterization tools, there are new opportunities to design liquid crystalline phases to achieve complex architectures and enable new applications in opto-electronics, multifunctional textiles, and conductive films. This review article provides a brief review of the liquid crystal phase behavior of one dimensional nanocylinders and two dimensional nanoplatelets, a discussion of investigations on the effects of size and shape dispersity on phase behavior, and outlook for exploiting size and shape dispersity in designing materials with controlled architectures.

Keywords: lyotropic liquid crystal; nanomaterial; mesogen; phase behavior

---

1. Introduction

The 130-year history of liquid crystal science is replete with examples of the discovery of new materials leading to advances in both fundamental science and breakthroughs in product development. Liquid crystals are mesophases that possess the fluidity of a liquid but the order of a solid. The building blocks of these phases are anisotropic materials with some rigidity known as mesogens. From 1888 through the 1950s, theoretical knowledge of liquid crystals advanced alongside the field of colloid science through empirical polarized microscopy investigations of naturally occurring materials such as cholesteryl benzoate, vanadium pentoxide, and tobacco mosaic virus (TMV). These observations led to the classification systems and seminal theories that still provide the foundation for modern liquid crystal research. Between the 1960s and 1970s, advances in chemical synthesis led to the production of materials such as 4-cyano-4′-pentylbiphenyl (5CB) and p-polyphenylene terephthalamide (PPTA). Notably, 5CB, a small molecule with a rigid core, is an example of a material that exhibits thermotropic liquid crystalline phase behavior. The discovery of thermotropic liquid crystals' ability to switch from disordered to ordered states with changes in temperature was the foundation for liquid crystal display technology. PPTA is a polymeric macromolecule whose aromatic rings impart rigidity. PPTA in sulfuric acid forms a lyotropic liquid crystal (LLC), a type of liquid crystal where the phase transitions are the result of changes in concentration. Fiber spinning of PPTA/H_2SO_4 LLCs results in the highly aligned, high strength fibers known as Kevlar®. From the 1990s to date, scientists have been developing the ability to synthesize new anisotropic nanomaterials and more scalably isolate natural nanomaterials such as DNA, *fd* virus, and cellulose nanocrystals (CNC). Characterization of these materials' intrinsic properties and the desire to preserve their anisotropic properties in macroscale solid structures, such as films and fibers, have resulted in a "new era for liquid crystal research" [1] that is focused on creating, understanding, and processing LLCs containing nanoscale mesogens. In many regards, however, the current nanotechnology-enabled lyotropic liquid crystal renaissance is based on a rediscovery of the field's roots [2]. In the 1920s and 1930s, decades before scanning electron microscopes (SEMs) became widely used or Norio Taniguchi coined the term "nanotechnology", Zocher reported the LLC phase behavior of aqueous vanadium pentoxide dispersions [3,4]. In 1935, Stanley first reported on the LLC phase formation of tobacco mosaic virus (TMV) [5]. The experiments conducted on TMV inspired the development of Onsager's 1949 theory on LLCs which still serves as the foundation of the field [6]. In modern nomenclature, a nanomaterial is an object with at least one dimension less than 100 nm; both TMV and vanadium pentoxide fulfill this definition, and since they form liquid crystalline phases, they can be termed nanomesogens.

In general, there are two primary motivations for studying nanomesogen LLCs. First, nanocylinders, such as carbon nanotubes (CNTs) and inorganic nanowires, are much closer to fulfilling Onsager's criterion for infinitely long rigid rods than the rod-like polymers that previously constituted the bulk of LLC research. The wide range of potential nanomesogen material sizes, densities, and surface chemistries provide new opportunities for advancing knowledge of the effects of size, shape, sedimentation, and thermodynamic interactions on phase behavior. Second, anisotropic nanomaterials provide new opportunities to produce bulk materials with anisotropic mechanical, electrical, thermal, and optical properties. The alignment in LLC phases facilitates manufacturing dispersions into highly aligned films, fibers, and electrical and optical devices that enable the nanoscale anisotropic properties to be manifested as bulk material properties. Processing nanomesogen LLCs is increasingly considered one of the most efficient methods for achieving aligned, large-scale, solid, assemblies of nanomesogens through manufacturing techniques such as solution spinning, film casting, and additive manufacturing. This combination of purely scientific and applied research opportunities has prompted several reviews highlighting how nanotechnology has enabled a renaissance in liquid crystal science and engineering [1,2,7,8].

The current review is focused on topics not yet widely discussed in the literature: the effects of size and shape dispersity on the phase behavior of lyotropic liquid crystals where one or more nanomaterials serve as the mesogen(s). From a scientific standpoint, many advances in these areas are relatively recent and there is still much to be understood. From an applications viewpoint, these are important topics because they can affect microstructure, processing, and performance properties of manufactured materials. Many nanomesogen syntheses or extraction schemes result in significant size polydispersity. Some, such as the polyol synthesis of silver nanowires, result in a mixture of shapes as well as sizes. Often, the unwanted shapes and sizes are removed through time-consuming separation techniques. However, it can also be desirable to exploit size or shape dispersity to tune the phase behavior, rheological properties, and performance properties such as thermal conductivity.

It is noted that in the interest of focusing the article, many important topics in nanomesogen LLC research are not described. This review only covers systems where a nanomesogen is in a non-mesogenic solvent; the equally important area of adding nanomaterials to known thermotropic or lyotropic liquid crystalline phases is not included. Additionally, the breadth of nanomaterial chemistries available means that detailed understanding requires considering the intersectionalities of size, shape, sedimentation, and non-athermal thermodynamic interactions. While sedimentation and non-athermal interactions are mentioned with respect to different systems, these topics are not discussed in detail. Also, while work from the 1980s and 1990s on nanoclays provide some theoretical insights, the complexities of sedimentation and thermodynamic interactions in dispersions of charged nanomesogens such as nanoclays, cellulose nanocrystals (CNC), MXenes, and other charged nanomesogen systems, are sufficiently complex to warrant a separate review.

2. Examples of Nanomesogens

Mesogens are materials that have the ability to form liquid crystalline phases; thus, nanomesogens are those for which at least one dimension is less than 100 nm. Such materials can be biological, naturally occurring, or synthesized; the main criteria are that they are anisotropic and can be classified as either "semi-flexible" or "rigid". The definition of rigidity varies in the literature but is typically defined as the persistence length being greater than the contour length. While there are many interesting mesogen shapes and sizes, this review is focused on one-dimensional (1D) and two- dimensional (2D) nanomaterial mesogens. In the colloid and polymer literature, 1D materials are referred to as "spherocylinders," "rigid-rods," or "rod-like." However, the term nanorod is commonly used for short solid cylinders and does not include the longer structures called nanowires or the hollow structures called nanotubes. Since the fundamental soft matter physics governing LLC phase behavior is based on their cylindrical shape, these materials will be collectively referred to as nanocylinders. Two-dimensional mesogens are referred to as sheets, plates, and platelets; for this review, the term nanoplatelet will primarily be used. Since zero-dimensional (0D) spherical nanoparticles often result from anisotropic nanomaterial syntheses and are sometimes intentionally added to some dispersions, they are also discussed in this review [9–12].

Investigation of nanomesogens began in the early 1900s, long before the prefix "nano-" became popular. Onsager's seminal 1949 theory for LLC phase behavior was based on experimental observations of tobacco mosaic virus (TMV), a relatively monodisperse, rigid, biological nanocylinder of length $L = 250$ nm and $D = 15$ nm (aspect ratio $L/D = 16.7$) [5,6,13]. An electron micrograph of TMV is shown in Figure 1. Biologically derived nanomesogens, such as *fd* virus, continue to serve as model experimental systems to aid theoretical advances [14]. Other biological nanomaterials such as cellulose nanocrystals (CNC) are more polydisperse, but their phase behavior is of interest for enabling the manufacture of anisotropic solid materials such as photonic films, high strength films, and fibers. Research on CNC liquid crystals began in the 1990s [15,16] and has recently gone through exciting advances spurred by increased material availability [17,18].

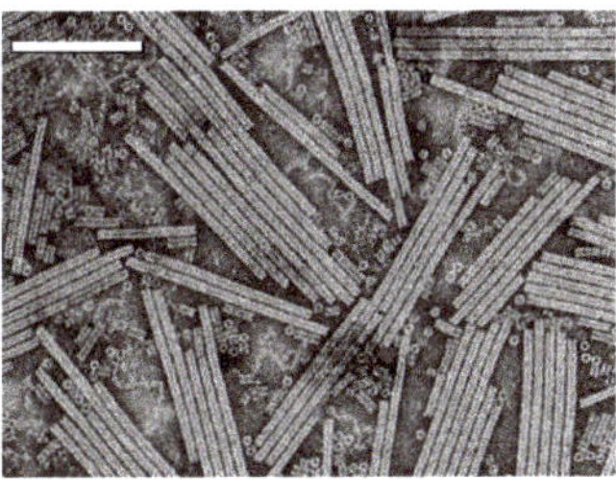

Figure 1. Electron micrograph of tobacco mosaic virus (TMV) on a substrate. Reprinted with permission from Fan, X.Z.; Pomerantseva, E.; Gnerlich, M.; Brown, A.; Gerasopoulos, K.; McCarthy, M.; Culver, J.; Ghodssi, R. Tobacco mosaic virus: A biological building block for micro/nano/biosystems. J. Vac. Sci. Technol. A 2013, 31, 050815 with the permission of AIP Publishing [19].

An even more historical area of LLC research that has been revived is mineral liquid crystals. Zocher first reported on vanadium pentoxide phase behavior in the 1920s. As of the early 2000s, there were very few mineral LLC systems that had been studied and most were nanoclays [20,21]. However, growing interest in nanomaterials such as MXenes and transition metal dichalcogenides (TMDCs) have not only prompted investigation of these materials but sparked a "fresh-look" [22] at the phase diagrams of nanoclays such as laponite and goethite [22–31].

Other actively researched nanomesogens are inorganic metallic semiconducting materials including oxides. Key examples include silver, gold, cadmium selenide (CdSe), germanium, zinc, tungsten, silica, zinc oxide, and manganese oxide. Research in this area has accelerated with improvements in synthesis techniques and interest in bottom-up assembly of electronic devices through additive manufacturing methods. Many of these inorganic 1D nanomesogens can be synthesized with aspect ratios ranging from five to over a thousand [12,32–38]. CdSe can be considered a model material for studying the phase behavior of inorganic 1D nanomesogens because it was one of the first reported systems and can be produced with excellent monodispersity [34]. However, the synthesis of many other materials results in significant size, and sometimes shape, polydispersity. While time-consuming protocols, such as isopycnic centrifugation, can be used to improve the geometric consistency, both polydispersity and sedimentation effects make inorganic nanomesogens a fascinating class of materials for studying phase behavior. In addition, they are of significant interest for use in display technology, wearable electronics, electrochemical devices, and other opto-electronic applications [7].

Some of the most rapid advances in nanomesogen LLCs have been made in the subfield of carbon nanomaterial liquid crystals. It is now known that the sharpness and durability of Damascus swords were partially due to the presence of carbon nanotubes [39] and that small quantities of carbon nanomaterials can be obtained by simple processes such as lighting a candle or exfoliating pencil markings with sticky tape. However, in the year 2000, a mere gram of carbon nanotubes was considered a large quantity, graphene had yet to be "discovered", and many doubted that carbon nanotubes could be dispersed at sufficiently high concentrations to form LLCs. However, during the last twenty years, liquid crystalline phases of pristine single-walled carbon nanotubes (SWNTs), multi-walled carbon nanotubes (MWNTs), and graphene in superacids have been discovered and applied to the production of aligned films and fibers [40–47]. In addition, favorable interactions with biomolecules such as DNA and hyaluronic acid as well as with γ-butyrolactone have resulted in LLC phase formation [48–51]. In addition, oxidized carbon nanomaterials such as graphene oxide (GO) can form liquid crystals in water and other solvents [41,52–61]. The combination of intrinsic properties and range of available solvents make carbon nanomaterials desirable mesogens for a range of applications including high strength materials, electrical conduits, electromagnetic interference shielding, thermal management, electrical devices, and biomedical materials.

3. A Very Brief Overview of Lyotropic Liquid Crystalline Phase Behavior

Based on mesogen orientational and positional arrangement, liquid crystals are categorized into four general classes: nematic, chiral nematic, smectic, or discotic. It should be noted that these arrangements are not static; a key criterion of liquid crystallinity is the ability to flow. In the nematic phase, mesogens possess a long-range orientational order but not positional order (Figure 2a). Their long axes align along a preferred direction, called director, with an order parameter S = (3cos^2q − 1)/2, where q is the angle between the director and long axis of the mesogen. The chiral nematic, or cholesteric, mesophase is similar to the nematic phase but the director orientation twists along a perpendicular axis (Figure 2b). In contrast, in smectic mesophases such as the smectic A mesophase shown in Figure 2c, mesogens have both orientational and positional order where the layers of mesogens can freely slide over one another [7,62]. Two-dimensional platelet mesogens can also form a columnar or discotic phase shown in Figure 2d, where platelets stack into aligned columns [63].

There are two primary frameworks for discussing LLC phase behavior: Onsager theory and Flory theory. Flory theory is perhaps more widely known. Its applicability to polymer systems made it the foundation for research on so-called "rod-like" polymer mesogens, and it is a frequently covered topic in polymer and thermodynamic textbooks. However, many researchers have realized that the high aspect ratios and extreme persistence length to length ratios (28,500 for High Pressure Carbon Monoide [HiPCo] SWNT) better fit Onsager's assumptions of infinite length and high rigidity. Since Onsager theory was based on observations of TMV nanocylinders, it should not be surprising that it provides the foundation for understanding nanomesogen phase behavior. Onsager's original theory was based on the highly idealized scenario of monodisperse spherocylinders interacting through only hard rod repulsion; numerous subsequent refinements have been made into a robust framework that can account for polydispersity and thermodynamic interactions [45,64]. A cylindrical rod is a geometric object of circular cross-section whose length L is much greater than its diameter D ($L \gg D$). If one imagines compressing a cylindrical rod, it will form a disk (or platelet) with $D \gg L$; however, it should be noted that many works use the convention that L is the lateral dimension of the platelet and D is the thickness. Therefore, the general elements of theories for the phase behaviors of 1D cylinders dispersed in a solvent are also applicable to 2D platelets; some caveats are discussed in Section 4.2. For clarity, the majority of discussion in this section is focused on 1D cylinders. In the absence of attractive or repulsive thermodynamic interactions when mesogens are dispersed in a solvent, their motion and alignment are solely dependent on their aspect ratio (L/D) and their concentration. Figure 3 depicts the general behavior. At low concentrations, the cylinders are in a rheologically dilute phase where they can rotate and translate freely. Increasing the concentration initially constrains rotational movement (semi-dilute phase), and then confines both translational and rotational motion to straw-like volumes in the isotropic concentrated phase. Further increases in concentration result in some of the cylinders aligning because the resulting increase in translational entropy is greater than the corresponding decrease in rotational entropy. This occurs at a critical volume fraction ϕ_I and results in a biphasic system where an anisotropic liquid crystalline phase exists in equilibrium with an isotropic phase. Further increases in concentration increase the fraction of the system that is liquid crystalline until at a second critical volume fraction ϕ_{LC} the system becomes fully liquid crystalline [6,7,47,64–67]. For monodisperse cylinders interacting through hard-rod repulsion, the concentration of cylinders in each phase remains constant as the total concentration is increased. However, this is not the case for polydisperse or charged cylinders. In fact, the complexities of polydispersity, sedimentation, and thermodynamic interactions result in complex behaviors such as size fractionation, gelation, multiple forms of phase coexistence, and various forms of aggregation including the formation of crystal solvates. It should be noted that nanomesogen densities range from approximately 1 g/cm^3 for organic materials to 10 g/cm^3 for inorganic nanomaterials, and solvent densities can similarly range from ~0.7 g/cm^3 to over 2 g/cm^3. Even in the same solvent, organic and inorganic mesogens with the same aspect ratio will require significantly different masses of material to reach the same volume fraction. Therefore, while dispersions are often prepared by weighing the nanomaterial, it is important to consider the corresponding volume fraction which can be obtained from:

$$\phi = \frac{m_{nc}/\rho_{nc}}{m_s/\rho_s + m_{nc}/\rho_{nc}} \tag{1}$$

where m is the mass, ρ is the density, and the subscripts nc and s refer to the nanocylinder and solvent, respectively. This relation can be extended for multiple components. It should be noted that the densities of nanomaterials are often assumed from bulk values and can vary between sources; therefore, it is helpful when authors note what densities they used in their calculations.

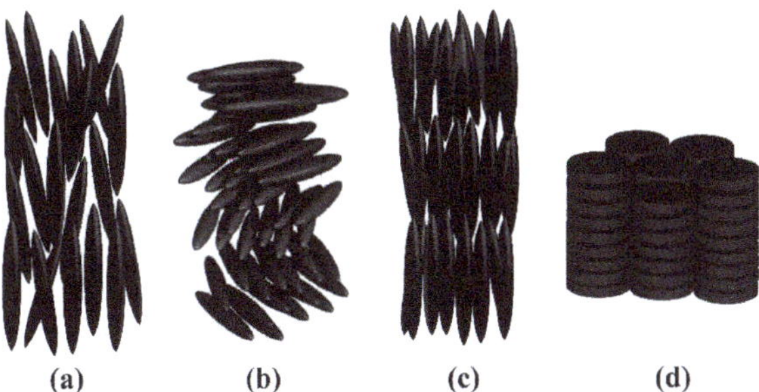

Figure 2. Schematic diagram representing the orientation of nanorods in (**a**) nematic, (**b**) cholesteric, (**c**) smectic A, and (**d**) discotic liquid crystalline phases.

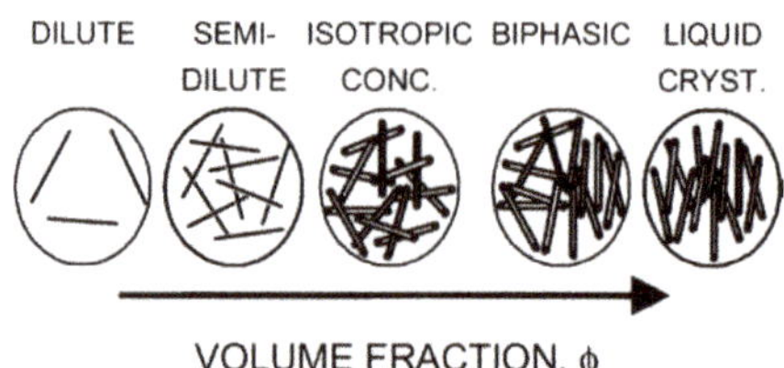

Figure 3. Phase behavior of dispersed rigid rods. Reprinted with permission from Davis, V.A.; Ericson, L.M.; Parra-Vasquez, A.N.G.; Fan, H.; Wang, Y.; Prieto, V.; Longoria, J.A.; Ramesh, S.; Saini, R.K.; Kittrell, C., et al. Phase Behavior and Rheology of SWNTs in Superacids. Macromolecules 2004, 37, 154–160. Copyright (2004) American Chemical Society [47].

For systems that closely obey Onsager's restrictions of monodispersity and only hard rod interactions, the phase boundaries are simply dependent on the aspect ratio, $\phi_I = 3.34/(L/D)$ and $\phi_{LC} = 4.43/(L/D)$. However, for real systems, the phase behavior can be markedly more complex; this can be shown on a phase diagram with volume fraction on the x-axis and solvent quality or a related parameter on the y-axis. Figure 4 shows a representative phase diagram for a polydisperse cylindrical mesogen in a non-athermal solvent [68]. Both polydispersity and differences in mesogen–mesogen versus mesogen–solvent interactions can result in complex behaviors including multiple liquid crystalline phases, aggregation, kinetic arrest (gelation), and aggregates such as crystal solvates which possess the order of a liquid crystal but lack the ability to flow. As described in Section 4, polydispersity can lead to the coexistence of multiple liquid crystal phases. In addition, when the solvent is less favorable the width of the coexistence region can be quite broad, while for more favorable solvents it is a narrow region known as the biphasic chimney. For rod-like polymers, the solvent quality can be improved (mesogen–solvent interactions can be made more favorable) by simply increasing temperature. This equates to decreasing the well-known Flory–Huggins interaction parameter χ. However, due to nanomesogens' typically greater length and rigidity, temperature is often an ineffective method of changing thermodynamic interactions and phase behavior of nanomesogen dispersions. For a given mesogen, the solvent quality can only be varied by changing the solvent itself. This is often achieved by using a mixture of liquids such as DMSO/water for silica nanorods [32] and sulfuric/chlorosulphonic acid for carbon nanotubes [69].

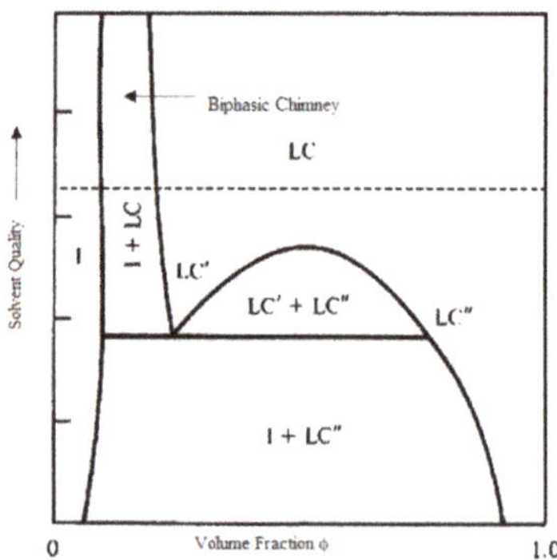

Figure 4. Generic phase diagram showing the biphasic chimney for favorable solvents, a broad coexistence region for less favorable solvents, and the potential for coexistence of multiple liquid crystalline phases. I: Isotropic, LC': one type of liquid crystal phase, LC": another type of liquid crystal phase. Adapted with permission from Zhang, S.J.; Kinloch, I.A.; Windle, A.H. Mesogenicity drives fractionation in lyotropic aqueous suspensions of multiwall carbon nanotubes. Nano Letters 2006, 6, 568–572. Copyright (2006) American Chemical Society [68].

4. Effects of Size Dispersity

Ever since Onsager's initial theory, experimental and computational researchers alike have wrestled with the impacts of size dispersity on phase behavior and the related topics of rheological behavior, processability, and properties of solid materials assembled from lyotropic liquid crystalline phases. Even after decades of refinement, most polymer syntheses result in a range of molecular weight distributions (and therefore length distributions). This affects not only phase behavior in fluid phase processing methods such as fiber spinning, but also chain alignment, crystallization processes, and mechanical properties. In polymers, the molecular weight (size) distribution is measured using techniques such as rheology, gel permeation chromatography, and light scattering. The various moments of the distribution are then represented in terms of the weight average M_w, number average M_n, and z-average M_z, which all reflect different moments of the distribution. Many polymer suppliers provide the polydispersity index (PDI) = M_w/M_n. For nanomaterials, size (as well as shape) polydispersity can be the result of synthesis or sonication, which is widely used to disperse 1D nanomaterials and exfoliate 2D nanomaterials. Nanomaterial sizes can be measured using methods such as atomic force microscopy (AFM), light scattering, and rheology; each of these methods has its advantages and disadvantages. While the most commonly used method is AFM, it rarely reflects the true size distribution impacting phase behavior due to the potential for aggregation in dispersions, biases resulting from sample preparation, larger objects being more visible and more likely to be manually measured, and the time required to count the thousands of objects required to obtain statistically meaningful results instead of the 100 or fewer objects typically counted. Software for automatic measurement can remove some of these issues [70], but as a method that requires drying the sample on a substrate it is never certain AFM truly reflects the size of the nanomaterial as it exists in the dispersion. For example, AFM samples prepared from dilute dispersions may show individual nanomaterials, but they may have a larger size in the dispersion of interest due to bundling or aggregation. Light scattering and rheology are both ensemble methods, but light scattering cannot capture larger objects and issues of optical opacity can be a challenge. In terms of rheology, Ubbeholde type capillary viscometers may produce erroneous results for rods because of flow alignment. Rotational rheometers can be limited by instrument sensitivity and the need for the material to meet stringent criteria such as behaving as Brownian rods [71]. Regardless of the method used, the size distributions can be quantified using similar metrics as the ones used for polymers (e.g., weight average length L_w, number average length L_n).

A more complicated issue is the nature of nanomaterial size dispersity. A mesogen's length to diameter ratio (*L/D*), known as the aspect ratio, is a fundamental determinant of LLC phase behavior. Different batches of a material can have very different aspect ratios. Even for the same batch, the aspect ratio can vary depending on the dispersion preparation method and changes due to aging. For example, length distribution of SWNTs in a dispersion is heavily dependent on sonication power, time, initial length, and stiffness [72]. Similarly, for CNC, length distribution depends on the source (e.g., wood, bacteria, tunicate), hydrolysis time, temperature and types of acid used to hydrolyze the cellulose to obtain CNC [73–75]. Another issue is deviations from a true cylindrical or disk shape. Some 1D nanomaterials, such as cellulose nanocrystals (CNC), have been found to generally obey the physics for dispersed spherocylinders, but they actually have three distinct dimensions and are perhaps more accurately described as lathes. The lateral dimension of some 2D nanomaterials such as the families of graphene-based and MXene materials are not smooth circular disks; this poses uncertainty in accounting for shape irregularity or their ability to wrinkle. Some researchers use more ideal materials such as *fd* virus [14,76] for comparison of theoretical developments with experiments. On the other hand, experimentalists working with polydisperse 1D and 2D systems are often reluctant to compare their results to theory, or if they do, they limit their comparisons to general concepts. For example, many experimental papers use average dimensions to calculate theoretical phase boundaries based on Onsager's original theory and then simply state that any discrepancies are the result of polydispersity and nanomesogen–solvent interactions. The sections below highlight the evolution of understanding the effects of size dispersity on phase behavior. They also describe that while monodispersity is typically viewed as desirable, many researchers are using the phase behavior of polydisperse mesogens to achieve complex phase behaviors that cannot be obtained in monodisperse systems.

4.1. Length Dispersity of 1D Materials

In general, rods of polydisperse lengths result in size fractionation, widening of the biphasic region [45,77,78], changes in the order parameter, and repressed smectic phase formation [77]. Diameter distribution also has some influence, but diameters are typically more uniform and length effects dominate the change in aspect ratio and phase behavior. The issue of length distribution is not new; even Onsager considered the potential impact of polydispersity. He correctly predicted that in a binary mixture of rod lengths, there would be a higher concentration of longer rods in the anisotropic phase. This prediction was confirmed by Oster who investigated mixtures of monodisperse TMV and end-to-end aggregated TMV [79]. Subsequent theoretical and experimental work on a wide variety of systems including 1D nanomesogens has confirmed the existence of size fractionation. In fact, some researchers have exploited this thermodynamically driven separation technique to obtain dispersions with lower polydispersity that could be processed into more uniform solid materials. For example, Honarato-Rios et al. found that by simply leaving aqueous biphasic dispersions of sulfated CNC in a separation funnel, the system would fractionate, and that repeating this process resulted in much lower polydispersity that enabled the production of more uniform optical films [80]. Similarly, Vroege et al. allowed highly polydisperse goethite nanorods to fractionate during sedimentation. Although polydisperse samples are generally deemed incapable of forming smectic liquid crystals, the resulting fractionation enabled the formation of different phases at different heights of the capillary. The nematic phase formed after approximately one week, while further fractionation resulting in the smectic phase did not occur for one month (Figure 5).

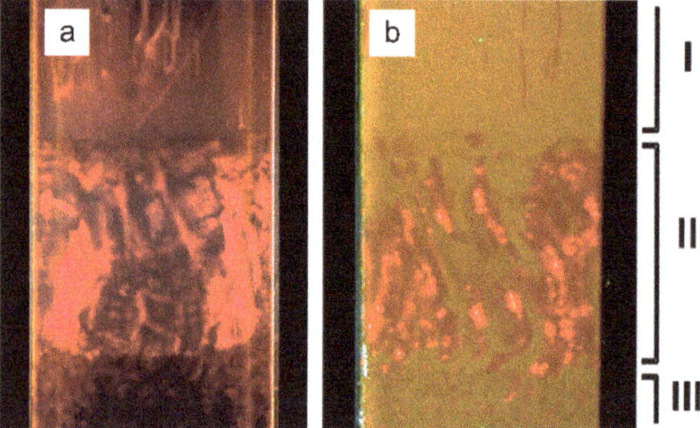

Figure 5. Capillary tube containing phase-separated goethite dispersions (16% volume fraction) displaying three distinctly different regions in addition to an isotropic region higher in the capillary (not shown). Images are taken using (**a**) transmitted light between crossed polarizers: Schlieren patterns typical for a nematic phase are observed in region I. (**b**) Reflected light: bright red Bragg reflections from a smectic phase are seen in region II. The columnar phase (predominant in region III) is not distinguished by special optical characteristics. Reprinted with permission from Vroege, G.J.; Thies-Weesie, D.M.E.; Petukhov, A.V.; Lemaire, B.J.; Davidson, P. Smectic Liquid-Crystalline Order in Suspensions of Highly Polydisperse Goethite Nanorods. Advanced Materials 2006, 18, 2565–2568. Copyright (2006) WILEY-VCH Verlag GmbH & Co. KGaA, Weinheim [23].

The example above also highlights how polydispersity can enable the coexistence of multiple liquid crystalline phases. Other work has also shown the complicated effects that vary with rod length [81]. Both Flory and Onsager based calculations have shown that significant size dispersity can enable the formation of multiple liquid crystalline phases (including multiple nematic phases) without violating the Gibbs phase rule [82–85]. Experimentally, triphasic phenomenon has been observed in bidisperse systems of schizophyllan in water, polydisperse aqueous dispersion of boehmite rods, and bimodal solutions of imogolite [86–88].

Early work on evaluating the effects of size dispersity considered the relatively simple case of binary mixtures of two rods of the same density and diameter, but two different lengths L_b and L_a with m = L_b/L_a > 1 [64,81]. If $\phi_{2,a}$ and $\phi_{2,b}$ are the volume fractions in the dispersion of rods of lengths L_b and L_a, then the overall rod volume fraction is $\phi_2 = \phi_{2,a} + \phi_{2,b}$. Defining $z = \phi_{2,b}/\phi_2$, the weight average length of the rods is:

$$\overline{L} = L_a(1 - z) + L_b z \tag{2}$$

A reduced volume fraction can then be defined as $\phi_r = \phi_2 \overline{L}/d$, where d is the diameter of the rod. This simple approach results in the well-known finding that polydispersity increases the width of the biphasic region by decreasing ϕ_I, and increasing ϕ_{LC} [64]. This effect is increased for greater values of z. It should be noted that this method results in only a small decrease in ϕ_I and a large increase in ϕ_{LC} [64], but experimental investigation of several nanomesogen systems have shown dramatic decreases in ϕ_I and negligible increases in ϕ_{LC}.

Most synthesized systems and systems extracted from larger materials (e.g., CNC, graphene) are not bidisperse but have continuous log-normal, Weibull, or Schultz distributions. For example, while typical commercial SWNT have a relatively narrow range of diameters, SWNT with an average length on the order of 700 nm, can contain a Weibull or log-normal distribution of lengths ranging from 100 nm to several microns. Even though continuous length distributions are common, it was not until the early 2000s that Onsager theory was adapted to account for such distributions. In 2002 and 2003, Speranza and Sollich used a simplified version of Onsager theory obtained by truncating the angular dependence of the excluded volume to calculate the effects of polydispersity on phase behavior. They used a moment free energy method to investigate a unimodal Schulz distribution, a bidisperse distribution, a bimodal mixture of two Schulz distributions, and log-normal distributions with a particular focus on the effects of "fat-tails" in the distribution [89–91]. In 2003, Wensink and Vroege published Onsager based results for Schulz and "fat-tailed" log-normal distributions using a Gaussian trial function ansatz. Both approaches resulted in biphasic regions greater than for monodisperse rods and three-phase I–N–N coexistence regions if the ratio of long and short rod lengths was sufficiently large. Green extended Wensink and Vroege's approach by adding a square well potential to account for inter-rod attraction that could be varied based on the favorability of the dispersion of the solvent, a parameter termed solvent quality [45,69]. This approach enabled the first polydisperse nanomesogen phase diagram where the theoretical phase boundaries matched those obtained by experiment (Figure 6). As previously noted, unlike rod-like polymer liquid crystals, the size and rigidity of many nanomaterials cause temperature to be an ineffective method for tuning nanomaterial–nanomaterial and nanomaterial–solvent interactions and the resulting phase behavior. As a result, the phase diagram can only be obtained by changing the dispersant [32,69].

4.2. Impact of Polydispersity on Phase Behavior of 2D Nanomesogens

As previously noted, 2D nanoplatelets are geometrically equivalent to short rods with lateral dimensions greater than their thicknesses. However, as pointed out by Onsager, applying his theory to infinitely short rods results in errors because of the truncation of the virial equation of state after the second virial term is justifiable for long rods, but not for disks [92]. The phase transitions for plates are often expressed in terms of dimensionless number densities nD^3 which can be related to volume fraction ϕ through $nD^3 = 4\phi D^3/\pi D^2 L$. Applying Onsager theory to uniform disks interacting only through hard repulsion results in $n_{iso}D^3 = 6.8$ and $n_{nem}D^3 = 7.7$ for an aspect ratio $D/L = 10$ and $n_{iso}D^3 = 5.3$ and $n_{nem}D^3 = 6.8$ for infinitely thin disks. However, Monte Carlo simulations using up to the fifth virial coefficient result in markedly lower values $n_{iso}D^3 = 3.81$ and $n_{nem}D^3 = 3.87$ for $D/L = 10$ and $n_{iso}D^3 = 4.04$ and $n_{nem}D^3 = 4.12$ for infinitely thin disks [92].

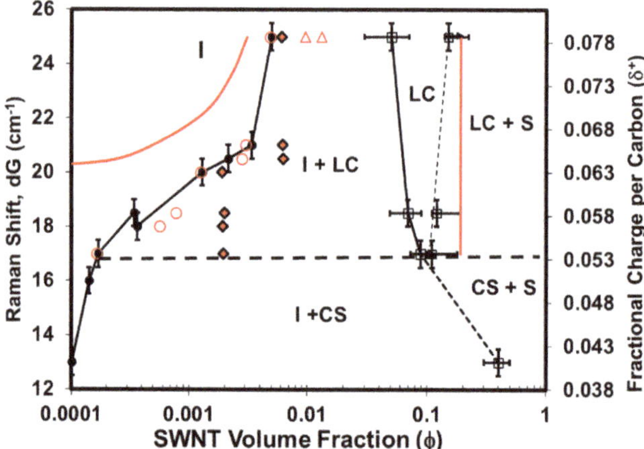

Figure 6. Phase diagram of single-walled carbon nanotubes (SWNTs) in mixtures of varying acid strengths showing isotropic (I), liquid crystal (LC), crystal solvate (CS), and solid (S) phases along with biphasic regions. Solvent quality is quantified by fractional charge per carbon, measured by the shift dG of the Raman G peak of SWNTs (514 nm laser). Onsager predictions for ϕ_I and ϕ_{LC} in a system of monodisperse hard rods are denoted by open red triangles. Black symbols denote experimental results and red symbols refer to model predictions. Circles designate ϕ_I from experiment (filled circles) and model (open circles). Black and red diamonds indicate the initial system concentration before phase separation. Red lines are the model predictions of the isotropic and liquid–crystalline stability limits (cloud curves). Black dotted lines connect the experimental data points to denote phase boundaries. Modified with permission from Davis, V.A.; Parra-Vasquez, A.N.G.; Green, M.J.; Rai, P.K.; Behabtu, N.; Prieto, V.; Booker, R.D.; Schmidt, J.; Kesselman, E.; Zhou, W., et al. True solutions of single-walled carbon nanotubes for assembly into macroscopic materials. Nature Nanotechnology 2009, 4, 830–834. Nature Publishing Group [69].

Phase behavior of real 2D systems is complicated by several factors. While some are nearly uniform in size and have a regular shape, they may not be circular. For example, gibbsite is hexagonal. Sedimentation of these and other inorganic, or primarily inorganic, nanomaterials also complicates phase behaviors. As previously mentioned, materials such as graphene and MXene, which are produced by exfoliation, have high polydispersities in nominal size and highly irregular cross sections. These materials can also wrinkle, further changing their aspect ratio. In their study of aqueous graphene oxide (GO) dispersions, Jalili et al. obtained experimental values for the liquid crystalline phase transitions that were approximately five times higher than the predicted values and attributed this to wrinkling of the sheets [60]. Finally, many 2D nanomaterials such as nanoclays are charged and may exhibit complex phase behaviors as a function of ionic strength including isotropic, nematic, columnar, and gels [22,25,26,30,31,93].

There have been fewer investigations on the effects of polydispersity on the phase behavior of 2D nanomaterials than on 1D nanomaterials. Much of the early work was performed on nanoclays with more recent work on "newer" 2D nanomesogens only beginning to emerge. In general, the volume fraction of polydisperse platelets (ϕ) is

$$\phi \;=\; \frac{3\sqrt{3}}{8}\,\frac{L(1+\sigma)}{D(1+3\sigma^2)}\left(nD^3\right) \tag{3}$$

where <L> is the average sheet thickness, <D> is the average lateral size, σ is the polydispersity defined as the average diameter divided by the standard deviation of the diameter, and n is the number density of sheets per volume (N/V) where the number densities for the isotropic and nematic phases are designated as n_{iso} and n_{nem}, respectively [30,60,92,94,95]. In van der Kooij and Lekkerkerker's study of polymer stabilized gibbsite with $<D>/<L> = 11$, they obtained $n_{iso} <D>^3 = 2.5$ and $n_{nem}<D^3> = 2.7$. They attributed this discrepancy from the theoretical values of $n_{iso} D^3 = 3.81$ and $n_{nem} D^3 = 3.87$ for $D/L = 10$ to three factors: polydispersity, gibbsite being hexagonally shaped, and gibbsite being slightly attractive [92].

Ahmad et al. explored the effects of changes in GO size, with increasing ultra-sonication time and the corresponding impacts on phase behavior. The GO diameter decreased from ~8 μm to ~0.1 μm over 9000 s (2.5 h) of ultrasonication (Figure 7a) [96]. Four different GO sheet sizes were used to investigate the effects of size on the concentrations (in wt%) at which the system became biphasic and liquid crystalline (Figure 7b) [96]. As expected, larger sheet sizes underwent the phase transitions at lower concentrations. Exploration of the effects of electro-optical sheet size on sensitivity showed that 2 wt% of the 0.5 μm GO provided the fastest response time [96].

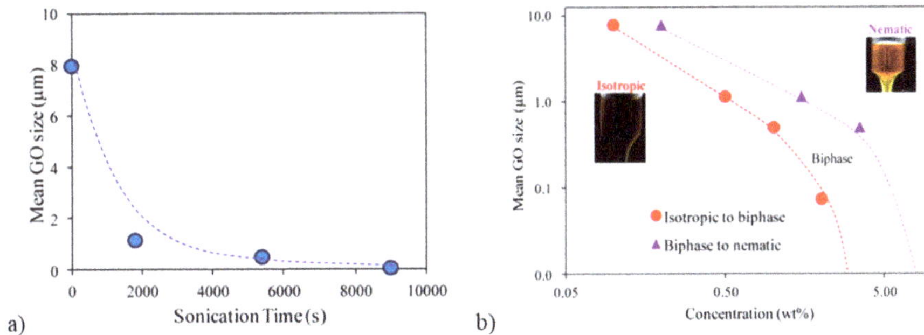

Figure 7. (**a**) Mean graphene oxide (GO) sheet size as a function of sonication time. (**b**) Comparison of isotropic to biphasic and biphasic to nematic phase boundaries of four different mean GO sheet sizes. Adapted and reprinted with permission from Ahmad, R.T.M.; Hong, S.-H.; Shen, T.-Z.; Song, J.-K. Optimization of particle size for high birefringence and fast switching time in electro-optical switching of graphene oxide dispersions. Opt. Express 2015, 23, 4435–4440 with the permission of AIP Publishing [96].

One of the most significant recent works highlighting the advantages of polydisperse systems is Jalili et al.'s exploration of the effects of size on aqueous GO dispersions' phase behavior, rheology, and ability to be wet spun into fibers [60,94]. For their ultra-large graphene oxide sheets with an average aspect ratio of 45,000, the dimensionless number densities were found experimentally to be $n_{iso}D^3 = 2.7$ and $n_{nem}D^3 = 4.3$ [60]. This corresponds to the biphasic region occurring theoretically between 0.05–0.09 mg/mL; however, polarized optical microscopy (POM) results portrayed that the nematic phase begins at 0.25 mg/mL [60]. They demonstrated the formation of liquid crystalline ultra-large GO with $<L> = 37 \pm 23$ μm and $<D> = 0.81$ nm or an aspect ratio of 45,000 [60], which is much larger than the ~1,600 aspect ratio typical for GO. They associated liquid crystalline phase formation with spinnability of dispersions into fibers [94]. They categorized aqueous 2.5 mg/mL GO dispersions with different average sheet sizes in terms of whether they were fully nematic ($\phi_{nem} = 1$), biphasic ($\phi_{nem} < 1$), or isotropic ($\phi_{nem} \sim 0$). Figure 8a shows representative images of sheets from each category [94]. Average sheets size of 3.2 ± 6.9 μm and higher were more suitable for wet spinning due to their nematic microstructure. Interestingly, they found that mixing a small quantity of an isotropic dispersion of large (37 ± 23 μm) GO with an isotropic dispersion of small (0.2 ± 0.15 μm) GO resulted in a nematic liquid crystal (Figure 8b) due to the change in excluded volume. These dispersions were suitable for fiber spinning; this result highlights how intentional use of polydispersity can result in liquid crystalline phase formation and enable the production of aligned solid materials.

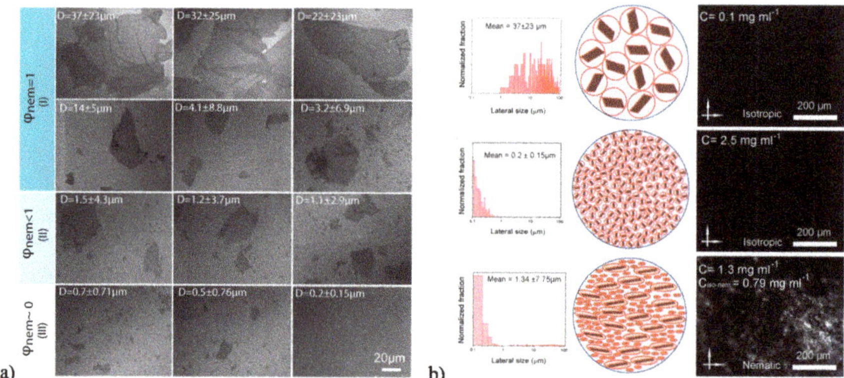

Figure 8. (**a**) SEM images examining the effect of ultra-large GO sheet size (D is lateral dimension) on wet-spinning capability, where (I), (II), and (III), represent spinnable, slightly spinnable, and non-spinnable. (**b**) Isotropic and non-spinnable ultra-large sheet (top) and small sheet GO (middle) can be combined to form a nematic liquid crystalline dispersion suitable for fiber spinning. Modified with permission from Jalili, R.; Aboutalebi, S.H.; Esrafilzadeh, D.; Konstantinov, K.; Razal, J.M.; Moultona, S.E.; Wallace, G.G. Formation and processability of liquid crystalline dispersions of graphene oxide. Material Horizons 2014, 1, 87–91, The Royal Society of Chemistry [94].

4.3. Outlook for Understanding and Exploiting Polydispersity

The few results described in this section highlight that polydispersity has a significant impact on LLC phase behavior and while computer models based on Onsager's original theory show similar findings in many cases, the approach and definition of the distribution function can yield different results. The results also show how polydisperse systems can lead to desirable phase behaviors and facilitate processing methods such as fiber spinning. The current challenge is three-fold: (1) to further refine these theories to account for complex thermodynamic interactions, including effects of ionic strength, (2) be able to account for shape irregularity in 2D systems, and (3) increased control over nanomesogen synthesis and purification processes to enable producing materials with the size distribution needed to yield a desired phase behavior.

5. Effects of Shape Dispersity

While less inherent than nanomaterial size dispersity, shape dispersity is an important topic of increasing interest. Synthesis schemes can result in multiple shapes. For example, inorganic nanocylinder syntheses can result in the formation of conical materials, cubes, or spheres. Shape polydisperse dispersions can be considered as hypercomplex fluids that exhibit emergent properties unachievable by other means [97]. These properties can also be intentionally tuned to stabilize dispersions, modify rheological characteristics, enable multi-component dispersions of materials that are typically produced from different shapes, or achieve desired performance properties in fibers, coatings, and devices. For example, including 1D or 2D nanomesogens in dispersions of 0D nanoparticles can increase the thermal or electrical conductivity of devices by bridging microstructural gaps [35]. However, even when the same material is used, dispersions containing multiple shapes exhibit complex phase behaviors that need to be better understood for their potential to be exploited in engineering applications. This section reviews theoretical and experimental findings for binary mixtures of 0D nanoparticles, 1D nanocylinders, and 2D nanoplatelets. The additional complexities of mixing different types of materials with different densities and thermodynamic interactions of ternary and higher mixtures have been largely unaddressed in the literature but provide an intriguing area for future study.

5.1. Nanocylinders and Nanoparticles (1D–0D)

As well described in the broader colloid literature, polymers, surfactants, and biomolecules are often used to stabilize dispersions by retarding van der Waals attraction either by inducing electrostatic repulsion or steric hindrance. However, this effect is very concentration dependent. At higher concentrations, these materials can actually cause phase separation due to depletion attraction. Depletion forces were first described by Asakura and Oosawa in 1954 [98]. In essence, when small particles of diameter σ are added to a dispersion of larger objects, they are excluded (or depleted) from the gaps between larger objects when $h < \sigma$ where h is the separation distance [99]. This results in the presence of only solvent between the larger objects and an osmotic pressure on the larger objects. The particles can be solid materials or soft materials such as polymers, surfactants, or globular proteins. For example, the globular protein lysozyme ($D < 90$ Angstrom) is highly effective at dispersing nanocylindrical SWNT ($D \sim 1$ nm, $L \sim 500$ nm) in water. This is due to lysozyme's ability to adsorb on the SWNT sidewalls via π–π interactions. However, increasing the concentration of free lysozyme which is not bound to SWNT results in the formation of large SWNT aggregates due to depletion forces driving the SWNT so close together that the van der Waals attraction between them causes large aggregates [100]. However, in cases where there are not strong thermodynamic attractions between 1D colloidal materials, depletion can actually stabilize liquid crystal phase formation by increasing the local concentration above that needed for liquid crystal phase formation. The concept of depletion is a recurring theme in both experimental and computational studies of nanomesogen dispersions. It is noted that while the addition of thin nanocylinders or platelets to dispersions of spheres can result in depletion [99,101,102], this section is primarily focused on the effects of adding spheres on the potential liquid crystalline phase formation in dispersions of nanocylinders.

In 2000, Dogic et al. published results of studying the phase behavior of spherocylinder-sphere (1D–0D) mixtures with Monte Carlo simulations [103]. In their model, each type of material was monodisperse, the spherocylinders were perfectly aligned, the orientational distribution function of the spherocylinders could not be disrupted by the presence of the spheres, and there were only hard interactions. Figure 9 illustrates the possible phases that can form from a nanocylinder–nanoparticle mixture. The miscible nematic phase includes a large excluded volume (gray areas) between the nanocylinders and nanoparticles (Figure 9a), the lamellar phase inhibits motion of the nanoparticles between the spherocylinders (Figure 9b), and the immiscible nematic or demixed system consists of spherocylinder-rich phase and sphere-rich phase (Figure 9c). Using a second virial approximation, the free energy difference between the uniformly mixed and layered states can be expressed as

$$\delta F \;=\; a_1^2\left(S_{11} - 2\frac{a_1}{a_2}S_{12} + \left(\frac{a_1}{a_2}\right)^2 S_{22}\right) \;=\; 0 \tag{4}$$

where S_{11} is the sphere–sphere interaction term, S_{22} is the rod–rod interaction term, S_{12} is the rod–sphere interaction term, and a_1 and a_2 are the amplitudes of fluctuation. The ratio of the amplitudes from Equation (4) is expressed as

$$\frac{a_1}{a_2} \;=\; -\frac{S_{12}(\eta_c, k_c)}{S_{11}(\eta_c, k_c)} \tag{5}$$

where η_c represents a specific total volume fraction and k_c represents a specific wave vector. A positive amplitude ratio is indicative of spheres and spherocylinders comprising the same layer, while a negative amplitude ratio is indicative of their intercalation [103].

Which phase forms depends on four parameters: the spherocylinder aspect ratio $(L/D)_{sc}$, the ratio of the spherocylinder and sphere diameters (D_{sc}/D_{sp}), the total volume fraction of spheres and spherocylinders (η), and the partial volume fraction of spheres (ρ_{sp}) [103]. Greater spherocylinder aspect ratio $(L/D)_{sc}$ increases the tendency for the lamellar phase to form at a lower volume fraction of spheres. Second, if the spherocylinder diameter is greater than the sphere diameter $(D_{sc}/D_{sp} > 1)$, decreasing the sphere diameter increases stability of the lamellar phase. Third, for low aspect ratio spherocylinders and larger diameter spheres (smaller L/D_{sc} and $D_{sc}/D_{sp} < 1$), the spheres promote the formation of a demixed nematic because they cannot fit between the spherocylinders. However, for higher aspect ratio of spherocylinders the transition from the lamellar to demixed phase requires greater sphere diameter. Figure 10 shows two examples resulting from the work of Dogic et al. [103].

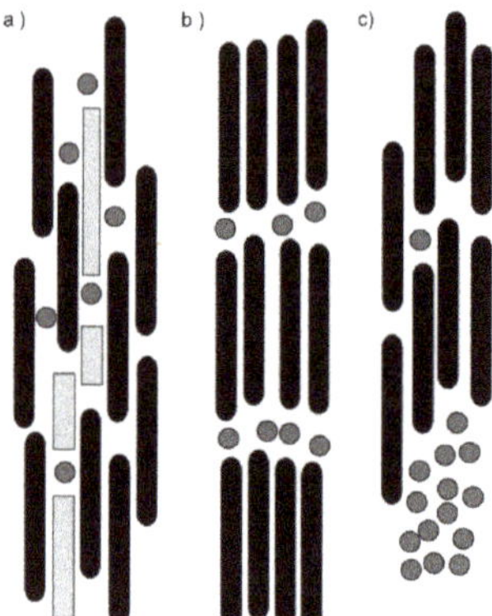

Figure 9. Illustrations of the (**a**) miscible (nematic) phase, (**b**) lamellar (layered) phase, and (**c**) immiscible phase. Reprinted with permission from Dogic, Z.; Frenkel, D.; Fraden, S. Enhanced stability of layered phases in parallel hard spherocylinders due to addition of hard spheres. Physical Review E 2000, 62, 3925–3933. Copyright (2000) by the American Physical Society [103].

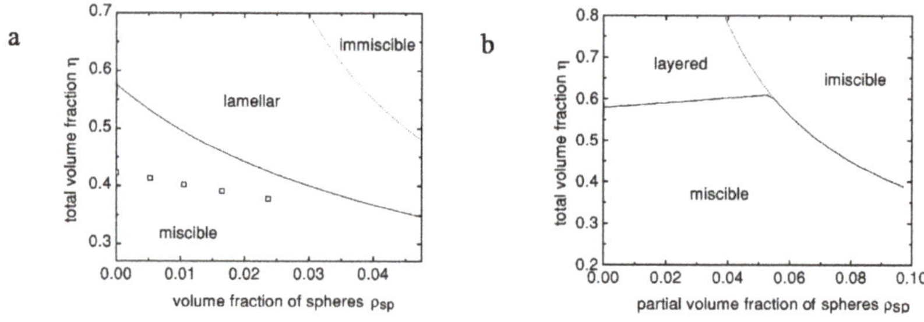

Figure 10. Phase diagrams of mixtures of spherocylinders $(L/D)_{sc} = 20$ and $D_{sc}/D_p = 1$, squares are results of computer simulations at which the lamellar transition is observed (**a**) small or (**b**) example effects of large diameter spheres for $(L/D)_{sc} = 10$ and $D_{sc}/D_{sp} = 0.15$. Reprinted with permission from Dogic, Z.; Frenkel, D.; Fraden, S. Enhanced stability of layered phases in parallel hard spherocylinders due to addition of hard spheres. Physical Review E 2000, 62, 3925–3933. Copyright (2000) by the American Physical Society [103].

Adams et al.'s investigation of the phase behavior of *fd* virus (a nanocylinder) and polymer nanoparticles generally agreed with the results of Dogic et al.'s simulations [103,104]. Adams et al. first explored adding polystyrene spheres of $D_s = 22$ nm at 5 vol%, to nematic dispersions of *fd* virus $(L_{sc} = 880, D_{sc} = 6.6)$ in Tris buffer (20 mg ml^{-1}, 10 mM Tris). This resulted in bulk demixing into a nematic nanocylinder-rich and isotropic nanocylinder-poor phase. The spheres partitioned into the isotropic phase and formed ellipsoidal shaped droplets whose long axes were parallel to the nematic director (rod alignment axis). Sphere concentrations above 700 nM resulted in complete demixing of the spheres into aggregates, which also elongated along the nematic director. The introduction of larger (100 nm) nanoparticles resulted in the complex phase behavior shown in Figure 11 [104]. Similar behavior was observed using a broad range of other diameters, but the required concentrations for each phase varied. The columnar phase was not expected and only present for 60 nm < D_s < 120 nm. In addition, in contrast to theory, the lamellar phase did not form for D_s > 250 nm, and low concentrations of 300 nm nanoparticles resulted in the spheres rapidly associating into chain-like structures.

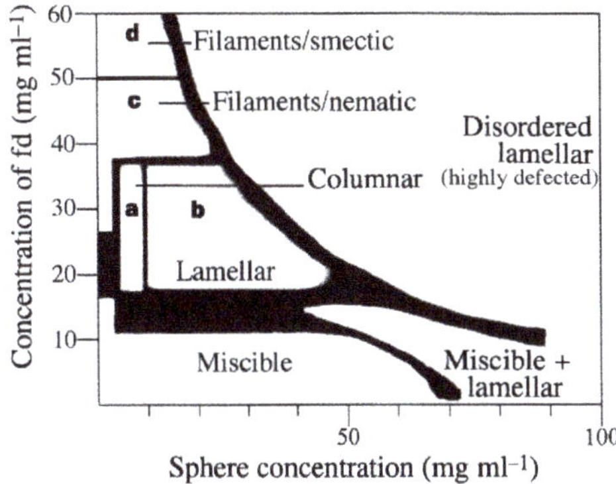

Figure 11. Phase diagram for *fd* virus and 100 nm polystyrene spheres. Reprinted with permission from Springer Nature: Nature, Entropically driven microphase transitions in mixtures of colloidal rods and spheres, Adams, M.; Dogic, Z.; Keller, S.L.; Fraden, Copyright (1998) [104].

TMV has also been used to understand the phase behavior of mixtures of 1D and 0D nanomesogens. For example, Urakami et al. performed Monte Carlo simulations on rods and spheres in an effort to understand changes in aqueous TMV phase behavior with the addition of chondroitin sulfate (Chs) and polyethylene oxide (PEO) where the Chs was treated as large spheres and PEO was treated as small spheres [105]. Their results generally agreed with the previously described results of Dogic et al. [103], as well as experimental studies [105]. Adams and Fraden [76] studied mixtures of TMV and the roughly spherical macromolecules bovine serum albumin (BSA), and polyethylene oxide (PEO) in 50 mM sodium borate buffer (ionic strength 7 mM) with a pH of 8.5. The spherical molecules acted as depletants and the researchers found a complex phase diagram that included isotropic, nematic, lamellar, and crystalline phase. These results also generally agreed with Dogic et al. [103].

More recent work on developing phase diagrams for 1D–0D mixtures has included inorganic nanomaterials. For example, Bakker et al. studied silica nanorods and nanoparticles [9]. Their results indicated the spontaneous formation of a binary smectic liquid crystalline phase, which they notate as Sm_2. This smectic phase was first observed by Adams and Fraden [76] with the *fd* virus. Bakker et al. expanded on this finding to show that stabilization was the result of entropy alone. Both experiments and simulations were used to understand the formation of this binary liquid crystalline phase. Experimentation on the sedimentation–diffusion equilibria of the mixture showed hard particle separation when in equilibrium, where most of the nanocylinders occupied the bottom portion and most of the nanoparticles occupied the top portion. The middle of the mixture consisted of both nanocylinders and nanoparticles, where a Sm_2-phase was ultimately formed as small domains. It was confirmed that entropy alone stabilized the domains when the Sm_2-phase remained stable after having a small AC-electric field applied to it and then removed.

In other work, the Davis group studied liquid crystalline phase behavior and aligned coatings from silver nanomaterials produced via the microwave-assisted polyol method [12,106]. As shown in Figure 12a, the synthesis resulted in a mixture of nanowires with $<L_{sc}>$ = 4.8 μm and $<D_{sc}>$ = 170 nm ($<L/D>_{sc}$ ~ 80) where *sc* indicates spherocylinders. However, while the relative mass and volume fraction of the nanowires was quite high, the number fraction of the nanowires was only 0.05; the remainder of the material was roughly spherical aggregates with D_{sp} = 170 nm where *sp* indicates the roughly spherical aggregates [106]. In ethylene glycol (EG), the system transitioned from isotropic to biphasic at a total silver concentration of $\phi_I = 0.10$ vol% and formed a fully demixed nematic at $\phi_{LC} = 0.42$ vol% total silver. The lyotropic phase behavior was evaluated using a combination of cross-polarized microscopy, rheology, differential scanning calorimetry and changes in the microstructure resulting from applied shear [12,106]. The authors hypothesized that the larger diameter of the spherical aggregates enabled formation of a demixed nematic at an order of magnitude lower concentration than the value of 4.8 vol% predicted by Onsager theory for monodisperse spherocylinders [12]. In cross-polarized optical microscopy images of the dispersions, the nanocylinders tended to sit above the spheres. Shearing and drying the sample increases this segregation and results in aligned silver nanowires sitting on a bed of packed nanoparticles (Figure 12b). The hypothesis of demixed nematic phase formation in the silver system led Green to develop a modified Onsager theory for mixtures of nanoparticles and length-polydisperse nanocylinders [107]. The experimental work by Murali et al. [12] was used to estimate the input data in the modeling parameters. However, the theoretical results were not in agreement with the experimental ones [107]. Instead, Green's model found that in mixtures with high nanocylinder length polydispersity, adding a small amount of monodisperse nanoparticles will increase length fractionation during phase separation, enabling ease of obtaining a near-monodisperse nanocylinder dispersion. Similar to Dogic et al. [103], Green found that D_{sc}/D_{sp} was a critical determinant of phase behavior highlighting that this parameter can be tuned to achieve desired phase behaviors.

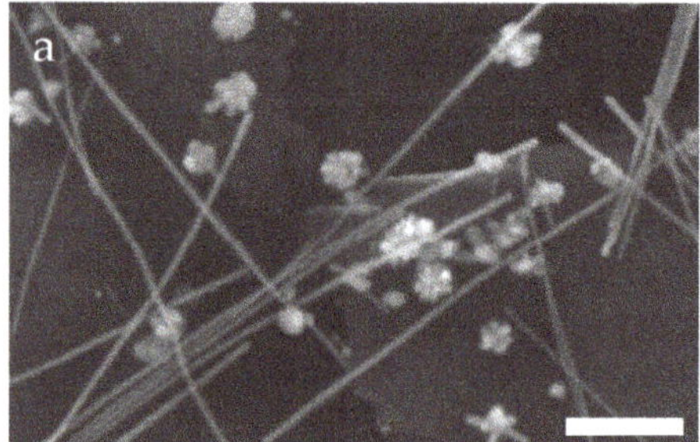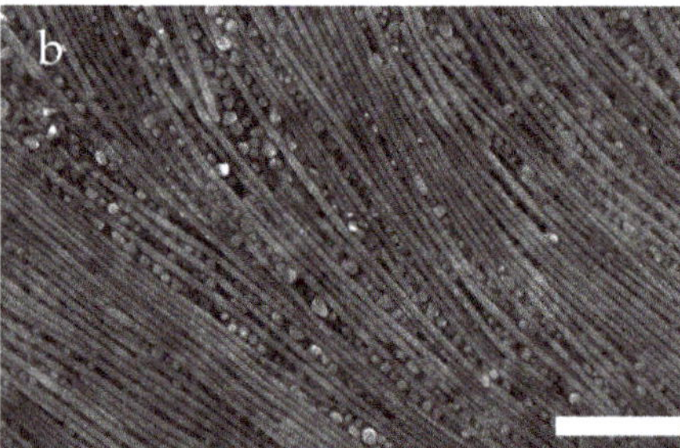

Figure 12. Scanning electron micrographs of silver nanomaterials produced by the microwave assisted polyol method (**a**) nanowires and spherical nanoparticle aggregates, (**b**) after shear the demixed nematic forms more aligned wires on a surface of packed nanoparticles. Scale bars are 1 μm. Adapted with permission from Murali, S.; Xu, T.; Marshall, B.D.; Kayatin, M.J.; Pizarro, K.; Radhakrishnan, V.K.; Nepal, D.; Davis, V.A. Lyotropic liquid crystalline self-assembly in dispersions of silver nanowires and nanoparticles. Langmuir 2010, 26, 11176–11183 American Chemical Society [12].

In summary, fundamental colloidal theories and simulation methods are very beneficial for gaining insights into the phase behavior of nanocylinder–nanoparticle systems. However, even for relatively ideal systems such as *fd* virus and polystyrene nanoparticles, the experimental phase diagrams can be even richer than those predicted from simulations [9,103,105,107–110]. The complexities of shape disperse mesogens are further complicated by length polydispersity and attractive interactions. For inorganic materials, sedimentation resulting from the mesogen density being much greater than that of the solvent likely also plays a role. Further theoretical, computational, and experimental work are needed to fully understand these issues and dictate dispersion microstructure through dispersion formulation.

5.2. Nanoplatelets and Nanoparticles (2D–0D)

The addition of 0D nanoparticles to 2D nanoplatelets is of interest because it can retard gelation [10] and facilitate processing. It has been studied using simulations based on the Parsons–Lee theory [111], density functional theory [112] and the free-volume scaled-particle approach [113]. Oversteegen and

Lekkerkerker [113] used a free-volume theory that employed a hypothetical reservoir in which the particles and platelets mixture was contained and considered the available system volume for the colloidal mixture. The scaled-particle theory was used to determine the free-volume fraction available to the platelets in the systems through $\alpha = e^{\frac{-W}{k_b T}}$ where α is the free-volume fraction, and $k_b T$ is the thermal energy. W is the insertion work determined by using particle size limits which ultimately reduce to finding the excluded volume of a platelet around a particle. The calculations performed by Oversteegen and Lekkerkerker [113] suggested that isotropic–isotropic demixing into particle-rich and platelet-rich regions would occur when the radius ratio of the particles to platelets is $R_{particle}/R_{platelet} < 2.44$. Harnau and Dietrich [112] obtained a similar result of $R_{particle}/R_{platelet} = 2$ using density functional theory (DFT). A geometry-based DFT was later used by de las Heras and Schmidt to study more complex and liquid crystalline phase behavior of colloidal particles and infinitely thin platelet mixtures with varying sizes of materials [114]. Unlike the previously mentioned work, de las Heras and Schmidt accounted for the excluded-volume interactions which play a large role in microstructure and phase behavior at low concentrations. They defined $F = F_{id} + F_{exc}$ where F_{id} is the ideal part and F_{exc} is the excess part. The platelets to particle size ratio was found to be the main parameter controlling the system's phase behavior. When the ratio is $0.2 \leq R_{platelet}/R_{particles} < 10$ only an isotropic–nematic phase separation was seen. When the particle volume fraction is low, the excluded-volume interaction between the platelets drives the isotropic–nematic phase transition. If the particles volume fraction is high, the interaction between the particles and platelets becomes the driving force, and strong demixing between a particle-rich isotropic phase and a platelet-rich nematic phase occurs. When the $R_{particle}/R_{platelet} > 10$, then demixing of two nematic phases occurs.

Chen et al. experimentally explored the phase behavior of mixtures of monolayer zirconium phosphate (ZrP) platelets and silica spheres. The polydispersities of the platelets and spheres were 0.13 and 0.17, respectively. It is known that polydispersity increases the width of the I–N coexistence region for platelets. However, as noted by the authors, the effects of polydispersity in a mixed shape system was an open question which had not yet been resolved by theory or experiment [10]. The authors showed that the presence of the silica spheres retarded gelation, and higher concentrations of spheres reduced the time required to achieve I–N coexistence [10]. Sedimentation resulted in changes in the dispersion over the thirty days of study. They also observed I_1–I_2 and N_1–I_1–I_2 phase coexistence (Figure 13a). They placed their experimental results on a graph showing results from other works (Figure 13b). In conclusion, they noted, "The exact boundary between N–N and I–I demixing still awaits for further experimental and theoretical investigations." [10].

This knowledge gap was partially addressed less than one year later by Aliabadi et al.'s investigation of mixtures of hard platelets and particles [111]. In contrast to much of the previous literature, they assumed the plates had finite thickness L instead of being infinitely thin. Using the Parsons–Lee theory the free energy F can be found from:

$$\frac{\beta F}{V} = \sum_{i = p,s} \rho_i (ln\rho_i - 1 + \sigma[f_i]) + \frac{4 - 3\eta}{8(1 - \eta)^2} \times \sum_{i,j = p,s} \rho_i \rho_j V_{exc}^{ij}[f_i, f_j] \tag{6}$$

where β is the inverse temperature, V is the system volume, ρ is the density of the respective component, η is the volume fraction of the system, $\sigma[f_i]$ is the orientational entropy term, and $V_{exc}^{ij}[f_i, f_j]$ is the excluded volume entropy term. They expressed the relevant size ratios as $d = \sigma/D$ and $k = L/D$ where σ is the spherical particle diameter, D is the lateral diameter of the platelet, and L is the thickness of the platelet. They explored $0.001 < k < 0.1$ to cover the range of thin materials such as laponite as well as thicker platelets. Phase diagrams of the fraction of the spheres η_s versus the fraction of platelets η_p are shown in Figure 14a–c for $\sigma/D = 1.3, 0.2,$ or 0.03 (note this is equivalent to $R_{sphere}/R_{platelet}$ or $R_{particle}/R_{platelet}$ used in other works). While the differences in the axes' scales used make direct visual comparison difficult, noteworthy differences are the existence of a narrow I_1–I_2 phase and N_1–I_1–I_2 phase for $\sigma/D = 1.3$ and N_1–N_2 and N_1–N_2–I phases for $\sigma/D = 0.03$. Figure 14d further highlights the complexity of the phase diagram by plotting d versus k. The authors noted that polydispersity, non-uniform charge distribution, and sedimentation could affect the results for experimental systems but included some experimental points on the phase diagram to help validate their results. In particular, they noted that their findings agreed with Chen et al.'s experimental results on demixing [10,111].

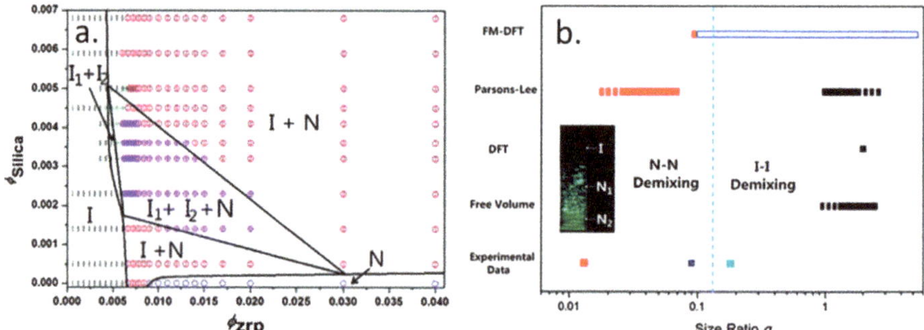

Figure 13. (**a**) Phase diagram as a function of ZrP plate and silica sphere concentrations. (**b**) The distribution of nematic–nematic demixing and isotropic–isotropic demixing found from several different theoretical frameworks as well as experimental data, as a function of size ratio q where $q = R_{particle}/R_{platelet}$. The dark blue squares give the bar size ratio in experiment and the light blue squares gives the effect size ratio accounting for ion clouds. The red square is the experimental observation of I–N$_1$–N$_2$ demixing at $q \sim 0.013$. The long empty rectangle box stands for neither I$_1$–I$_2$ nor the N$_1$–N$_2$ demixing, while the solid red rectangle stands for N$_1$–N$_2$ demixing, and the solid black rectangle stands for the I$_1$–I$_2$ demixing. Dashed rectangles indicate that the limit valve has yet to be determined. The vertical dotted line indicates the possible boundary between I$_1$–I$_2$ and N$_1$–N$_2$ demixing. Reproduced with permission from Chen, M.; Li, H.; Chen, Y.; Mejia, A.; Wang, X.; Cheng, Z. Observation of isotropic–isotropic demixing in colloidal platelet-sphere mixtures. Soft Matter 2015, 11, The Royal Society of Chemistry [10].

Another issue in nanoplatelet–nanoparticle phase behavior briefly explored by Chen et al. [10] is temporal changes resulting from sedimentation. This was studied in more detail by Kleshchaok et al. for gibbsite nanoplatelets and silica nanoparticles [115]. The gibbsite was D_{plate} = 232.5 nm, L_{plate} = 8.4 nm and the silica had diameter $D_{particle}$ = 16.8 nm. The particles and platelets had the same charge sign to induce strong depletion attraction between them. In previous work, the authors had shown that this system did not form a liquid crystalline phase on short time scales [24]. Samples of either 4 vol% or 8 vol% gibbsite and ranging through 2, 3, 5, 7, and 8 vol% of silica were prepared and left to sediment for a year. The results are shown in Figure 15. A columnar phase formed after only one month in the samples with <7 vol% silica particles. The samples with ≥7 vol% silica particles had no columnar phase formed after one year of sedimentation. The size of the columnar phase region was dependent on particle volume fraction; it decreased with increasing particle concentration. Sui addressed the issue of sedimentation in platelet–particle colloidal mixtures from a theoretical perspective and used a minimal energy model to further study the sedimentation dynamics of particle–platelet colloidal mixtures, specifically the shape stratification that occurs in nematic phases of the particle–platelet mixtures [116]. Similar to the results of previous work, Sui found that both the volume fraction of particles and the size ratio of the two types of mesogens have large effects on the nematic stratification structure. When the size ratio is $R_{particle}/R_{platelet} \leq 0.3$, the nematic platelet-rich regions falls to the bottom of the sedimented mixture; however, when $R_{particle}/R_{platelet} \geq 0.4$, the nematic platelet-rich region floats on top as a result of excluded volume interactions that push the platelets upward above sedimenting particles. Between these two values, which the phase forms depend on the concentration of spheres [116].

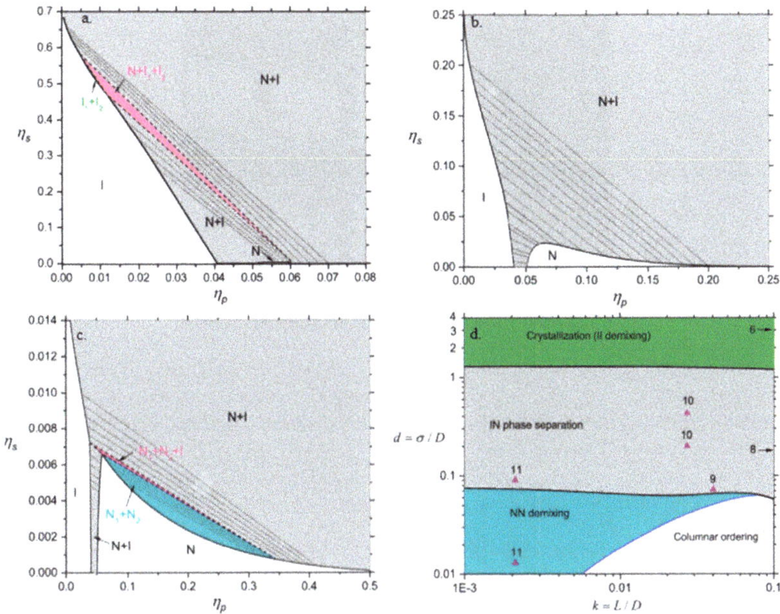

Figure 14. Effects of spheres on the phase behavior of platelets. η_s = fraction of spheres, η_p = fraction of platelets, σ = sphere diameter, D = platelet lateral diameter. In (a–c), $k = L/D = 0.01$ and the coexisting isotropic (nematic) phases are labeled as I_1 and I_2 (N_1 and N_2). The two-phase and three-phase regions are indicated by gray (I–N), green (I_1–I_2), cyan (N_1–N_2), and pink (I–N_1–N_2 and I_1–I_2–N). (**a**) σ/D = 1.3, (**b**) σ/D = 0.2, and (**c**) σ/D = 0.03. (**d**) Phase diagrams of sphere-plate mixtures shown in the diameter ratio–aspect ratio (σ/D–L/D) plane. Regions of the phase diagrams are coloured differently: (1) The phase diagram is dominated by strong fractionation and reentrance of isotropic–nematic phase transition (gray), (2) the region of isotropic–nematic and isotropic–crystal transition, where the isotropic–isotropic demixing is not stable (green), (3) the region of nematic–nematic demixing and isotropic–nematic transition (cyan) and (4) those systems where the nematic–nematic demixing is unlikely to occur and is replaced by isotropic–columnar transition (white). The molecular parameters of some experimental systems are highlighted by triangle symbols and the numbers refer to their references. Modified from Aliabadi, R.; Moradi, R.; Varga, S. Tracking three-phase coexistences in binary mixtures of hard plates and spheres. J. Chem. Phys. 2016, 144, 074902, with the permission of AIP Publishing [111].

Understanding of the phase behavior of 2D–0D mixtures has improved significantly in the last five years. While there are some discrepancies based on the approach and whether the platelets are assumed to have finite thickness or be infinitely thin, there is general agreements that the ratio of the particle to platelet diameter is a key determinant of phase behavior. In addition, the absolute and relative concentrations have an influence. A key issue with experimental studies is the time scale of investigation since changes can occur over a year, and perhaps even longer. In addition, there are still many unanswered questions about the role of polydispersity and inter-particle interactions in both theoretical and experimental studies. Continuing to refine the understanding of these systems will facilitate controlled microstructure formation in dispersions that are suitable for coatings with desired microstructure and properties.

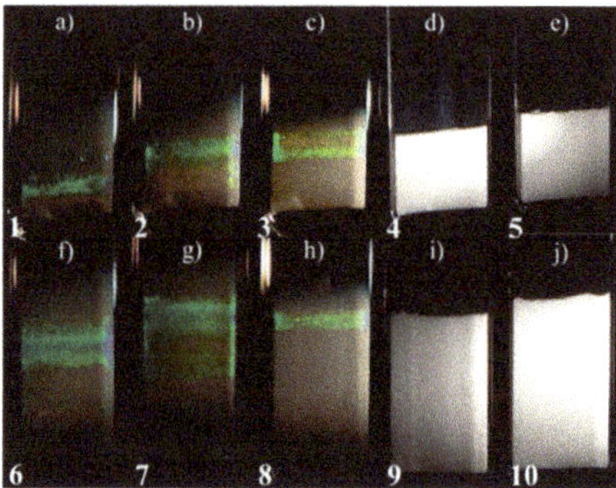

Figure 15. Samples of gibbsite/silica suspensions after one year of sedimentation. Samples (**a–e**) are 4 vol% gibbsite and 2, 3, 5, 7, and 8 vol% of silica particles, respectively. Samples (**f–j**) have 8 vol% gibbsite and 2, 3, 5, 7, and 8 vol% of silica particles, respectively. Reprinted with permission from Kleshchanok, D.; Meijer, J.-M.; Petukhov, A.V.; Portale, G.; Lekkerkerker, H.N.W. Sedimentation and depletion attraction directing glass and liquid crystal formation in aqueous platelet/sphere mixtures. *Soft Matter* 2012, *8*, 191–197, The Royal Society of Chemistry.

5.3. Nanoplatelets and Nanocylinders (2D–1D)

There is growing interest in using systems such as CNT/graphene [117], CNT/MXene [118,119], silver nanowire/graphene [120], silver nanowire/MXene [121,122] and CNC/GO [123,124], and CNC/MXene [125] to create multifunctional materials. For example, Jalili et al. highlighted that combining liquid crystals of ultra-large GO in organic solvents with SWNT enabled multifunctional papers that had higher moduli than those containing only GO. This was due to the SWNT being able to bridge the gaps between the GO platelets [56]. In spite of this interest, the phase behavior and microstructure of mixtures of nanoplatelets and nanocylinders have been largely unaddressed.

Based on simulations, 2D–1D mixtures can form a biaxial phase (two directors oriented perpendicular to each other) which is not found in the pure components [126]. However, van der Kooij and Lekkerkerker found the biaxial phase was unstable in their investigation of boehmite nanocylinders with $L/D = 10$ and gibbsite nanoplatelets with $L/D = 15$. Up to a total volume fraction of 0.3, macroscopic phase separation was complete after 24 h without biaxial phase formation. At higher concentrations of $0.3 \leq \phi_{total} \leq 0.5$, they found the very rich phase behavior shown in Figure 16. The macroscopic phase separation took 2 weeks, but their reproducibility was suggestive of thermodynamic equilibrium. The formation of up to five phases shows that due to its polydispersity, the system should not simply be viewed as a two-component system; for a system of two mesogens and the solvent, the Gibbs phase rule would have been violated. The authors noted that it is difficult to predict the phase behavior of cylinder-platelet systems due to the entropic contributions from orientation, excluded volume, and mixing [127].

Woolston and Duijneveldt found that combining montmorillonite (MMT) nanoplatelets with sepiolite (Sep) nanocylinders also showed complex phase behavior [128]. The aqueous MMT alone formed a gel, and aqueous Sep formed a nematic, but the addition of Sep to MMT could result in three-phase coexistence where the bottom layer consisted of a nematic Sep cylinder-rich phase, the middle layer consisted of an MMT platelet-rich phase, and the top layer was a dilute isotropic layer [128]. For all MMT concentrations above 1 vol%, however, the system gelled and the MMT concentration required for gelation decreased as the Sep concentration was increased to 4 vol%. The authors highlighted that their findings could be used in dispersion formulations to strengthen or weaken a gel phase; however, the effects of sepiolite on MMT gel strength were not rheologically quantified.

In contrast, ten Brinke et al. focused on the rheological properties of clay dispersions in a two-part work investigating platelets (gibbsite), cylinders (boehmite), and laths (hectorite) individually and then in binary mixtures [28,29]. Laths are distinct from platelets as they allow three different dimensions for the particle axis. The rheological behavior of the materials was characterized by oscillatory, transient, and steady shear flow pre-treated with a pre-shear and recovery protocol and modeled via a simple viscoelastic model [29]. Interestingly, these materials behave as elasto-viscous high viscosity solids at low strains and shear-thinning low viscosity liquids at high strains. However, there is a complex intermediate range defined as a "yield space" that accounts for sample microstructure of the sample as well as stress–strain–time history [29]. The mixtures generally demonstrated similar behavior to the pure components. In addition, the rheological properties were greater in magnitude when a small amount (0.25 wt%) of the second component is added to the first component in the dispersion. The shear storage and viscous moduli, effective viscosity, and yield stress were larger for the mixtures of the laths and platelets than the laths and cylinders [29]. The authors highlighted that a detailed model for explaining their results does not yet exist [29]. However, their results provide an important baseline and motivation for understanding 2D–1D systems with the goal of being able to formulate dispersions that possess both desirable microstructures and rheological properties suitable for processing solid coatings, fibers, or objects using a given manufacturing technique.

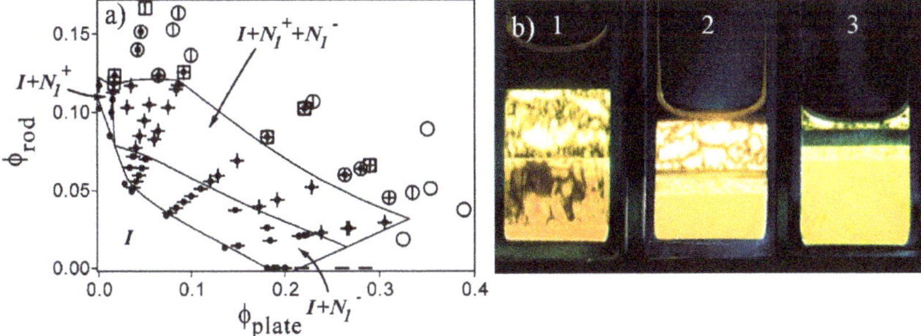

Figure 16. (**a**) Phase diagram for boehmite/gibbsite (nanocylinder/nanoplatelet). N^+ indicates a nematic phase of predominantly nanocylinders, and N^- indicates a nematic phase of predominantly nanoplatelets. Filled circles indicate isotropic, vertical lines N_1^+, dashes N_1^-, open squares N_1^+, open circles N_2^-. (**b**) Polarized light images of concentrated dispersions of concentrations (boehmite, gibbsite) and top to bottom. (1) (0.06, 0.26) $I + N_1^+ + N_1^- + N_2^-$, (2) (0.10, 0.22) $I + N_1^+ + N_2^+ N_1^- + N_2^-$, (3) (0.07, 0.29) $N_1^+ + N_2^+ N_1^- + N_2^-$. Reprinted with permission from van der Kooij, F.M.; Lekkerkerker, H.N.W. Liquid-crystalline phase behavior of a colloidal rod-plate mixture. Phys. Rev. Lett. 2000, 84, 781–784. Copyright 2000 by the American Physical Society [127].

5.4. Outlook for Understanding and Exploiting Shape Dispersity

Dispersions of nanomesogens of different shapes, and often different chemical compositions, are increasingly being studied to achieve desired thermal, mechanical, or electrical properties. They are also being studied to achieve the rheological properties required for processing by a given manufacturing method. Fundamental study of the phase behavior of shape disperse systems will lead to better understanding of both rheological properties and the ability to achieve solid materials with desirable microstructures. The behavior of 1D–0D systems is generally understood, particularly for monodisperse athermal systems. However, practical systems often also include size dispersity and thermodynamic interactions that further complicate phase behavior. Therefore, a concerted experimental and theoretical effort is needed. This is particularly necessary for 1D–2D and 2D–0D systems.

6. Conclusions and Future Directions

The last twenty years have resulted in a significant shift in lyotropic liquid crystalline science from focusing almost exclusively on polymeric mesogens to focusing on nanomesogens. Much has been achieved with respect to characterizing the phase behavior of individual systems, and computational studies have enabled significant advances in understanding the profound effects of size and shape dispersity on liquid crystalline phase behavior. While traditionally viewed as an annoyance, recent work suggests that polydispersity actually represents an opportunity to achieve a broader range of microstructures. There has been significant progress in evaluating size and shape dispersity separately, but much work needs to be done to understand real systems consisting of non-athermal interactions, size polydispersity, and shape polydispersity. This review provided highlights of experimental and theoretical work related to the phase behavior of polydisperse nanocylinders and nanoplatelets as well as binary mixtures of nanocylinder, nanoplatelets, and nanoparticles. A remarkable theme throughout the literature is the amazing robustness of the Onsager framework and how the initial investigations of TMV and vanadium pentoxide started a field that is still being explored today.

To date, traditional methods such as cross-polarized microscopy, rheology, and small angle x-ray scattering (SAXS) have been used to characterize nanomesogen systems. While these methods will continue to play a critical role in studying nanomesogen phase behavior, future advances in evaluating shape, size, and material disperse systems will be facilitated by the ongoing evolution of manufacturing technologies and characterization methods. Advanced scattering methods such as grazing incidence small angle x-ray and neutron scattering (GISAXS and GISANS) promise new insights into microstructure. Integrated spectroscopic and morphological methods such as Raman spectroscopy–atomic force microscopy (Raman–AFM) will enable detailed insights into how chemistry affects ordering in multicomponent systems. As new methodologies for dispersion processing such as ink jet printing, direct ink writing, and aerosol jet printing continue to evolve alongside traditional methods of film casting and solution spinning, the intersection of phase behavior and rheological properties need to be further studied to enable the production of materials with controlled microstructures. Phase behavior affects rheological properties, and quiescent dispersion microstructures can be deformed by both processing shear and solvent removal during solidification. Therefore, methods that examine the evolution of microstructure under shear such as rheo-optics, rheo-SALS (small angle light scattering), rheo-SAXS (small angle X-ray scattering), and rheo-SANS (small angle neutron scattering) will be critical to advancing both the fundamental science and applications of size and shape disperse lyotropic liquid crystals.

Increasing interest in multifunctional materials for electronics, energy storage, optics, sensors, mechanical structures, wearables, and biomedical devices will continue to drive research on both individual nanomaterials and mixtures of nanomaterials. Many of these will benefit from the anisotropic properties that can be achieved by exploiting lyotropic liquid crystalline phase behavior. Increased comprehension of how to tailor microstructure through size and shape dispersity is needed to enable the rational design of dispersions instead of the current trial and error approach. While significant advances have been made in the last ten years, the current models largely neglect the complexities of real systems that possess size polydispersity, shape polydispersity, and can have significant electrostatic or van der Waals interactions. In addition, the phase behavior of dispersions containing more than two components has yet to be addressed. Significant advancements and progress toward functional materials assembled from complex dispersions of nanomesogen building blocks will require ongoing theoretical advances and close collaboration between the theoretical and experimental soft matter communities.

Author Contributions: Conceptualization, V.A.D.; Writing–Original Draft Preparation, F.H., M.B.W., S.K.A., V.A.D.; Writing–Review and Editing, F.H., M.B.W., S.K.A., V.A.D.; Funding Acquisition, V.A.D. All authors have read and agreed to the published version of the manuscript.

Funding: The authors acknowledge funding from the US National Science Foundation Award (Grant CBET-2005413).

Conflicts of Interest: The authors declare no conflict of interest.

References

1. Lagerwall, J.P.F.; Scalia, G. A new era for liquid crystal research: Applications of liquid crystals in soft matter nano-, bio- and microtechnology. *Curr. Appl. Phys.* **2012**, *12*, 1387–1412. [CrossRef]
2. Lekkerkerker, H.N.W.; Vroege, G.J. Liquid crystal phase transitions in suspensions of mineral colloids: new life from old roots. *Philos. Trans. R. Soc. A Math. Phys. Eng. Sci.* **2013**, *371*, 20120263. [CrossRef] [PubMed]
3. Livage, J.; Pelletier, O.; Davidson, P. Vanadium pentoxide sol and gel mesophases. *J. Sol-Gel Sci. Technol.* **2000**, *19*, 275–278. [CrossRef]
4. Sonin, A.S. Inorganic lyotropic liquid crystals. *J. Mater. Chem.* **1998**, *8*, 2557–2574. [CrossRef]
5. Stanley, W.M. Isolation of a crystalline protein possessing the properties of tobacco mosaic virus. *Science* **1935**, *81*, 644. [CrossRef]
6. Onsager, L. The effects of shape on the interaction of colloidal particles. *Ann. N.Y. Acad. Sci.* **1949**, *51*, 627–659. [CrossRef]
7. Davis, V.A. Liquid crystalline assembly of nanocylinders. *J. Mater. Res.* **2011**, *26*, 140–153. [CrossRef]
8. Solomon, M.J.; Spicer, P.T. Microstructural regimes of colloidal rod suspensions, gels, and glasses. *Soft Matter* **2010**, *6*, 1391–1400. [CrossRef]
9. Bakker, H.E.; Dussi, S.; Droste, B.L.; Besseling, T.H.; Kennedy, C.L.; Wiegant, E.I.; Liu, B.; Imhof, A.; Dijkstra, M.; van Blaaderen, A. Phase diagram of binary colloidal rod-sphere mixtures from a 3D real-space analysis of sedimentation–diffusion equilibria. *Soft Matter* **2016**, *12*, 9238–9245. [CrossRef]
10. Chen, M.; Li, H.; Chen, Y.; Mejia, A.; Wang, X.; Cheng, Z. Observation of isotropic-isotropic demixing in colloidal platelet-sphere mixtures. *Soft Matter* **2015**, *11*, 5775–5779. [CrossRef]
11. Kleshchanok, D.; Holmqvist, P.; Meijer, J.-M.; Lekkerkerker, H.N.W. Lyotropic Smectic B phase formed in suspensions of charged colloidal platelets. *J. Am. Chem. Soc.* **2012**, *134*, 5985–5990. [CrossRef] [PubMed]
12. Murali, S.; Xu, T.; Marshall, B.D.; Kayatin, M.J.; Pizarro, K.; Radhakrishnan, V.K.; Nepal, D.; Davis, V.A. Lyotropic liquid crystalline self-assembly in dispersions of silver nanowires and nanoparticles. *Langmuir* **2010**, *26*, 11176–11183. [CrossRef] [PubMed]
13. Lauffer, M.A. The size and shape of tobacco mosaic virus particles. *J. Am. Chem. Soc.* **1944**, *66*, 1188–1194. [CrossRef]
14. Tang, J.; Fraden, S. Isotropic-cholesteric phase transition in colloidal suspensions of filamentous bacteriophage fd. *Liq. Cryst.* **1995**, *19*, 459–467. [CrossRef]
15. Revol, J.F.; Bradford, H.; Giasson, J.; Marchessault, R.H.; Gray, D.G. Helicoidal self-ordering of cellulose microfibrils in aqueous suspension. *Int. J. Biol. Macromol.* **1992**, *14*, 170–172. [CrossRef]
16. Revol, J.F.; Godbout, L.; Dong, X.M.; Gray, D.G.; Chanzy, H.; Maret, G. Chiral nematic suspensions of cellulose crystallites—Phase-separation and magnetic-field orientation. *Liq. Cryst.* **1994**, *16*, 127–134. [CrossRef]
17. Reid, M.S.; Villalobos, M.; Cranston, E.D. Benchmarking cellulose nanocrystals: From the laboratory to industrial production. *Langmuir* **2017**, *33*, 1583–1598. [CrossRef]
18. Schütz, C.; Bruckner, J.R.; Honorato-Rios, C.; Tosheva, Z.; Anyfantakis, M.; Lagerwall, J.P. From equilibrium liquid crystal formation and kinetic arrest to photonic bandgap films using suspensions of cellulose nanocrystals. *Crystals* **2020**, *10*, 199. [CrossRef]
19. Fan, X.Z.; Pomerantseva, E.; Gnerlich, M.; Brown, A.; Gerasopoulos, K.; McCarthy, M.; Culver, J.; Ghodssi, R. Tobacco mosaic virus: A biological building block for micro/nano/biosystems. *J. Vac. Sci. Technol. A* **2013**, *31*, 050815. [CrossRef]
20. Gabriel, J.C.; Davidson, P. Mineral liquid crystals from self-assembly of anisotropic nanosystems. In *Colloid Chemistry I. Topics in Current Chemistry*; Antonietti, M., Ed.; Springer: Berlin/Heidelberg, Germany, 2003; Volume 226, pp. 119–172. [CrossRef]
21. Gabriel, J.C.P.; Davidson, P. New trends in colloidal liquid crystals based on mineral moieties. *Adv. Mater.* **2000**, *12*, 9–20. [CrossRef]
22. Ruzicka, B.; Zaccarelli, E. A fresh look at the laponite phase diagram. *Soft Matter* **2011**, *7*, 1268–1286. [CrossRef]
23. Vroege, G.J.; Thies-Weesie, D.M.E.; Petukhov, A.V.; Lemaire, B.J.; Davidson, P. Smectic liquid-crystalline order in suspensions of highly polydisperse goethite nanorods. *Adv. Mater.* **2006**, *18*, 2565–2568. [CrossRef]

24. Kleshchanok, D.; Meijer, J.-M.; Petukhov, A.V.; Portale, G.; Lekkerkerker, H.N.W. Attractive glass formation in aqueous mixtures of colloidal gibbsite platelets and silica spheres. *Soft Matter* **2011**, *7*, 2832–2840. [CrossRef]

25. Mourad, M.C.D.; Byelov, D.V.; Petukhov, A.V.; Matthijs de Winter, D.A.; Verkleij, A.J.; Lekkerkerker, H.N.W. Sol–gel transitions and liquid crystal phase transitions in concentrated aqueous suspensions of colloidal gibbsite platelets. *J. Phys. Chem. B* **2009**, *113*, 11604–11613. [CrossRef] [PubMed]

26. Mourad, M.C.D.; Wijnhoven, J.E.G.J.; van't Zand, D.D.; van der Beek, D.; Lekkerkerker, H.N.W. Gelation versus liquid crystal phase transitions in suspensions of plate-like particles. *Philos. Trans. R. Soc. A Math. Phys. Eng. Sci.* **2006**, *364*, 2807–2816. [CrossRef]

27. Prestidge, C.A.; Ametov, I.; Addai-Mensah, J. Rheological investigations of gibbsite particles in synthetic bayer liquors. *Coll. Surf. A Physiochem. Eng. Asp.* **1999**, *157*, 137–145. [CrossRef]

28. ten Brinke, A.J.W.; Bailey, L.; Lekkerkerker, H.N.W.; Maitland, G.C. Rheology modification in mixed shape colloidal dispersions. Part I: Pure components. *Soft Matter* **2007**, *3*, 1145–1162. [CrossRef]

29. ten Brinke, A.J.W.; Bailey, L.; Lekkerkerker, H.N.W.; Maitland, G.C. Rheology modification in mixed shape colloidal dispersions. Part ii: Mixtures. *Soft Matter* **2008**, *4*, 337–348. [CrossRef]

30. van der Beek, D.; Lekkerkerker, H.N.W. Liquid crystal phases of charged colloidal platelets. *Langmuir* **2004**, *20*, 8582–8586. [CrossRef]

31. Verhoeff, A.A.; Wensink, H.H.; Vis, M.; Jackson, G.; Lekkerkerker, H.N.W. Liquid crystal phase transitions in systems of colloidal platelets with bimodal shape distribution. *J. Phys. Chem. B* **2009**, *113*, 13476–13484. [CrossRef]

32. Xu, T.; Davis, V.A. Liquid crystalline phase behavior of silica nanorods in dimethyl sulfoxide and water. *Langmuir* **2014**, *30*, 4806–4813. [CrossRef] [PubMed]

33. Wang, J.M.; Khoo, E.; Lee, P.S.; Ma, J. Synthesis, assembly, and electrochromic properties of uniform crystalline WO_3 nanorods. *J. Phys. Chem. C* **2008**, *112*, 14306–14312. [CrossRef]

34. Li, L.-S.; Alivisatos, A.P. Semiconductor nanorod liquid crystals and their assembly on a substrate. *Adv. Mater.* **2003**, *15*, 408–411. [CrossRef]

35. Yang, X.; He, W.; Wang, S.; Zhou, G.; Tang, Y.; Yang, J. Effect of the different shapes of silver particles in conductive ink on electrical performance and microstructure of the conductive tracks. *J. Mater. Sci. Mater. Electron.* **2012**, *23*, 1980–1986. [CrossRef]

36. Dessombz, A.; Chiche, D.; Davidson, P.; Panine, P.; Chanéac, C.; Jolivet, J.-P. Design of liquid-crystalline aqueous suspensions of rutile nanorods: Evidence of anisotropic photocatalytic properties. *J. Am. Chem. Soc.* **2007**, *129*, 5904–5909. [CrossRef] [PubMed]

37. Jana, N.R.; Gearheart, L.A.; Obare, S.O.; Johnson, C.J.; Edler, K.J.; Mann, S.; Murphy, C.J. Liquid crystalline assemblies of ordered gold nanorods. *J. Mater. Chem.* **2002**, *12*, 2909–2912. [CrossRef]

38. Nikoobakht, B.; Wang, Z.L.; El-Sayed, M.A. Self-assembly of gold nanorods. *J. Phys. Chem. B* **2000**, *104*, 8635–8640. [CrossRef]

39. Reibold, M.; Paufler, P.; Levin, A.A.; Kochmann, W.; Patzke, N.; Meyer, D.C. Materials: Carbon nanotubes in an ancient Damascus sabr. *Nature* **2006**, *444*, 286. [CrossRef]

40. Akbari, A.; Sheath, P.; Martin, S.T.; Shinde, D.B.; Shaibani, M.; Banerjee, P.C.; Tkacz, R.; Bhattacharyya, D.; Majumder, M. Large-area graphene-based nanofiltration membranes by shear alignment of discotic nematic liquid crystals of graphene oxide. *Nat. Commun.* **2016**, *7*, 10891. [CrossRef]

41. Dan, B.; Behabtu, N.; Martinez, A.; Evans, J.S.; Kosynkin, D.V.; Tour, J.M.; Pasquali, M.; Smalyukh, I.I. Liquid crystals of aqueous, giant graphene oxide flakes. *Soft Matter* **2011**, *7*, 11154–11159. [CrossRef]

42. Liu, Y.; Xu, Z.; Gao, W.; Cheng, Z.; Gao, C. Graphene and other 2D colloids: Liquid crystals and macroscopic fibers. *Adv. Mater.* **2017**, *29*, 1606794-n/a. [CrossRef] [PubMed]

43. Xu, Z.; Gao, C. Graphene in macroscopic order: Liquid crystals and wet-spun fibers. *Acc. Chem. Res.* **2014**, *47*, 1267–1276. [CrossRef] [PubMed]

44. Zakri, C.; Blanc, C.; Grelet, E.; Zamora-Ledezma, C.; Puech, N.; Anglaret, E.; Poulin, P. Liquid crystals of carbon nanotubes and graphene. *Philos. Trans. R. Soc. A Math. Phys. Eng. Sci.* **2013**, *371*. [CrossRef] [PubMed]

45. Green, M.J.; Parra-Vasquez, A.N.G.; Behabtu, N.; Pasquali, M. Modeling the phase behavior of polydisperse rigid rods with attractive interactions with applications to single-walled carbon nanotubes in superacids. *J. Chem. Phys.* **2009**, *131*, 041401. [CrossRef] [PubMed]

46. Behabtu, N.; Lomeda, J.R.; Green, M.J.; Higginbotham, A.L.; Sinitskii, A.; Kosynkin, D.V.; Tsentalovich, D.; Parra-Vasquez, A.N.G.; Schmidt, J.; Kesselman, E. Spontaneous high-concentration dispersions and liquid crystals of graphene. *Nat. Nanotechnol.* **2010**, *5*, 406–411. [CrossRef]

47. Davis, V.A.; Ericson, L.M.; Parra-Vasquez, A.N.G.; Fan, H.; Wang, Y.; Prieto, V.; Longoria, J.A.; Ramesh, S.; Saini, R.K.; Kittrell, C.; et al. Phase behavior and rheology of SWNTs in superacids. *Macromolecules* **2004**, *37*, 154–160. [CrossRef]

48. Moulton, S.E.; Maugey, M.; Poulin, P.; Wallace, G.G. Liquid crystal behavior of single-walled carbon nanotubes dispersed in biological hyaluronic acid solutions. *J. Am. Chem. Soc.* **2007**, *129*, 9452–9457. [CrossRef]

49. Barisci, J.N.; Tahhan, M.; Wallace, G.G.; Badaire, S.; Vaugien, T.; Maugey, M.; Poulin, P. Properties of carbon nanotube fibers spun from DNA-stabilized dispersions. *Adv. Funct. Mater.* **2004**, *14*, 133–138. [CrossRef]

50. Bergin, S.D.; Nicolosi, V.; Giordani, S.; de Gromard, A.; Carpenter, L.; Blau, W.J.; Coleman, J.N. Exfoliation in ecstasy: Liquid crystal formation and concentration-dependent debundling observed for single-wall nanotubes dispersed in the liquid drug γ-butyrolactone. *Nanotechnology* **2007**, *18*, 455705. [CrossRef]

51. Ao, G.; Nepal, D.; Aono, M.; Davis, V.A. Cholesteric and nematic liquid crystalline phase behavior of double-stranded DNA stabilized single-walled carbon nanotube dispersions. *ACS Nano* **2011**, *5*, 1450–1458. [CrossRef]

52. Song, W.; Kinloch, I.A.; Windle, A.H. Nematic liquid crystallinity of multiwall carbon nanotubes. *Science* **2003**, *302*, 1363. [CrossRef]

53. Song, Y.S.; Youn, J.R. Influence of dispersion states of carbon nanotubes on physical properties of epoxy nanocomposites. *Carbon* **2005**, *43*, 1378–1385. [CrossRef]

54. Gudarzi, M.M. Colloidal stability of graphene oxide: Aggregation in two dimensions. *Langmuir* **2016**, *32*, 5058–5068. [CrossRef]

55. Paredes, J.I.; Villar-Rodil, S.; Martinez-Alonso, A.; Tascon, J.M.D. Graphene oxide dispersions in organic solvents. *Langmuir* **2008**, *24*, 10560–10564. [CrossRef] [PubMed]

56. Jalili, R.; Aboutalebi, S.H.; Esrafilzadeh, D.; Konstantinov, K.; Moulton, S.E.; Razal, J.M.; Wallace, G.G. Organic solvent-based graphene oxide liquid crystals: A facile route toward the next generation of self-assembled layer-by-layer multifunctional 3D architectures. *ACS Nano* **2013**, *7*, 3981–3990. [CrossRef] [PubMed]

57. Ahmad, R.T.M.; Hong, S.-H.; Shen, T.-Z.; Masud, A.R.; Song, J.-K. Effect of solvents on the electro-optical switching of graphene oxide dispersions. *Appl. Phys. Lett.* **2016**, *108*, 251903. [CrossRef]

58. Naficy, S.; Jalili, R.; Aboutalebi, S.H.; Gorkin, R.A.; Konstantinov, K.; Innis, P.C.; Spinks, G.M.; Poulin, P.; Wallace, G.G. Graphene oxide dispersions: Tuning rheology to enable fabrication. *Mater. Horiz.* **2014**, *1*, 326–331. [CrossRef]

59. Liu, Y.; Chen, C.M.; Liu, L.Y.; Zhu, G.R.; Kong, Q.Q.; Hao, R.X.; Tan, W. Rheological behavior of high concentrated dispersions of graphite oxide. *Soft Mater.* **2015**, *13*, 167–175. [CrossRef]

60. Jalili, R.; Aboutalebi, S.H.; Esrafilzadeh, D.; Shepherd, R.L.; Chen, J.; Aminorroaya-Yamini, S.; Konstantinov, K.; Minett, A.I.; Razal, J.M.; Wallace, G.G. Scalable one-step wet-spinning of graphene fibers and yarns from liquid crystalline dispersions of graphene oxide: Towards multifunctional textiles. *Adv. Funct. Mater.* **2013**, *23*, 5345–5354. [CrossRef]

61. Aboutalebi, S.H.; Gudarzi, M.M.; Zheng, Q.B.; Kim, J.K. Spontaneous formation of liquid crystals in ultralarge graphene oxide dispersions. *Adv. Funct. Mater.* **2011**, *21*, 2978–2988. [CrossRef]

62. Andrienko, D. Introduction to liquid crystals. *J. Mol. Liq.* **2018**, *267*, 520–541. [CrossRef]

63. Veerman, J.A.C.; Frenkel, D. Phase behavior of disklike hard-core mesogens. *Phys. Rev. A* **1992**, *45*, 5632–5648. [CrossRef] [PubMed]

64. Khokhlov, A.R. Theories based on the Onsager approach. In *Liquid Crystallinity in Polymers*; Ciferri, A., Ed.; VCH Publishers: New York, NY, USA, 1991; pp. 97–129.

65. Doi, M.; Edwards, S.F. *The Theory of Polymer Dynamics*; Oxford University Press: Oxford, UK, 1986.

66. Roij, R.V. The isotropic and nematic liquid crystal phase of colloidal rods. *Eur. J. Phys.* **2005**, *26*, S57–S67. [CrossRef]

67. Larson, R.G. *The Structure and Rheology of Complex Fluids*; Oxford University Press: New York, NY, USA, 1999.

68. Zhang, S.; Kinloch, I.A.; Windle, A.H. Mesogenicity drives fractionation in lyotropic aqueous suspensions of multiwall carbon nanotubes. *Nano Lett.* **2006**, *6*, 568–572. [CrossRef] [PubMed]

69. Davis, V.A.; Parra-Vasquez, A.N.G.; Green, M.J.; Rai, P.K.; Behabtu, N.; Prieto, V.; Booker, R.D.; Schmidt, J.; Kesselman, E.; Zhou, W.; et al. True solutions of single-walled carbon nanotubes for assembly into macroscopic materials. *Nat. Nanotechnol.* **2009**, *4*, 830–834. [CrossRef] [PubMed]

70. Ziegler, K.J.; Rauwald, U.; Gu, Z.; Liang, F.; Billups, W.; Hauge, R.H.; Smalley, R.E. Statistically accurate length measurements of single-walled carbon nanotubes. *J. Nanosci. Nanotechnol.* **2007**, *7*, 2917–2921. [CrossRef]

71. Parra-Vasquez, A.N.G.; Stepanek, I.; Davis, V.A.; Moore, V.C.; Haroz, E.H.; Shaver, J.; Hauge, R.H.; Smalley, R.E.; Pasquali, M. Simple length determination of single-walled carbon nanotubes by viscosity measurements in dilute suspensions. *Macromolecules* **2007**, *40*, 4043–4047. [CrossRef]

72. Pagani, G.; Green, M.J.; Poulin, P.; Pasquali, M. Competing mechanisms and scaling laws for carbon nanotube scission by ultrasonication. *Proc. Natl. Acad. Sci. USA* **2012**, *109*, 11599. [CrossRef]

73. Moon, R.J.; Martini, A.; Nairn, J.; Simonsen, J.; Youngblood, J. Cellulose nanomaterials review: Structure, properties and nanocomposites. *Chem. Soc. Rev.* **2011**, *40*, 3941–3994. [CrossRef]

74. Elazzouzi-Hafraoui, S.; Nishiyama, Y.; Putaux, J.-L.; Heux, L.; Dubreuil, F.; Rochas, C. The shape and size distribution of crystalline nanoparticles prepared by acid hydrolysis of native cellulose. *Biomacromolecules* **2008**, *9*, 57–65. [CrossRef]

75. Beck-Candanedo, S.; Roman, M.; Gray, D.G. Effect of reaction conditions on the properties and behavior of wood cellulose nanocrystal suspensions. *Biomacromolecules* **2005**, *6*, 1048–1054. [CrossRef] [PubMed]

76. Adams, M.; Fraden, S. Phase behavior of mixtures of rods (tobacco mosaic virus) and spheres (polyethylene oxide, bovine serum albumin). *Biophys. J.* **1998**, *74*, 669–677. [CrossRef]

77. Bates, M.A.; Frenkel, D. Influence of polydispersity on the phase behavior of colloidal liquid crystals: A monte carlo simulation study. *J. Chem. Phys.* **1998**, *109*, 6193–6199. [CrossRef]

78. Woolston, P.; van Duijneveldt, J.S. Isotropic-nematic phase transition of polydisperse clay rods. *J. Chem. Phys.* **2015**, *142*, 184901. [CrossRef] [PubMed]

79. Donald, A.M.; Windle, A.H.; Hanna, S. Theories of liquid crystallinity in polymers. In *Liquid Crystalline Polymers*, 2nd ed.; Windle, A.H., Donald, A.M., Hanna, S., Eds.; Cambridge University Press: Cambridge, UK, 2006; pp. 133–228. [CrossRef]

80. Honorato-Rios, C.; Lehr, C.; Schütz, C.; Sanctuary, R.; Osipov, M.A.; Baller, J.; Lagerwall, J.P.F. Fractionation of cellulose nanocrystals: Enhancing liquid crystal ordering without promoting gelation. *NPG Asia Mater.* **2018**, *10*, 455–465. [CrossRef]

81. Lekkerkerker, H.N.W.; Coulon, P.; Van Der Haegen, R.; Deblieck, R. On the isotropic-liquid crystal phase separation in a solution of rodlike particles of different lengths. *J. Chem. Phys.* **1984**, *80*, 3427–3433. [CrossRef]

82. Flory, P.J.; Abe, A. Statistical thermodynamics of mixtures of rodlike particles. 1. Theory for polydisperse systems. *Macromolecules* **1978**, *11*, 1119–1122. [CrossRef]

83. Flory, P.J. Phase equilibria in solutions of rod-like particles. *Proc. R. Soc. Lond. Ser. A Math. Phys. Sci.* **1956**, *234*, 73–89. [CrossRef]

84. Vroege, G.J.; Lekkerkerker, H.N.W. Theory of the isotropic-nematic-nematic phase separation for a solution of bidisperse rodlike particles. *J. Chem. Phys.* **1993**, *97*, 3601–3605. [CrossRef]

85. Donald, A.M.; Windle, A.H. *Liquid Crystalline Polymers*; Cambridge University Press: Cambridge, UK, 1992.

86. Itou, T.; Teramoto, A. Multi-phase equilibrium in aqueous solutions of the triple-helical polysaccharide, schizophyllan. *Polym. J.* **1984**, *16*, 779–790. [CrossRef]

87. Kajiwara, K.; Donkai, N.; Hiragi, Y.; Inagaki, H. Lyotropic mesophase of imogolite, 1. Effect of polydispersity on phase diagram. *Die Makromol. Chem.* **1986**, *187*, 2883–2893. [CrossRef]

88. Buining, P.A.; Lekkerkerker, H.N.W. Isotropic-nematic phase separation of a dispersion of organophilic boehmite rods. *J. Chem. Phys.* **1993**, *97*, 11510–11516. [CrossRef]

89. Speranza, A.; Sollich, P. Simplified onsager theory for isotropic–nematic phase equilibria of length polydisperse hard rods. *J. Chem. Phys.* **2002**, *117*, 5421–5436. [CrossRef]

90. Speranza, A.; Sollich, P. Isotropic-nematic phase equilibria in the Onsager theory of hard rods with length polydispersity. *Phys. Rev. E* **2003**, *67*. [CrossRef] [PubMed]

91. Speranza, A.; Sollich, P. Isotropic-nematic phase equilibria of polydisperse hard rods: The effect of fat tails in the length distribution. *J. Chem. Phys.* **2003**, *118*, 5213–5223. [CrossRef]

92. van der Kooij, F.M.; Lekkerkerker, H.N.W. Formation of nematic liquid crystals in suspensions of hard colloidal platelets. *J. Chem. Phys. B* **1998**, *102*, 7829–7832. [CrossRef]

93. Gabriel, J.-C.P.; Sanchez, C.; Davidson, P. Observation of nematic liquid-crystal textures in aqueous gels of smectite clays. *J. Chem. Phys.* **1996**, *100*, 11139–11143. [CrossRef]

94. Jalili, R.; Aboutalebi, S.H.; Esrafilzadeh, D.; Konstantinov, K.; Razal, J.M.; Moultona, S.E.; Wallace, G.G. Formation and processability of liquid crystalline dispersions of graphene oxide. *Mater. Horiz.* **2014**, *1*, 87–91. [CrossRef]

95. Pusey, P.N.; Fijnaut, H.M.; Vrij, A. Mode amplitudes in dynamic light scattering by concentrated liquid suspensions of polydisperse hard spheres. *J. Chem. Phys.* **1982**, *77*, 4270–4281. [CrossRef]

96. Ahmad, R.T.M.; Hong, S.-H.; Shen, T.-Z.; Song, J.-K. Optimization of particle size for high birefringence and fast switching time in electro-optical switching of graphene oxide dispersions. *Opt. Express* **2015**, *23*, 4435–4440. [CrossRef]

97. Dogic, Z.; Sharma, Z.; Zakhary, M.J. Hypercomplex liquid crystals. *Annu. Rev. Condens. Matter Phys.* **2014**, *5*, 137–157. [CrossRef]

98. Asakura, S.; Oosawa, F. On interaction between two bodies immersed in a solution of macromolecules. *J. Chem. Phys.* **1954**, *22*, 1255–1256. [CrossRef]

99. Mao, Y.; Cates, M.E.; Lekkerkerker, H.N.W. Depletion force in colloidal systems. *Phys. A Stat. Mech. Its Appl.* **1995**, *222*, 10–24. [CrossRef]

100. Horn, D.W.; Ao, G.; Maugey, M.; Zakri, C.; Poulin, P.; Davis, V.A. Dispersion state and fiber toughness: Antibacterial lysozyme-single walled carbon nanotubes. *Adv. Funct. Mater.* **2013**, *23*, 6082–6090. [CrossRef]

101. Koenderink, G.H.; Vliegenthart, G.A.; Kluijtmans, S.G.J.M.; van Blaaderen, A.; Philipse, A.P.; Lekkerkerker, H.N.W. Depletion-induced crystallization in colloidal rod–sphere mixtures. *Langmuir* **1999**, *15*, 4693–4696. [CrossRef]

102. Asakura, S.; Oosawa, F. Interaction between particles suspended in solutions of macromolecules. *J. Polym. Sci.* **1958**, *33*, 183–192. [CrossRef]

103. Dogic, Z.; Frenkel, D.; Fraden, S. Enhanced stability of layered phases in parallel hard spherocylinders due to addition of hard spheres. *Phys. Rev. E* **2000**, *62*, 3925–3933. [CrossRef]

104. Adams, M.; Dogic, Z.; Keller, S.L.; Fraden, S. Entropically driven microphase transitions in mixtures of colloidal rods and spheres. *Nature* **1998**, *393*, 349–352. [CrossRef]

105. Urakami, N.; Imai, M. Dependence on sphere size of the phase behavior of mixtures of rods and spheres. *J. Chem. Phys.* **2003**, *119*, 2463–2470. [CrossRef]

106. Xu, T.; Davis, V.A. Rheology and shear-induced textures of silver nanowire lyotropic liquid crystals. *J. Nanomater.* **2015**, *2015*, 9. [CrossRef]

107. Green, M.J. Isotropic–nematic phase separation and demixing in mixtures of spherical nanoparticles with length-polydisperse nanorods. *J. Polym. Sci. Part B Polym. Phys.* **2012**, *50*, 1321–1327. [CrossRef]

108. Lüders, A.; Siems, U.; Nielaba, P. Dynamic ordering of driven spherocylinders in a nonequilibrium suspension of small colloidal spheres. *Phys. Rev. E* **2019**, *99*, 022601. [CrossRef] [PubMed]

109. Wu, L.; Malijevský, A.; Avendaño, C.; Müller, E.A.; Jackson, G. Demixing, surface nematization, and competing adsorption in binary mixtures of hard rods and hard spheres under confinement. *J. Chem. Phys.* **2018**, *148*, 164701. [CrossRef] [PubMed]

110. Wu, L.; Malijevský, A.; Jackson, G.; Müller, E.A.; Avendaño, C. Orientational ordering and phase behaviour of binary mixtures of hard spheres and hard spherocylinders. *J. Chem. Phys.* **2015**, *143*, 044906. [CrossRef]

111. Aliabadi, R.; Moradi, R.; Varga, S. Tracking three-phase coexistences in binary mixtures of hard plates and spheres. *J. Chem. Phys.* **2016**, *144*, 074902. [CrossRef]

112. Harnau, L.; Dietrich, S. Bulk and wetting phenomena in a colloidal mixture of hard spheres and platelets. *Phys. Rev. E* **2005**, *71*, 011504. [CrossRef]

113. Oversteegen, S.M.; Lekkerkerker, H.N.W. Phase diagram of mixtures of hard colloidal spheres and discs: A free-volume scaled-particle approach. *J. Chem. Phys.* **2004**, *120*, 2470–2474. [CrossRef]

114. de las Heras, D.; Schmidt, M. Bulk fluid phase behaviour of colloidal platelet-sphere and platelet-polymer mixtures. *Philos. Trans. R. Soc. A Math. Phys. Eng. Sci.* **2013**, *371*. [CrossRef]

115. Kleshchanok, D.; Meijer, J.-M.; Petukhov, A.V.; Portale, G.; Lekkerkerker, H.N.W. Sedimentation and depletion attraction directing glass and liquid crystal formation in aqueous platelet/sphere mixtures. *Soft Matter* **2012**, *8*, 191–197. [CrossRef]

116. Sui, J. Stratification in the dynamics of sedimenting colloidal platelet–sphere mixtures. *Soft Matter* **2019**, *15*, 4714–4722. [CrossRef]

117. Lu, H.; Zhang, J.; Luo, J.; Gong, W.; Li, C.; Li, Q.; Zhang, K.; Hu, M.; Yao, Y. Enhanced thermal conductivity of free-standing 3D hierarchical carbon nanotube-graphene hybrid paper. *Compos. Part A Appl. Sci.* **2017**, *102*, 1–8. [CrossRef]

118. Zhao, M.Q.; Ren, C.E.; Ling, Z.; Lukatskaya, M.R.; Zhang, C.; Van Aken, K.L.; Barsoum, M.W.; Gogotsi, Y. Flexible MXene/carbon nanotube composite paper with high volumetric capacitance. *Adv. Mater.* **2015**, *27*, 339–345. [CrossRef] [PubMed]

119. Chen, H.; Yu, L.; Lin, Z.; Zhu, Q.; Zhang, P.; Qiao, N.; Xu, B. Carbon nanotubes enhance flexible MXene films for high-rate supercapacitors. *J. Mater. Sci.* **2020**, *55*, 1148–1156. [CrossRef]

120. Ricciardulli, A.G.; Yang, S.; Wetzelaer, G.J.A.; Feng, X.; Blom, P.W. Hybrid silver nanowire and graphene-based solution-processed transparent electrode for organic optoelectronics. *Adv. Funct. Mater.* **2018**, *28*, 1706010. [CrossRef]

121. Tang, H.; Feng, H.; Wang, H.; Wan, X.; Liang, J.; Chen, Y. Highly conducting MXene–silver nanowire transparent electrodes for flexible organic solar cells. *ACS Appl. Mater. Interfaces* **2019**, *11*, 25330–25337. [CrossRef]

122. Chen, W.; Liu, L.-X.; Zhang, H.-B.; Yu, Z.-Z. Flexible, transparent and conductive Ti3C2Tx MXene-silver nanowire films with smart acoustic sensitivity for high-performance electromagnetic interference shielding. *ACS Nano* **2020**. [CrossRef]

123. Chen, L.; Hou, X.; Song, N.; Shi, L.; Ding, P. Cellulose/graphene bioplastic for thermal management: Enhanced isotropic thermally conductive property by three-dimensional interconnected graphene aerogel. *Compos. Part A Appl. Sci.* **2018**, *107*, 189–196. [CrossRef]

124. Chen, Q.; Liu, P.; Sheng, C.; Zhou, L.; Duan, Y.; Zhang, J. Tunable self-assembly structure of graphene oxide/cellulose nanocrystal hybrid films fabricated by vacuum filtration technique. *RSC Adv.* **2014**, *4*, 39301–39304. [CrossRef]

125. Tian, W.; VahidMohammadi, A.; Reid, M.S.; Wang, Z.; Ouyang, L.; Erlandsson, J.; Pettersson, T.; Wågberg, L.; Beidaghi, M.; Hamedi, M.M. Multifunctional nanocomposites with high strength and capacitance using 2D MXene and 1D nanocellulose. *Adv. Mater.* **2019**, *31*, 1902977. [CrossRef]

126. van Roij, R.; Mulder, B. Demixing in a hard rod-plate mixture. *J. Phys. II* **1994**, *4*, 1763–1769. [CrossRef]

127. van der Kooij, F.M.; Lekkerkerker, H.N.W. Liquid-crystalline phase behavior of a colloidal rod-plate mixture. *Phys. Rev. Lett.* **2000**, *84*, 781–784. [CrossRef] [PubMed]

128. Woolston, P.; van Duijneveldt, J.S. Three-phase coexistence in colloidal rod–plate mixtures. *Langmuir* **2015**, *31*, 9290–9295. [CrossRef] [PubMed]

Review

Experimental Conditions for the Stabilization of the Lyotropic Biaxial Nematic Mesophase

Erol Akpinar [1] and Antônio Martins Figueiredo Neto [2,*

[1] Faculty of Arts and Sciences, Department of Chemistry, Bolu Abant Izzet Baysal University, Golkoy, Bolu 14030, Turkey; akpinar_e@ibu.edu.tr
[2] Instituto de Física, Universidade de São Paulo, Rua do Matão, 1371, São Paulo-SP 05508-090, Brazil
* Correspondence: afigueiredo@if.usp.br

Received: 1 February 2019; Accepted: 15 March 2019; Published: 19 March 2019

Abstract: Nematic phases are some of the most common phases among the lyotropic liquid crystalline structures. They have been widely investigated during last decades. In early studies, two uniaxial nematic phases (discotic, N_D, and calamitic, N_C) were identified. After the discovery of the third one, named biaxial nematic phase (N_B) in 1980, however, some controversies in the stability of biaxial nematic phases began and still continue in the literature. From the theoretical point of view, the existence of a biaxial nematic phase is well established. This review aims to bring information about the historical development of those phases considering the early studies and then summarize the recent studies on how to stabilize different nematic phases from the experimental conditions, especially, choosing the suitable constituents of lyotropic mixtures.

Keywords: Lyotropic liquid crystals; uniaxial nematic phase; biaxial nematic phase; stabilization of nematic phases; micelle; surfactants

---

1. Introduction

Lyotropic liquid crystalline phases may be encountered in mixtures of amphiphilic molecules and a solvent, with or without a salt, in particular regions of the multidimensional phase diagram (relative concentrations and temperature) of these mixtures [1,2]. From the symmetry point of view, two types of lyotropic nematic phases exist: uniaxial (two of them were identified, the discotic, N_D, and the calamitic, N_C, phases, $D_{\infty h}$ symmetry), and one biaxial (N_B, D_{2h} symmetry). In the uniaxial phases we define a director represented by a vector $\vec{n}$ that corresponds to the optical axis of the phase. The $\vec{n}$ and $-\vec{n}$ states are indistinguishable. In the case of the N_B phase, there are three orthogonal two-fold symmetry axes ($\vec{l}$, $\vec{m}$ and $\vec{n}$, where $\vec{n} = \vec{l} \times \vec{m}$) and two optical axes [3–5]. These phases are diamagnetic and high-intensity magnetic fields ($H \sim 10kG$) are used to orient them in actual experiments.

After the discovery of the first lyotropic mixture showing the biaxial nematic phase [1], intensive studies have begun to characterize its properties and structure [6]. These studies were mostly conducted to understand whether the biaxial phases are thermodynamically stable or whether a mixture of the two uniaxial phases. Many experimental and theoretical studies have confirmed the first hypothesis.

In the phase diagrams of lyotropic mixtures presenting the three nematic phases, the N_B phase is mainly located between the other two uniaxial phases [1,2,7–15]. The transition from the uniaxial (N_D or N_C) to the biaxial phase is of second order, as predicted by the Landau-type mean-field theory [16,17]. Since the N_B phase domain (at least in some phase diagrams) is located in between those from the N_D and N_C phases, some researchers claimed that the N_B phase was thermodynamically unstable and even constituted by a mixture of both uniaxial phases [18–22]. Assuming that the N_D and N_C phases consist of disc-like and cylindrical-like micelles, respectively, the N_B phase should be composed by a mixture of these types of micelles. However, it was theoretically proved that disc-like and cylindrical-like objects could not coexist, i.e., a mixture of both objects (in the case of nematic phases, micelles) is not stable and phase separation occurs [23].

This review article deals with the recent studies related to the stabilization of different lyotropic nematic phases from the experimental point of view. Mainly, we will review some important studies obtained in our research laboratories, giving to experimentalists hints of mixture compositions that favor the presence of the N_B phase in the phase diagram.

2. Background

The first lyotropic nematic phase was reported towards the end of the 1960s [24]. In the following years, many studies were conducted to understand the properties of this phase. In these early studies, the existence of two types of nematic phases was determined and these phases were classified as Type I and Type II, according to the orientation of the phase director with respect to an external applied magnetic field [25–27]. Since the lyotropic mixture investigated was composed by amphiphilic molecules with carbonic chains, it was verified that the director of the Type I (Type II) phase aligns parallel (perpendicular) to the magnetic field direction. In the following years, Type I (Type II) phase was named N_C (N_D) [28]. At that time, it seemed natural to assume that Type I and Type II phases were composed of cylindrical-like and disc-like micelles [28–32], respectively, as shown in Figure 1.

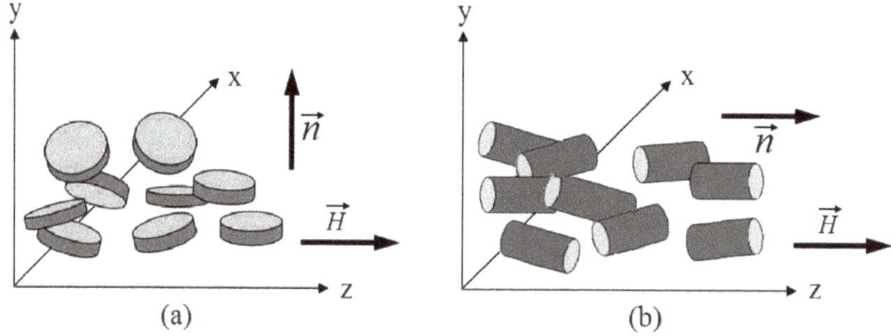

Figure 1. Orientations of (**a**) disc-like and (**b**) cylindrical-like micelles perpendicular and parallel, respectively, to magnetic field direction, $\vec{H}$, along z. $\vec{n}$ (phase director) represents the preferred alignment of the individual micelles with their local directors.

Freiser [16,33] theoretically predicted the existence of the N_B phase for the first time [5,34–38] after Taylor et al. [39] reported a biaxial smectic C phase in thermotropic liquid crystals. However, at that time, there were no experimental results to support this theoretical prediction. In 1980, Yu and Saupe [1] experimentally showed the existence of the N_B phase for the first time in the ternary mixture of potassium laurate (KL)/decanol (DeOH)/D_2O via Nuclear Magnetic Resonance (NMR) and microscopic conoscopy studies, as shown in Figure 2. They constructed the partial phase diagram of this ternary mixture where the N_B phase was an intermediate phase between the two uniaxial N_D and N_C phases. In 1985, Neto et al. [40] investigated the same ternary mixture in more details. They used the optical microscopy, laser conoscopy and X-ray diffraction techniques and found that the biaxial phase domain was relatively large (~15 °C) for appropriate relative concentrations of the mixture compounds.

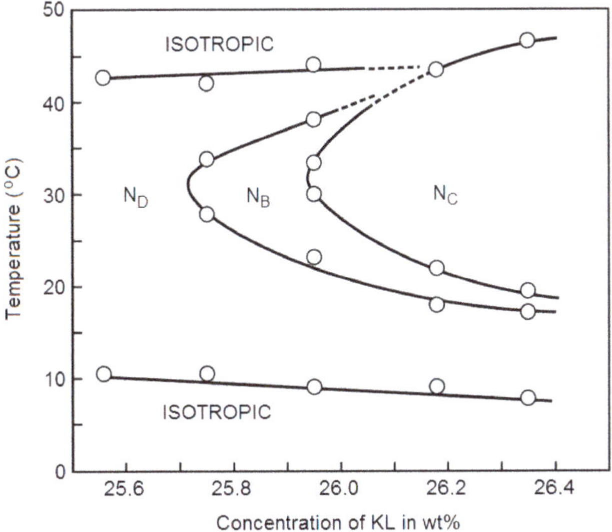

Figure 2. Phase diagram of KL/DeOH/water mixture which was established for the first time in the
Reference [1].

In 1982, Bartolino et al. [7] reported a partial phase diagram of the ternary mixture of sodium
decylsulfate (SdS)/DeOH/H_2O to contribute to the understanding of the biaxial nematic phases,
as shown in Figure 3a. They showed that the uniaxial N_D (N_C) phase had positive (negative)
birefringence, as shown in Figure 3b. Their results seem to show a discontinuity of the birefringence
in the biaxial domain, however, to prove this claim it should be necessary measurements of the
birefringences with more precision in temperature.

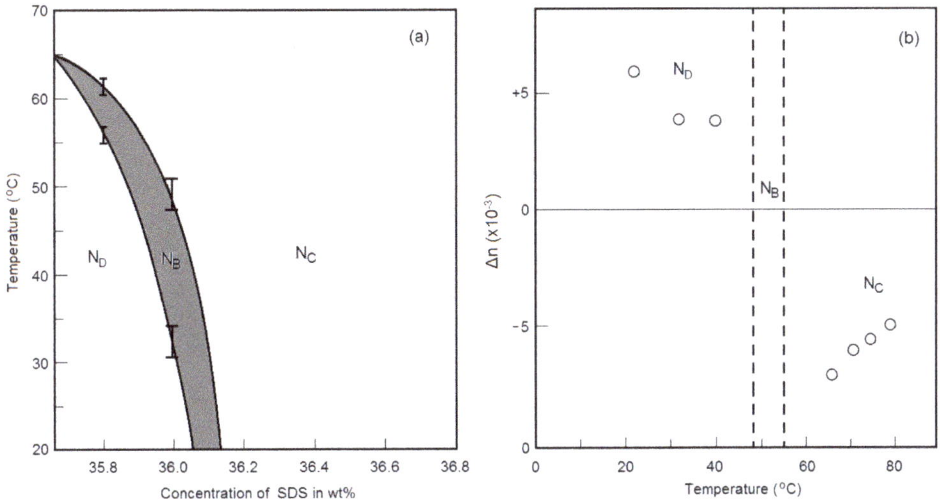

Figure 3. (a) Partial phase diagram of SDS/DeOH/water mixture given in the Reference [7]. (b) The
birefringences of the uniaxial nematic phases for the sample of 35.8% SDS.

Finding the first biaxial nematic phase Yu and Saupe not only verified Freiser's prediction on the
existing that phase, but also started interesting discussions about the nature and structure of the N_B

phase [1,36,41–47]. Theoretical and computer simulation studies of binary mixtures of rod-like and disk-like rigid particles were reported [19–21,48–57]. In 1984 Stroobants and Lekkerkerker proposed a model, based on the Onsager's theory, where rod-like and disc-like particles coexist (mixture of disc-like and cylindrical-like micelles, MCD model), giving rise to a biaxial nematic phase [21]. However, some theoretical and experimental studies showed that MCD-type models presented problems to describe the N_B phase [58]. For instance, in 2000, Kooij and Lekkerkerker reported an interesting experimental study about the liquid crystal phase behavior of suspensions with mixtures of rod- and plate-like units. They verified that this type of mixture separates, i.e., the coexistence of rod- and plate-like particles is not stable [23]. In another study, Palffy-Muhoray et al. [22] investigated the phase behavior of a binary mixture of uniaxial nematic liquid crystals (rods and disks), in the absence of the external field, employing the Maier-Saupe mean-field theory. They concluded that the N_B phase was not thermodynamically stable. In a recent study Martinez-Raton et al. showed that phase-separation of rod-plate mixtures and the stability of the biaxial phase depend on the aspect ratio and molar fraction of the rods [59]. Sharma et al. [60] showed that there could be a possibility to stabilize the biaxial phase in mixtures by increasing either the isotropic or the anisotropic parts of the interspecies interaction strength. The latter situation was also supported by other theoretical studies, i.e., biaxial phase may be favored in multi-component (polydisperse) mixtures [61] by the specific interactions (indeed, hydrogen bonding) between the unlike basic units [62,63]. However, for the symmetric [64–67] or asymmetric [23,68–73] binary rod-plate mixtures, considering the stability of the biaxial phase with respect to phase separation, the coexistence of rod-plate particles is not possible and the phase separation is favored. In summary, the separation of the rod and plate-like objects (in the case of the lyotropic systems, prolate or cylindrical-like and oblate or disc-like micelles) requires the rejection of the MCD-type model, as experimentally showed by Kooij and Lekkerkerker [23].

Although the discussions about the stability of the N_B phase were going on, experimentalists tried to find new lyotropic mixtures presenting the N_B phases. In 1985, Galerne et al. [8] introduced a new ternary mixture of rubidium laurate (RbL)/DeOH/H_2O. They investigated the defect lines (disclinations) in both uniaxial and biaxial nematic phases and observed that the disclination lines in the biaxial region made "zig-zags" while the uniaxial counterparts exhibited a "flexible" behavior, i.e., the disclinations observed in the biaxial phase seem to be different from those in the uniaxial phases. However, some time later the same authors show that the same zig-zag disclinations were observed in uniaxial phases with different mechanism than in the biaxial [74–76]. Santos and Galerne investigated the uniaxial discotic nematic phase of KL/DeOH/D_2O by Rayleigh-scattering technique [77] in the vicinity of the uniaxial to biaxial phase transition. They observed the dynamic of fluctuations of the biaxial order parameter close to the uniaxial to biaxial phase transition, at the uniaxial phase domain, and showed that the micelles in the uniaxial discotic nematic phase exhibit biaxial ordering.

In 1985, a reliable model was proposed by Neto and co-workers [41,78], assuming that the micelles in all the three nematic phases in lyotropics have orthorhombic symmetry. They investigated the uniaxial and biaxial nematic phases of the mixture KL/DeOH/D_2O by X-ray diffraction and observed that the micelles in all the nematic phases are locally biaxial, arranged in a pseudo-lamellar structure. This means that in the three nematic phases, the micelles are similar from the point of view of the local symmetry. Their model, so-called "intrinsically biaxial micelles model, IBM", is mainly based on two pillars: (a) micelle symmetry and (b) orientational fluctuations of the micelles. According to the IBM model, the micelles have orthorhombic symmetry, as shown in Figure 4a, in all three different nematic phases and different orientational fluctuations are responsible for the stabilization of the three nematic phases. Figure 4a shows a sketch a micelle as an object of orthorhombic symmetry, with typical dimensions A', B' and C'. The axes of the local coordinate system fixed in the micelle are α, β and γ. These orthorhombic micelles exhibit different orientational fluctuations for the stabilization of the N_D, N_B or N_C phases. The orientational fluctuations around the axis perpendicular to the largest micelle surface (along axis y) give rise to the N_D phase, as shown in Figure 4b. In this situation, the micelle size A' approximately equal to B' (A'~B'). By changing temperature to reach the transition from

N_D to N_B, the micelle size along the symmetry axis α increases, and the small-amplitude orientational fluctuations along the three axes (x, y and z) lead to the N_B phase. If the temperature is changed until obtaining the N_C phase, the micelle dimension along α continues to increase, which favors the orientational fluctuations around the local α micellar axis (along the z axis).

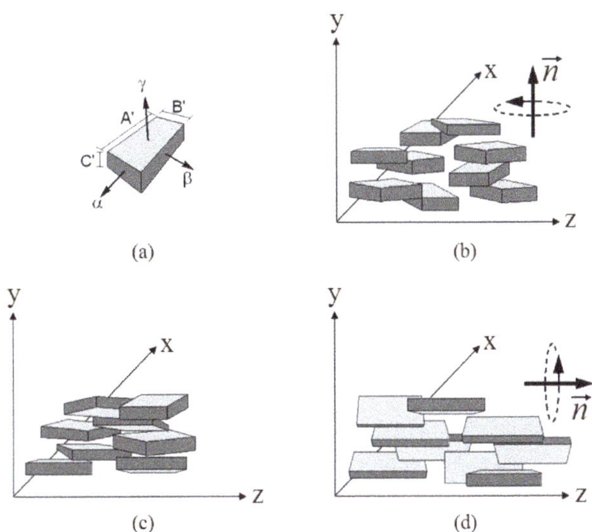

Figure 4. (**a**) Sketch of the orthorhombic micelle in the framework of the IBM model (Reference [79]). The detergent amphiphilic bilayer is represented by C′. According to the IBM model, based on the different orientational fluctuations, the formations of (**b**) ND, (**c**) NB, and (**d**) NC phases. $\vec{n}$ represents the director of the ND and NC phases."

A neutron scattering study reported by Hendrikx et al. [80] in 1986 supported the IBM model. In that study, authors experimentally showed that the micelles of the N_C phase are statistically biaxial.

Until 1989, since the lyotropic mixtures exhibiting the biaxial nematic phases contained alcohol and surfactant, Oliveira et al. [9] achieved to find a new mixture of KL/decylammonium chloride (DACl)/water, alcohol-free. This new mixture was important because, as suggested by Saupe et al. [81], if a mixture has surfactant and alcohol, there is a possibility of a slow esterification occur in the mixture. In the study of the [9], researchers investigated the mixtures of KL/DACl/water and KL/DeOH/water, i.e., alcohol-free and with alcohol mixtures, respectively, to compare the physical-chemical stability of both types of mixtures. Their long-term studies were based on the birefringence and X-ray diffraction measurements. It was shown that the phase-transition temperatures, birefringences and microscopic structures changed with time in the case of the mixtures including alcohol. However, for the alcohol-free mixtures, especially the X-ray diffraction results of the nematic phases, revealed that the alcohol-free mixtures were more stable than the one with alcohol.

In 1991, Vasilevyskaya et al. [82] investigated a new lyotropic mixture exhibiting the biaxial nematic phase SdS/DeOH/H$_2$O/Na$_2$SO$_4$. Their experimental study was based on the measurement of the birefringences of the nematic phases, as a function of the temperature. They observed that this new system had much smaller birefringence values with respect to the conventional lyotropic mixture KL/DeOH/H$_2$O [1]. As a result of their study, they concluded that the phase transitions from biaxial to uniaxial arise from different orientational fluctuations of the micelles around their axes. Their conclusion is similar to those from the IBM model.

Experimental and theoretical studies stated that the biaxial nematic phase is an intermediate phase between two uniaxial nematic ones. Ho et al. [83] reported an interesting study related to a new lyotropic mixture of sodium lauroyl sulfate/hexadecanol (HDeOH)/water, exhibiting a biaxial nematic phase in the case of diluted solutions. They obtained this biaxial phase after applying the sonication process on a gel-like phase. They showed, surprisingly that a biaxial phase could be found in a different place in the lyotropic phase diagram, not as an intermediate phase between two uniaxial nematics. Another interesting study was published by Quist [84] to investigate novel lyotropic mixture of sodium dodecylsulfate SDS/DeOH/H_2O via NMR spectroscopy. Based on the Landau theory, the phase transitions from uniaxial to biaxial nematic phase transitions are of second-order. However, Quist proposed that these phase transitions might be of first-order. The author attributed the first–order N_D-N_B and N_B-N_C phase transitions to a variation in the aggregate (micelle) shape. This result is really very interesting, because other experimental and theoretical studies showed that the uniaxial to biaxial transitions have to be of second-order. In addition, early [40] and recent [13,15] studies, especially obtained from the temperature dependence of the birefringence, showed that these phase transitions are of second-order, as predicted by mean-field theory [16,18].

From the diamagnetic-susceptibility anisotropy (Equation (1)) point of view, considering the diamagnetic susceptibilities along three two-fold symmetry axes being χ_{33}, χ_{22} and χ_{11}, the biaxial nematics are identified with positive and negative $\Delta\chi$, N_{B+} and N_{B-}.

$$\Delta\chi = \chi_{33} - \frac{\chi_{11} + \chi_{22}}{2}. \tag{1}$$

In the case of $\Delta\chi > 0$ ($\Delta\chi < 0$), the largest (smallest) diamagnetic susceptibility is larger (less) than the average of the diamagnetic susceptibilities of other axes and the biaxial phase aligns parallel (perpendicular) to the magnetic field direction [84,85]. Quist [84] reported a study with these two biaxial nematic phases, with positive and negative diamagnetic-susceptibility anisotropy in the phase diagram; however, they did not study an eventual phase transition between these N_{B+} and N_{B-} phases. Moreover, from the symmetry point of view, there is no transition between them. This type of "phase transition" was investigated by de Melo Filho et al. [85] in the novel lyotropic mixture of tetradecyltrimethylammonium bromide (TDTMABr)/DeOH/H_2O by measuring the deuterium quadrupolar splitting, via Nuclear Magnetic Resonance. In 2009, van den Pol et al. [36] reported a model-system to experimentally obtain the biaxial phase in colloidal dispersions of board-like particles. According to theory and simulations, the particles have the dimensions L/W ≈ W/T, where L, W and T are length, width and thickness, to obtain a biaxial nematic phase [48,50,86]. They used goethite particles as a model-system with L/W = 3.1 and W/T = 3.0, and investigated the properties of the model-system by small angle x-ray scattering, under the effect of external magnetic field. Their results indicated the evidence of the biaxial nematic phase. Indeed, their model system is similar to what is proposed by the IBM model in terms of the micelle shapes and, from this respect, the study in [36] supports the IBM model.

As reviewed above, despite the evidence of the stability of the biaxial nematic phase in lyotropics, some studies still reject the existence of this phase. Indeed, these controversies arise from the absence of enough number of lyotropic mixtures in the literature showing this phase and the missing thermotropic counterpart.

In the next sections, we will discuss how to stabilize the lyotropic biaxial nematic phase from the sample preparation point of view. In other words, we will show which experimental conditions have to be fulfilled to prepare a lyotropic mixture presenting this phase.

3. Recent Experimental Studies on the Stabilization of the Lyotropic Biaxial Nematic Phase

Lyotropic mixtures are prepared by dissolution of a surfactant molecule in water (mainly). In some cases, a cosurfactant and/or an electrolyte are/is added to that binary solution. Indeed, while the former provides both the decrease in the repulsions between similarly charged headgroups of the surfactants at the micelle surfaces and the increase in the attractive forces (van der Waals) between the surfactant alkyl chains in the interior of the micelles, the latter one just reduces the repulsions between the headgroups. Because of this, modifications at the micelle surfaces and in the micelles shape anisotropy play a key role to obtain not only uniaxial nematic phases but also the biaxial one. Thus, we will review our recent experimental results focusing, mainly, in the modifications at micelle surfaces and in the micelles shape anisotropy.

3.1. Effects of Alkyl-Chain Lengths of Surfactant and Cosurfactant

Surfactant molecules consist of two different parts, hydrophilic water-soluble head groups and hydrophobic water-insoluble alkyl tails or chains. By the same way, cosurfactants have water-insoluble parts and polar water-soluble parts attached. In general, long-chain alcohols, such as 1-decanol, are added to the lyotropic mixtures as cosurfactants.

In 2012, we studied the effect of the alkyl-chain length of the alcohols in the mixture of KL/K_2SO_4/alcohol/water [11]. We constructed a partial phase diagram by changing the number of carbon atoms in the alcohol-chain length from 8 (1-octanol) to 16 (hexadecanol), as shown in Figure 5. The laser conoscopy technique, which is a very accurate method to measure the birefringences of the nematic phases [87], was applied to determine the uniaxial to biaxial nematic phase transitions from the temperature dependences of the birefringences. The results indicated that there is a strong relation between the alkyl-chain length of the surfactant and that of the alcohol. As the alkyl-chain length of the alcohol gets longer (shorter), the N_C (N_D) phase is stable and at intermediate alkyl chain length there exists the possibility to obtain the N_B one. In general, at constant mixture composition, the N_B phase is stabilized when the $n = m \pm 2$, where m (n) is the number of carbons in alkyl chain of the main surfactant (co-surfactant) molecule.

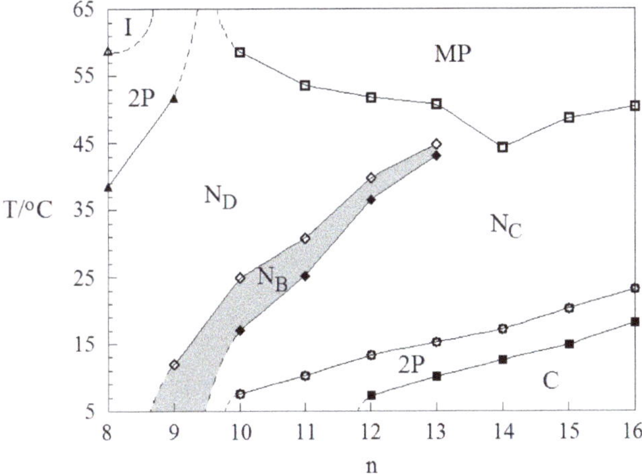

Figure 5. Partial phase diagram of KL/K_2SO_4/alcohol/water mixtures [11]. n is the number of carbon atoms in the alkyl chain of the alcohol molecules. I, 2P, MP and C represent an isotropic phase, two-phase region, multi-phase region and crystalline-like phase, respectively. Solid and dashed lines are only guides. The grey region shows the biaxial nematic phase domain.

In a more recent study [15], we have also investigated the relation between the alkyl-chain length of the surfactant and the alcohol in a different experimental design. In this case, we examined the effect of the surfactant alkyl-chain length on the stability of the different nematic phases, at constant alcohol chain-length and mixture composition. The mixtures chosen were potassium alkanoates/Rb_2SO_4/DeOH/water and sodium alkylsulfates/Na_2SO_4/DeOH/water. The experimental techniques employed were the laser conoscopy, polarizing optical microscopy and small-angle x-ray scattering. In both systems, it was observed that as the number of the carbon atoms in the surfactant alkyl-chain length increases (decreases), the stabilization of the N_D (N_C) phase is favored. Here, again, for the intermediate level of the relation between the alkyl-chain length of the surfactant and the alcohol, the N_B is stabilized. It was also observed that the $n = m \pm 2$ rule is still applicable, except that the concentration of the surfactant, which is the main component of the lyotropic mixtures, is bigger than that of alcohol.

3.2. Effect of Interactions between Head Groups and Ions of Electrolytes

Lyotropic systems are also known as micellar systems because their building blocks (or basic units) are micelles. The micellar systems consist of three main regions: (a) intermicellar region, (b) interfacial region and (c) micelle core. The intermicellar region includes water, some counterions of the surfactants and, if an electrolyte is added to the mixture, the ions of the electrolytes, and free surfactant molecules. The micelle cores consist of the hydrocarbon part, i.e., the alkyl chain of the surfactants. The interfacial region is a region at the micelle surfaces where the head groups of the surfactants, some counterions, ions of electrolytes and structured water are present. In the previous section, we have dealt with the modifications in the micelle core by examining the relative effect of the alkyl-chain lengths of the surfactant and of the alcohol. In this section, we discuss the effect of the modifications at the micelle surfaces, i.e., in the interfacial region.

The interaction of alkali ions with the head groups of the surfactant molecules on obtaining the different nematic phases, especially biaxial one, was investigated [88]. In this study, two different lyotropic mixtures were examined, KL/alkali sulfate/undecanol/water, as shown in Figure 6, and sodium dodecylsulfate (SDS)/alkali sulfate/dodecanol/water, as shown in Figure 7, where alkali sulfates are Li_2SO_4, Na_2SO_4, K_2SO_4, Rb_2SO_4 and Cs_2SO_4. In Figure 6 the optical birefringences are given. The refraction indices along the x, y and z axes of the laboratory coordinate system are n_x, n_y and n_z, and the birefringences are defined as: $\delta n = |n_z - n_y|$; $\Delta n = |n_x - n_z|$. The magnetic field is applied along the z axis. In the N_D phase, $n_x = n_z \neq n_y$, in the N_C phase $n_x = n_y \neq n_z$, and in the N_B phase $n_x \neq n_z \neq n_y$.

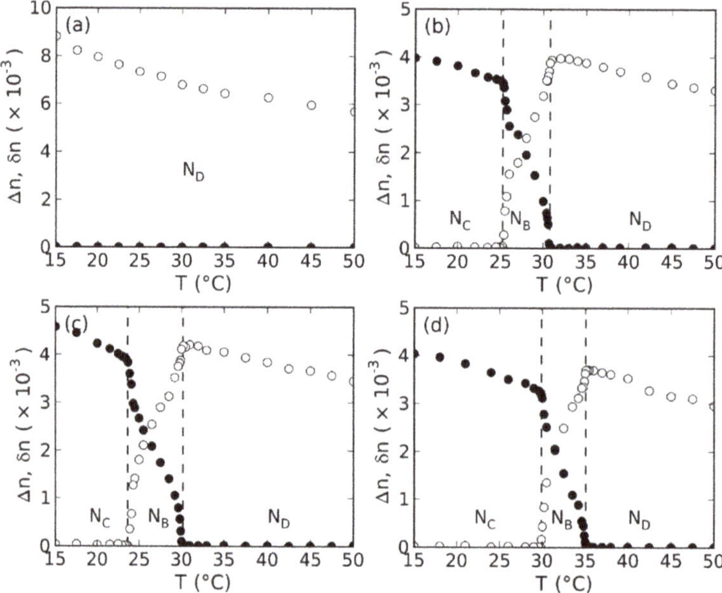

Figure 6. Temperature dependences of the birefringences of the mixtures KL/alkali sulfate/UndeOH/water, including (**a**) Na_2SO_4, (**b**) K_2SO_4, (**c**)Rb_2SO_4 and (**d**) Cs_2SO_4, separately. (○) and (•) represent δn and Δn, respectively [88].

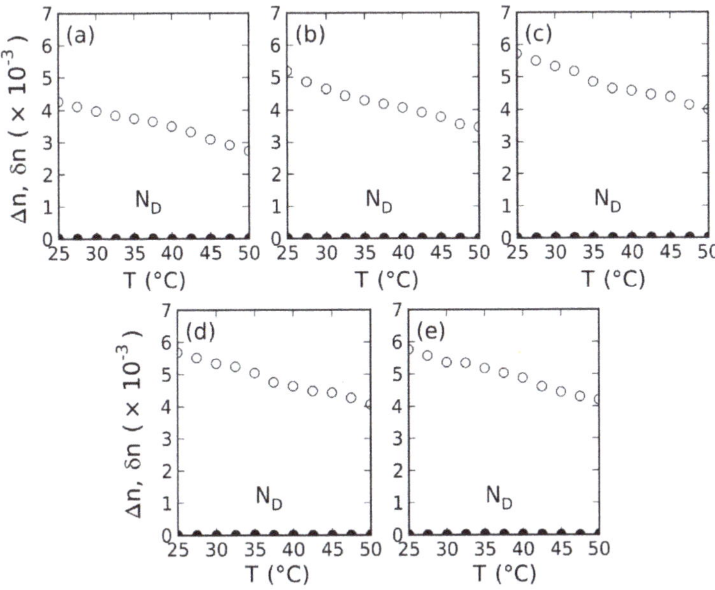

Figure 7. Temperature dependences of the birefringences of the mixtures SDS/alkali sulfate/DDeOH/water, including (**a**) Li_2SO_4, (**b**) Na_2SO_4, (**c**) K_2SO_4, (**d**)Rb_2SO_4 and (**e**) Cs_2SO_4, separately. (○) and (●) represent δn and Δn, respectively [88].

As seen in Figure 6, except for Na_2SO_4, three nematic phases were obtained for K_2SO_4, Rb_2SO_4 and Cs_2SO_4 and the biaxial temperature ranges were about the same, ~5–6 °C. The mixture with Na_2SO_4 has higher birefringence in the N_D region (about twice higher) with respect to those from the other salts. Although we did not observe the biaxial regions for SDS mixtures, as shown in Figure 7, we extracted some useful information from those samples, which were in good agreement with the results of the KL mixtures. While the sample with Li_2SO_4 and Na_2SO_4 exhibited maximum birefringence of ~4.0×10^{-3} and ~5.0×10^{-3}, respectively, K_2SO_4, Rb_2SO_4 and Cs_2SO_4 presented similar birefringence values of ~6.0×10^{-3}, in the N_D phases of the lyotropic host mixture SDS/DDeOH/water. The ~50% increase in the birefringence value, going from Li_2SO_4 to K_2SO_4, was really significant because the birefringences are related to the micelle size and shape anisotropy. The higher the birefringences the bigger the micelle shape anisotropy. Moreover, the amount (or degree) of structured water around the ions of the salts provides a different electrostatic capability to them against the head groups of the surfactants in the micelle interface that plays an important role on the stabilization of the different nematic phases. Indeed, this effect arises from the chaotropic and kosmotropic properties of the ions/counterions. Small ions have relatively high surface-charge density [89] and they exhibit high tendency to be hydrated by a large amount of free water molecules (water-structuring or kosmotropic ions [90–92]) with the negatively greater enthalpy of hydration with respect to bigger ions [90–92]. On the contrary, big ions show opposite behaviour with respect to small ones and they are less hydrated (water-breaking or choatropic ions [90–92]). The kosmotropic and chaotropic properties of the ions are in close relation with the hydrodynamic radius of the ions, R. Comparing two ions with different physical (ionic) radius, r_s and r_b ($r_s < r_b$), and same electrostatic charge, as stated above, smaller ions are surrounded by a higher amount of structured water molecules with respect to the bigger ions, as shown in Figure 8. If two ions with $r_s \neq r_b$ and $R_s \approx R_b$, it is expected that they show similar kosmotropic or chaotropic character. This can be seen in our experimental results given in Figures 6 and 7. While the sequential ordering of the ionic radii of the alkali ions is Li^+ (0.59) < Na^+ (0.99) < K^+ (1.37) < Rb^+ (1.52) < Cs^+ (1.67), their hydrodynamic radii sequence is Li^+ (3.40) >Na^+ (2.76) >K^+ (2.32) $\approx$ Rb^+ (2.28) $\approx$ Cs^+ (2.28), where the numbers between parentheses are the ionic and hydrodynamic radius of the ions in Å [93]. In Figures 6 and 7, it was observed that K_2SO_4, Rb_2SO_4 and Cs_2SO_4 affected the phase topology of the nematic-host phases in the similar way because their alkali ions have almost same hydrodynamic radii. At first sight, it seems that the birefringences of the nematic phases could be related to the hydrodynamic radius of ions. Although, for instance, K^+ ion has smaller hydrodynamic radius than Na^+, the mixture with K^+ has lower birefringences in the N_D phase with respect to the one with Na^+ in KL system (Figure 6a,b). However, in the case of SDS system, as shown in Figure 7b,c, there is an opposite situation for the same ions. Remembering that KL is kosmotropic and SDS is chaotropic type surfactants, the head groups of the surfactants KL and SDS interact with strongly kosmotropic Na^+ and chaotropic K^+ ions, respectively, to produce strongly bound ion-pairs at the micelle surfaces, which leads to the formation of larger micelles. X-ray diffraction studies indicated that the larger the micelles the higher the birefringences [15]. Consequently, it would be better to conclude that the larger value of the birefringences is attributed to the formation of tightly bound ion-pairs at the micelle surfaces.

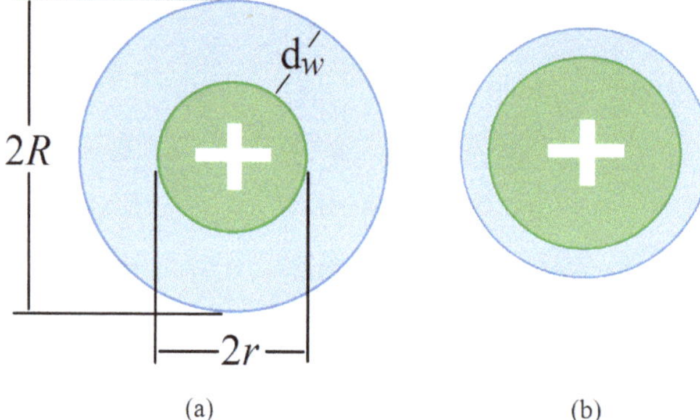

Figure 8. Comparative representation of ionic ($2r$) and hydrodynamic ($2R$) diameter of small kosmotropic (**a**) and the bigger chaotropic ions (**b**) with the same electrostatic charge. d_w is the thickness of the water layer.

The results discussed in [88] may be considered as preliminary. For this reason, complementary investigations were needed to understand the effect of the specific interactions between the ionic species at the micelle surfaces on the stabilization of the lyotropic nematic phases, especially the biaxial. Sodium salts of some Hofmeister ions were added [12] to the host lyotropic mixture of dodecyltrimethylammonium bromide (DTMABr)/dodecanol (DDeOH)/water. The results showed that the interactions between the positively charged head group of DTMABr, trimethylammonium, and the Hofmeister series kosmotropic (SO_4^{2-}, F^-, Cl^-), and chaotropic (Br^-, NO_3^-, I^-, ClO_4^-) anions play an important role on stabilizing the lyotropic nematic phases. The relatively strong interactions of the trimethylammonium head group, which has a chaotropic property, with the most chaotropic anions ClO_4^- promoted the stabilization of just the N_D phase. The higher the kosmotropic (chaotropic) character of the anions the bigger the N_C (N_D) temperature domain.

Although the results of [12] provided useful results about the stabilization the different nematic phases, it was necessary to generalize the role of specific interactions on the micelle's surfaces in terms of the kosmotropic and chaotropic properties of both the surfactant head groups and the counterions/ions. For this purpose, three surfactant molecules were selected, two of them with chaotropic head groups (one positively charged, the other negatively charged) and one with a negatively charged kosmotropic head group. The chaotropic surfactants were sodium dodecylsulfate (SDS) and tetradecyltrimethylammonium bromide (TTMABr), and the kosmotropic one potassium laurate (KL). Three mixtures were prepared, SDS/alkali salt/dodecanol/water, TTMABr/sodium salt/DeOH/water and KL/alkali salt/DeOH/water. The alkali salts (sodium salts) were Li_2SO_4,

Na_2SO_4, K_2SO_4, Rb_2SO_4 and Cs_2SO_4 (Na_2SO_4, NaF, NaCl, NaBr, $NaNO_3$, $NaClO_4$ and NaSCN). The results [14] indicated that when the surfactant head groups and the ions bound to them at the micelle's surface exhibit the same property in terms of the chaotropic or kosmotropic character, the ionic species shows highly-strongly bound ion pairs (Figure 9a), which lead to the stabilization of the N_D phase. If they form highly-loosely bound ion pairs (Figure 9b), the N_C phase is stabilized. For the ion pairs formed by the intermediate level of the interactions between the head groups and the ions, the N_B is favored, as shown in Figure 9c.

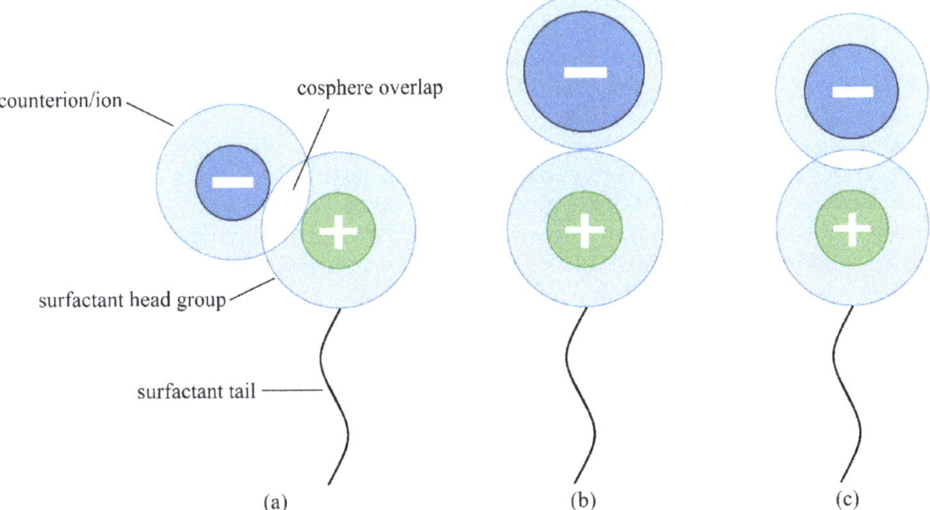

Figure 9. Formation of the ion pairs, considering the interactions between the positively charged kosmotropic surfactant head group and (**a**) kosmotropic, (**b**) chaotropic, (**c**) weakly chaotropic or kosmotropic counterion/ion. This phenomenon was described by Moriera and Froozabadi in Ref [94]. For a surfactant with chaotropic head group, similar situation is in question, except that it produces tightly bound ions pairs with a chaotropic ions and loosely bound ion pairs with a kosmotropic ions, as in the (**a**) and (**b**), respectively.

At this point, it would be better to discuss how the surfactant molecules aggregate in the micelles as a result of the formation of tightly/loosely bound ion-pairs at the micelle surfaces. When an ion is strongly bound to the head group of a surfactant, it screens the Coulombic repulsions between the head groups and the surfactant molecules are packed well in the micelles. In this case, the micelle surface curvature is flatter with respect to the one appeared with loosely interacting head group-ion pairs. This situation applies to strong kosmotrope-kosmotrope or chaotrope-chaotrope interactions between the ionic species. As the surfactant head groups and the ions presence the opposite kosmotrope/chaotrope characters, the repulsions between the head groups are screened less and the micelle surface curvature starts to increase. Thus, in the case of lyotropic nematic phases, from N_C to N_D phase passing through the N_B phase by changing the temperature, the surfactant head groups are packed well at the micelle surfaces leading to the growth in the micelle size and the micelle surface curvature decreases.

3.3. Effect of Localization of Weak Electrolytes at Micelle's Surface

Another effect of the modifications on the micelle's surface in the stabilization of uniaxial and biaxial nematic phases is the location of the dopant molecules (e.g., strong or weak electrolytes) on the micelle's surface. It is known that when a strong electrolyte is added to a lyotropic mixture, its oppositely charged ions with respect to the surfactant head groups produce highly bound ion pairs as a result of strong Coulombic attractions on the micelle's surface. This gives rise to the packing of the surfactant molecules within the micelles and the growth of the micelle size in the A'-B' plane of the orthorhombic micelles (Figure 4a). This process leads to the stabilization of N_D and/or N_B phase. In the case of weak electrolytes, as expected, the interactions of polar parts of these electrolytes with the surfactant head groups are not as strong as those of ions from the strong electrolytes. At first sight, it might be assumed that this situation is due to the difference in the solubilities of strong and weak electrolytes. We chose some weak electrolytes with –OH and –COOH polar parts [13] to investigate this hypothesis. The experimental results indicated that there exists no direct relation between the solubilities of those weak electrolytes in water and their effectiveness for stabilizing the uniaxial and biaxial nematic phases. In contrast to the solubility, the acidity constants, K_a, of the dopant molecules (i.e., weak electrolytes) emerged as an important parameter in obtaining different types of nematic phase. If a dopant molecule with $pK_a < 7$ is chosen, it is highly possible to stabilize three nematic phases. However, in the case of $pK_a > 7$ and exactly the same amount of dopant with pKa < 7, the mixture with the dopant molecule exhibit the same nematic phase type that the mixture without the dopant. In other words, the dopant with $pK_a > 7$ does not change the type of the nematic phase of the host mixture. This situation was attributed to the different location of the dopant molecules in the micelles and their effectiveness of screening the repulsions between the ionic head groups of the surfactants.

4. Conclusions

These recent studies have provided advances on the understanding of the biaxial nematic phase stabilization. We may conclude that the choosing the optimum mixture constituents is a key point to stabilize different nematic phases. Parameters as the relative alkyl chain lengths of surfactant and alcohols and occurrence of strong or weak interactions between the ionic species at the micelle surfaces have to be considered in the preparation of a lyotropic mixture aiming the obtaining of particular nematic phases. The following conclusions may be helpful for experimentalists to obtain the lyotropic mixture presenting uniaxial and/or biaxial nematic phases:

- Considering the relative alkyl chain length of both surfactant (m) and alcohol (n), the higher (smaller) the value of the m (n), when compared with the value of n (m), the larger the phase domains of the N_D and N_B (N_C) [11,15]. Indeed, this situation is a result of the molecular segregation of the surfactant and alcohol molecules in the micelles. Remember that according to the IBM model proposed for the stabilization of three lyotropic nematic phases, micelles are assumed to have an orthorhombic symmetry and there exist two main parts in the micelles: the flattest part in the plane perpendicular to the surfactant long molecular axis (A'-B' plane in Figure 4a) and the curved part at the rims of the micelles. Our results [11,15] indicated that there is a possibility that more alcohol molecules tend to be located in the curved parts of the micelle when the value of n gets bigger. Consequently, if we have a lyotropic mixture with an alcohol as a cosurfactant exhibiting just N_D (N_C), and one wants to stabilize the N_B phase, we have to change the alcohol by another with shorter (longer) alkyl chain length with respect to that of the main surfactant.
- In the case of the specific interactions between head groups of the surfactants and the counterions/ions of strong electrolytes, in terms of their kosmotropic and chaotropic characters, choosing the surfactant and the electrolyte with slightly opposite (strongly same) character may help to stabilize a lyotropic mixture of the N_B (N_D) phase. Strongly opposite characters of both ionic species, head groups and the ions present in the mixture, stabilizes the N_C phase [12,14].
- The researchers have been using, in general, KL/DeOH/water mixtures to prepare lyotropic mixtures presenting the biaxial nematic phase. However, sometimes the reproducible results could not be obtained. As known, the purity of the KL is very important to obtain the reproducible experimental results. In early studies, the researchers reported in the synthesis of the KL molecule

"KL was synthesized by neutralization of lauric acid with KOH". Here, the term "neutralization" does not include the control of the pH. However, Berejnov et al. showed that the neutralization of all lauric acid to give KL is a crucial point to have reproducible results for any mixture of the surfactant KL [95]. To do so, the reaction pH is kept at ~10.8. Otherwise, if the pH is less than 10.8, some amount of lauric acid may remain in the solution without neutralizing, and then excess of lauric acid and DeOH presence in the lyotropic mixture may cause the esterification of lauric acid. Melnik and Saupe proposed a slow esterification reaction between the alcohol and KL; however, most probably this reaction occurs between the alcohol and lauric acid, instead of the soap [81]. Thus, we can say that the experimentalists have to consider the purity of the KL to have chemically stable lyotropic mixtures.

Consequently, if we have a chance to give "a receipt" to experimentalists for preparation of a lyotropic mixture presenting the biaxial phase, the followings may be considered:

- check the purities of the surfactant and other ingredients of the lyotropic mixture,
- consider the relative alkyl chain lengths of both surfactant and alcohol molecules,
- choose the surfactant head groups and counterions/ions of electrolytes with oppositely kosmotrope or chaotrope characters.

Funding: This study was funded by the Scientific and Technological Research Council of Turkey (TÜBİTAK) [grant number: 113Z469 and 114Z031], CNPq (Conselho Nacional de Desenvolvimento Científico e Tecnológico) (465259/2014-6), FAPESP (Fundação de Amparo à Pesquisa do Estado de São Paulo) (2014/50983-3 and 2016/24531-3), CAPES (Coordenação de Aperfeiçoamento de Pessoal de Nível Superior), INCT-FCx (Instituto Nacional de Ciência e Tecnologia de Fluidos Complexos), and NAP-FCx (Núcleo de Apoio à Pesquisa de Fluidos Complexos).

Conflicts of Interest: The authors declare no conflicts of interest.

References

1. Yu, L.J.; Saupe, A. Observation of a Biaxial Nematic Phase in Potassium Laurate-1-Decanol-Water Mixtures. *Phys. Rev. Lett.* **1980**, *45*, 1000–1003. [CrossRef]
2. Neto, A.M.F.; Salinas, S.R.A. *The Physics of Lyotropic Liquid Crystals: Phase Transitions and Structural Properties*, 1st ed.; Oxford University Press: New York, NY, USA, 2005.
3. Kim, Y.K.; Senyuk, B.; Shin, S.T.; Kohlmeier, A.; Mehl, G.H.; Lavrentovich, O.D. Surface alignment, anchoring transitions, optical properties, and topological defects in the thermotropic nematic phase of organo-siloxane tetrapodes. *Soft Matter* **2014**, *10*, 500–509. [CrossRef] [PubMed]
4. Kim, Y.K.; Cukrov, G.; Vita, F.; Scharrer, E.; Samulski, E.T.; Francescangeli, O.; Lavrentovich, O.D. Search for Microscopic and Macroscopic Biaxiality in the Cybotactic Nematic Phase of New Oxadiazole Bent-Core Mesogens. *Phys. Rev. E* **2016**, *93*, 062701. [CrossRef] [PubMed]
5. Nasrin, L.; Kabir, E.; Rahman, M. External Magnetic Field-Dependent Tricritical Points of Uniaxial-To-Biaxial Nematic Transition. *Phase Transit.* **2016**, *89*, 193–201. [CrossRef]
6. Luckhurst, G.R.; Sluckin, T.J. *Biaxial Nematic Liquid Crystals, Theory, Simulation, and Experiment*, 1st ed.; Wiley: Chichester, UK, 2015.
7. Bartolino, R.; Chiaranza, T.; Meuti, M.; Compagnoni, R. Uniaxial and Biaxial Lyotropic Nematic Liquid Crystals. *Phys. Rev. A* **1982**, *26*, 1116–1119. [CrossRef]
8. Galerne, Y.; Liébert, L. Zigzag Disclinations in Biaxial Nematic Liquid Crystals. *Phys. Rev. Lett.* **1985**, *55*, 2449–2451. [CrossRef] [PubMed]
9. Oliveira, E.A.; Liebert, L.; Neto, A.M.F. A New Soap/Detergent/Water Lyotropic Liquid Crystal with a Biaxial Nematic Phase. *Liq. Cryst.* **1989**, *5*, 1669–1675. [CrossRef]
10. Akpinar, E.; Reis, D.; Neto, A.M.F. Lyotropic Mixture Made of Potassium Laurate/1-Undecanol/K_2SO_4/Water Presenting High Birefringences and Large Biaxial Nematic Phase Domain: A Laser Conoscopy Study. *Eur. Phys. J. E Soft Matter Biol. Phys.* **2012**, *35*, 50. [CrossRef]

11. Akpinar, E.; Reis, D.; Neto, A.M.F. Effect of Alkyl Chain Length of Alcohols on Nematic Uniaxial-to-Biaxial Phase Transitions in a Potassium Laurate/Alcohol/K_2SO_4/Water Lyotropic Mixture. *Liq. Cryst.* **2012**, *39*, 881–888. [CrossRef]

12. Akpinar, E.; Reis, D.; Neto, A.M.F. Effect of Hofmeister Anions on the Existence of the Biaxial Nematic Phase in Lyotropic Mixtures of Dodecyltrimethylammonium Bromide/Sodium Salt/1-Dodecanol/Water. *Liq. Cryst.* **2015**, *42*, 973–981. [CrossRef]

13. Akpinar, E.; Otluoglu, K.; Turkmen, M.; Canioz, C.; Reis, D.; Neto, A.M.F. Effect of the Presence of Strong and Weak Electrolytes on the Existence of Uniaxial and Biaxial Nematic Phases in Lyotropic Mixtures. *Liq. Cryst.* **2016**, *43*, 1693–1708. [CrossRef]

14. Akpinar, E.; Turkmen, M.; Canioz, C.; Neto, A.M.F. Role of Kosmotrope-Chaotrope Interactions at Micelle Surfaces on the Stabilization of Lyotropic Nematic Phases. *Eur. Phys. J. E* **2016**, *39*, 107. [CrossRef] [PubMed]

15. Akpinar, E.; Canioz, C.; Turkmen, M.; Reis, D.; Neto, A.M.F. Effect of the Surfactant Alkyl Chain Length on the Stabilization of Lyotropic Nematic Phases. *Liq. Cryst.* **2018**, *45*, 219–229. [CrossRef]

16. Freiser, M.J. Ordered States of a Nematic Liquid. *Phys. Rev. Lett.* **1970**, *24*, 1041–1043. [CrossRef]

17. Shih, C.S.; Alben, R. Lattice model for biaxial liquid crystals. *J. Chem. Phys.* **1972**, *57*, 3055–3061. [CrossRef]

18. Alben, R. Liquid crystal phase transitions in mixtures of rodlike and platelike molecules. *J. Chem. Phys.* **1973**, *59*, 4299–4304. [CrossRef]

19. Rabin, Y.; McMullen, W.E.; Gelbart, W.M. Phase Diagram Behaviors for Rod/Plate Liquid Crystal Mixtures. *Mol. Cryst. Liq. Cryst.* **1982**, *89*, 67–76. [CrossRef]

20. Chen, Z.Y.; Deutch, J.M. Biaxial Nematic Phase, Multiphase Critical Point, and Reentry Transition in Binary Liquid Crystal Mixtures. *J. Chem. Phys.* **1984**, *80*, 2151–2162. [CrossRef]

21. Stroobants, A.; Lekkerkerker, H.N.W. Liquid Crystal Phase Transitions in a Solution of Rodlike and Disklike Particles. *J. Phys. Chem.* **1984**, *88*, 3669–3674. [CrossRef]

22. Muhoray, P.P.; Bruyn, J.R.; Dunmur, D.A. Phase Behavior of Binary Nematic Liquid Crystal Mixtures. *J. Chem. Phys.* **1985**, *82*, 5294–5295. [CrossRef]

23. Kooij, F.M.; Lekkerkerker, H. Liquid-Crystal Phases Formed in Mixed Suspensions of Rod- and Platelike Colloids. *Langmuir* **2000**, *16*, 10144–10149. [CrossRef]

24. Lawson, K.D.; Flautt, T.J. Magnetically Oriented Lyotropic Liquid Crystalline Phases. *J. Am. Chem. Soc.* **1967**, *89*, 5489. [CrossRef]

25. Radley, K.; Reeves, L.W. Studies of Ternary Nematic Phases by Nuclear Magnetic Resonance. Alkali Metal Decyl Sulfates/Decanol/D2O. *Can. J. Chem.* **1975**, *53*, 2998–3004. [CrossRef]

26. Radley, K.; Reeves, L.W. Effect of counterion substitution on the type and nature of nematic lyotropic phases from nuclear magnetic resonance studies. *J. Phys. Chem.* **1976**, *80*, 174–182. [CrossRef]

27. Forrest, B.J.; Reeves, L.W.; Robinson, C.J. Transition in the sign of the diamagnetic anisotropy of a lyotropic mesophase without a phase change. *Type O disk micelle systems. J. Phys. Chem.* **1981**, *85*, 3244–3247.

28. Charvolin, J.; Levelut, A.M.; Samulski, E.T. Lyotropic nematics: Molecular aggregation and susceptibilities. *J. Phys. Lett.* **1979**, *40*, 587–592. [CrossRef]

29. Radley, K.; Saupe, A. The Structure of Lyotropic Nematic Decylammoniumchloride and Bromide Systems by PMR of Monomethyltin Complexes and by Microscopic Studies. *Mol. Cryst. Liq. Cryst.* **1978**, *44*, 227–235. [CrossRef]

30. Amaral, L.Q.; Pimentel, C.A.; Tavares, M.R.; Vanin, J.A. Study of a Magnetically Oriented Lyotropic Mesophase. *J. Chem. Phys.* **1979**, *71*, 2940–2945. [CrossRef]

31. Amaral, L.Q.; Tavares, M.R. On the Possible Formation of Macromicelles in a Lyomesophase. *Mol. Cryst. Liq. Cryst. Lett.* **1980**, *56*, 203–208. [CrossRef]

32. Yu, L.J.; Saupe, A. Liquid Crystalline Phases of the Sodium Decylsulfate/Decanol/Water System. Nematic-Nematic and Cholesteric-Cholesteric Phase Transitions. *J. Am. Chem. Soc.* **1980**, *102*, 4879–4883. [CrossRef]

33. Freiser, M.J. Successive Transitions in a Nematic Liquid. *Mol. Cryst. Liq. Cryst.* **1971**, *14*, 165–182. [CrossRef]

34. Henriques, E.F.; Henriques, V.B. Biaxial Phases in Polydisperse Mean-Field Model Solution of Uniaxial Micelles. *J. Chem. Phys.* **1997**, *107*, 8036–8040. [CrossRef]

35. Severing, K.; Saalwaachter, K. *Phase Biaxiality in Nematic Liquid Crystals in: Thermotropic Liquid Crystals: Recent Advances*; Springer: Dordrecht, The Netherlands, 2007; Chapter 5; pp. 141–170.

36. Van den Pol, E.; Petukhov, A.V.; Thies-Weesie, D.M.; Byelov, D.V.; Vroege, G.J. Experimental Realization of Biaxial Liquid Crystal Phases in Colloidal Dispersions of Boardlike Particles. *Phys. Rev. Lett.* **2009**, *103*, 258301. [CrossRef] [PubMed]

37. Sauerwein, R.A.; Oliveira, M.J. Lattice Model for Biaxial and Uniaxial Liquid Crystals. *J. Chem. Phys.* **2016**, *144*, 194904. [CrossRef] [PubMed]

38. Liu, J.; Guan, R.; Dong, X.; Dong, Y. Molecular Properties of a Bent-Core Nematic Liquid Crystal A131 by Multi-Level Theory Simulations. *Mol. Simul.* **2018**, *44*, 1539–1543. [CrossRef]

39. Taylor, T.R.; Fergason, J.L.; Arora, S.L. Biaxial Liquid Crystals. *Phys. Rev. Lett.* **1970**, *24*, 359–362. [CrossRef]

40. Neto, A.M.F.; Liebet, L.; Galerne, Y. Temperature and Concentration Range of the Biaxial Nematic Lyomesophase in the Mixture Potassium Laurate/1-Decanol/D_2O. *J. Phys. Chem.* **1985**, *89*, 3737–3739. [CrossRef]

41. Galerne, Y.; Neto, A.MF.; Liebert, L. Microscopic Structure of the Uniaxial and Biaxial Lyotropic Nematics. *J. Chem. Phys.* **1987**, *87*, 1851–1856. [CrossRef]

42. Madsen, L.A.; Dingemans, T.J.; Nakata, M.; Samulski, E.T. Thermotropic Biaxial Nematic Liquid Crystals. *Phys. Rev. Lett.* **2004**, *92*, 145505. [CrossRef]

43. Luckhurst, G.R. Biaxial Nematic Liquid Crystals: Fact or Fiction? *Thin Solid Film.* **2001**, *393*, 40–52. [CrossRef]

44. Acharya, B.R.; Primak, A.; Kumar, S. Biaxial Nematic Phase in Bent-Core Thermotropic Mesogens. *Phys. Rev. Lett.* **2004**, *92*, 145506. [CrossRef]

45. Luckhurst, G.R. Liquid Crystals—A Missing Phase Found at Last? *Nature* **2004**, *430*, 413–414. [CrossRef]

46. Galerne, Y. Comment on Thermotropic Biaxial Nematic Liquid Crystals. *Phys. Rev. Lett.* **2006**, *96*, 219803. [CrossRef]

47. Madsen, L.A.; Dingemans, T.J.; Nakata, M.; Samulski, E.T. A Reply to the Comment by Yves Galerne. *Phys. Rev. Lett.* **2006**, *96*, 219804. [CrossRef]

48. Straley, J.P. Ordered Phases of a Liquid of Biaxial Particles. *Phys. Rev. A* **1974**, *10*, 1881–1887. [CrossRef]

49. Biscarini, F.; Chiccoli, C.; Pasini, P.; Semeria, F.; Zannoni, C. Phase Diagram and Orientational Order in a Biaxial Lattice Model: A Monte Carlo Study. *Phys. Rev. Lett.* **1995**, *75*, 1803–1806. [CrossRef]

50. Camp, P.J.; Allen, M.P. Phase Diagram of the Hard Biaxial Ellipsoid Fluid. *J. Chem. Phys.* **1997**, *106*, 6681. [CrossRef]

51. Taylor, M.P.; Herzfeld, J. Nematic and Smectic Order in a Fluid of Biaxial Hard Particles. *Phys. Rev. A* **1991**, *44*, 3742–3751. [CrossRef]

52. Vanakaras, A.G.; Bates, M.A.; Photinos, D.J. Theory and Simulation of Biaxial Nematic and Orthogonal Smectic Phases Formed by Mixtures of Boardlike Molecules. *Phys. Chem. Chem. Phys.* **2003**, *5*, 3700–3706. [CrossRef]

53. Berardi, R.; Muccioli, L.; Orlandi, S.; Ricci, M.; Zannoni, C. Computer Simulations of Biaxial Nematics. *J. Phys. Condens Matter* **2008**, *20*, 463101. [CrossRef]

54. Hashim, R.; Luckhurst, G.R.; Romano, S. Computer Simulation Studies of Anisotropic Systems. *Mol. Phys.* **1984**, *53*, 1535. [CrossRef]

55. Hashim, R.; Luckhurst, G.R.; Prata, F.; Romano, S. Computer Simulation Studies of Anisotropic Systems. XXII. An Equimolar Mixture of Rods and Discs: A Biaxial Nematic? *Liq. Cryst.* **1993**, *15*, 283–309. [CrossRef]

56. Pratibha, R.; Madhusudana, N.V. Evidence for two Coexisting Nematic Phases in Mixtures of Rod-Like and Disc-Like Nematogens. *Mol. Cryst Liq. Cryst.* **1985**, *1*, 111–116.

57. Pratibha, R.; Madhusudana, N.V. On the Occurrence of Point and Ring Defects in the Nematic-Nematic Coexistence Range of a Binary Mixture of Rod-like and Disc-like Mesogens, Molecular Crystals and Liquid Crystals Incorporating Nonlinear Optics. *Mol. Cryst. Liq. Cryst.* **1990**, *178*, 167–178.

58. Neto, A.M.F.; Galerne, Y. Yotropic Systems. In *Biaxial Nematic Liquid Crystals, Theory, Simulation, and Experiment*; Wiley: Chichester, UK, 2015; Chapter 11; pp. 285–304.

59. Martínez-Ratón, Y.; Gonzalez-Pinto, M.; Velasco, E. Biaxial Nematic Phase Stability and Demixing Behaviour in Monolayers of Rod-Plate Mixtures. *Phys. Chem. Chem. Phys.* **2016**, *18*, 24569. [CrossRef] [PubMed]

60. Sharma, S.R.; Muhoray, P.P.; Bergersen, B.; Dunmur, D.A. Stability of a Biaxial Phase in a Binary Mixture of Nematic Liquid Crystals. *Phys. Rev. A* **1985**, *32*, 3752. [CrossRef]

61. Martínez-Ratón, Y.; Cuesta, J.A. Enhancement by Polydispersity of the Biaxial Nematic Phase in a Mixture of Hard Rods and Plates. *Phys. Rev. Lett.* **2002**, *89*, 185701. [CrossRef]

62. Vanakaras, A.G.; Photinos, D.J. Theory of Biaxial Nematic Ordering in Rod-Disc Mixtures Revisited. *Mol. Cryst. Liq. Cryst.* **1997**, *299*, 65–71. [CrossRef]

63. Vanakaras, A.G.; McGrother, S.C.; Jackson, G.; Photinos, D.J. Hydrogen-Bonding and Phase Biaxiality in Nematic Rod-Plate Mixtures. *Mol. Cryst. Liq. Cryst.* **1998**, *323*, 199–209. [CrossRef]

64. Varga, S.; Galindo, A.; Jackson, G. Global Fluid Phase Behavior in Binary Mixtures of Rodlike and Platelike Molecules. *J. Chem. Phys.* **2002**, *117*, 7207. [CrossRef]

65. Varga, S.; Galindo, A.; Jackson, G. Phase Behavior of Symmetric Rod–Plate Mixtures Revisited: Biaxiality Versus Demixing. *J. Chem. Phys.* **2002**, *117*, 10412. [CrossRef]

66. Varga, S.; Galindo, A.; Jackson, G. Ordering Transitions, Biaxiality, and Demixing in The Symmetric Binary Mixture of Rod and Plate Molecules Described with the Onsager Theory. *Phys. Rev. E* **2002**, *66*, 011707. [CrossRef]

67. Galindo, A.; Haslam, A.J.; Varga, S.; Jackson, G.; Vanakaras, A.G.; Photinos, D.J.; Dunmur, D.A. The Phase Behavior of a Binary Mixture of Rodlike and Disclike Mesogens: Monte Carlo Simulation, Theory, and Experiment. *J. Chem. Phys.* **2003**, *119*, 5216–5225. [CrossRef]

68. Kooij, F.M.; Lekkerkerker, H.N.W. Liquid-Crystalline Phase Behavior of a Colloidal Rod-Plate Mixture. *Phys. Rev. Lett.* **2000**, *84*, 781. [CrossRef]

69. Kooij, F.M.; Vogel, M.; Lekkerkerker, H.N.W. Phase Behavior of a Mixture of Platelike Colloids and Nonadsorbing Polymer. *Phys. Rev. E* **2000**, *62*, 5397. [CrossRef]

70. Kooij, F.M.; Lekkerkerker, H.N.W. Liquid-Crystal Phase Transitions in Suspensions of Plate-Like Particles. *Philos Trans. R Soc. B* **2001**, *359*, 985–995. [CrossRef]

71. Kooij, F.M.; Beek, D.; Lekkerkerker, H.N.W. Isotropic−Nematic Phase Separation in Suspensions of Polydisperse Colloidal Platelets. *J. Phys. Chem. B* **2001**, *105*, 1696. [CrossRef]

72. Wensink, H.H.; Vroege, G.J.; Lekkerkerker, H.N.W. Isotropic-Nematic Phase Separation in Asymmetrical Rod-Plate Mixtures. *J. Chem. Phys.* **2001**, *115*, 7319. [CrossRef]

73. Wensink, H.H.; Vroege, G.J.; Lekkerkerker, H.N.W. Biaxial versus uniaxial nematic stability in asymmetric rod-plate mixtures. *Phys. Rev. E* **2002**, *66*, 041704. [CrossRef]

74. Galerne, Y.; Itoua, J.; Liebert, L. Zigzag disclinations in uniaxial nematic liquid crystals. *J. Phys.* **1988**, *49*, 681–687. [CrossRef]

75. Galerne, Y. Biaxial nematics. *Mol. Cryst. Liq. Cryst.* **1988**, *165*, 131–149. [CrossRef]

76. Galerne, Y. Vanishing disclination lines in lyotropic biaxial nematics. *Mol. Cryst. Liq. Cryst.* **1997**, *292*, 103–112. [CrossRef]

77. Santos, M.B.L.; Galerne, Y.; Durand, G. Critical Slowing Down of Biaxiality Fluctuations at the Uniaxial-to-Biaxial Phase Transition in a Lyotropic Disklike Nematic Liquid Crystal. *Phys. Rev. Lett.* **1984**, *53*, 787–790. [CrossRef]

78. Neto, A.M.F.; Galerne, Y.; Levelut, A.M.; Liebert, L. Pseudo-Lamellar Ordering in Uniaxial and Biaxial Lyotropic Nematics-A Synchrotron X-Ray Diffraction Experiment. *J. Phys. Lett.* **1985**, *46*, 409–505.

79. Akpinar, E.; Reis, D.; Neto, A.M.F. Anomalous Behavior in the Crossover Between the Negative and Positive Biaxial Nematic Mesophases in a Lyotropic Liquid Crystal. *ChemPhysChem* **2014**, *15*, 1463–1469. [CrossRef] [PubMed]

80. Hendrikx, Y.; Charvolin, J.; Rawiso, M. Uniaxial-Biaxial Phase Transition in Lyotropic Nematic Solutions: Local Biaxiality in the Uniaxial Phase. *Phys. Rev. B Condens Matter* **1986**, *33*, 3534–3537. [CrossRef] [PubMed]

81. Melnik, G.; Saupe, A. Microscopic Textures of Micellar Cholesteric Liquid Crystals. *Mol. Cryst. Liq. Cryst.* **1987**, *145*, 95–110. [CrossRef]

82. Vasilevskaya, A.S.; Kitaeva, E.L.; Sonin, A.S. Optically Biaxial Mesophases in Lyotropic Nematics. *Russ. J. Phys. Chem.* **1990**, *64*, 599–601.

83. Ho, C.C.; Hoetz, R.J.; El-Aasser, M.S. A Biaxial Lyotropic Nematic Phase in Dilute Solutions of Sodium Lauryl Sulfate-1-Hexadecanol-Water. *Langmuir* **1991**, *7*, 630–635. [CrossRef]

84. Quist, P.O. First Order Transitions to a Lyotropic Biaxial Nematic. *Liq. Cryst.* **1995**, *18*, 623–629. [CrossRef]

85. Filho, A.M.; Laverde, A.; Fujiwara, F.Y. Observation of Two Biaxial Nematic Mesophases in the Tetradecyltrimethylammonium Bromide/Decanol/Water System. *Langmuir* **2003**, *19*, 1127–1132. [CrossRef]

86. Alben, R. Phase transitions in a fluid of biaxial particles. *Phys. Rev. Lett.* **1973**, *30*, 778–781. [CrossRef]

87. Galerne, Y.; Marcerou, J.P. Temperature Behavior of the Order-Parameter Invariants in the Uniaxial and Biaxial Nematic Phases of a Lyotropic Liquid Crystal. *Phys. Rev. Lett.* **1983**, *51*, 2109–2111. [CrossRef]

88. Akpinar, E.; Reis, D.; Meto, A.M.F. Investigation of the Interaction of Alkali Ions with Surfactant Head Groups for the Formation of Lyotropic Biaxial Nematic Phase via Optical Birefringence Measurements. *Proc. Soc. Photo-Opt. Inst. Eng. (SPIE)* **2013**, *8642*, 864203.

89. Trompette, J.L.; Arurault, L.; Fontorbes, S.; Massot, L. Influence of the anion specificity on the electrochemical corrosion of anodized aluminum substrates. *Electrochim. Acta* **2010**, *55*, 2901–2910. [CrossRef]

90. Bridges, N.J.; Gutowski, K.E.; Rogers, R.D. Investigation of aqueous biphasic systems formed from solutions of chaotropic salts with kosmotropic salts (salt–salt ABS). *Green Chem.* **2007**, *9*, 77–183. [CrossRef]

91. Collins, K.D. Ions from the Hofmeister series and osmolytes: Effects on proteins in solution and in the crystallization process. *Methods* **2004**, *34*, 300–311. [CrossRef]

92. Collins, K.D.; Neilson, G.W.; Enderby, J.E. Ions in water: Characterizing the forces that control chemical processes and biological structure. *Biophys. Chem.* **2007**, *128*, 95–104. [CrossRef]

93. Ropers, M.H.; Czichocki, G.; Brezesinski, G. Counterion Effect on the Thermodynamics of Micellization of Alkyl Sulfates. *J. Phys. Chem. B* **2003**, *107*, 5281–5288. [CrossRef]

94. Moriera, L.; Firoozabadi, A. Molecular Thermodynamic Modeling of Specific Ion Effects on Micellization of Ionic Surfactants. *Langmuir* **2010**, *26*, 15177–15191. [CrossRef]

95. Berejnov, V.V.; Cabuil, V.; Perzynski, R.; Raikher, Y.L.; Lysenko, S.N.; Sdobnov, V.N. Lyotropic Nematogenic System Potassium Laurate–1-Decanol–Water: Method of Synthesis and Study of Phase Diagrams. *Cryst. Rep.* **2000**, *45*, 541–545. [CrossRef]

Article

Magnetic Field and Dilution Effects on the Phase Diagrams of Simple Statistical Models for Nematic Biaxial Systems

Daniel D. Rodrigues, André P. Vieira * and Silvio R. Salinas *

Instituto de Fisica, Universidade de Sao Paulo, Rua do Matao, 1371, Sao Paulo 05508-090, Brazil;
daniel.dias.rodrigues@usp.br
* Correspondence: apvieira@if.usp.br (A.P.V.); ssalinas@if.usp.br (S.R.S.)

Received: 19 June 2020; Accepted: 20 July 2020; Published: 22 July 2020

Abstract: We use a simple statistical model to investigate the effects of an applied magnetic field and of the dilution of site elements on the phase diagrams of biaxial nematic systems, with an emphasis on the stability of the Landau multicritical point. The statistical lattice model consists of intrinsically biaxial nematogenic units, which interact via a Maier–Saupe potential, and which are characterized by a discrete choice of orientations of the microscopic nematic directors. According to previous calculations at zero field and in the absence of dilution, we regain the well-known sequence of biaxial, uniaxial, and disordered structures as the temperature is increased, and locate the Landau point. We then focus on the topological changes induced in the phase diagram by the application of an external magnetic field, and show that the Landau point is destabilized by the presence of an applied field. On the other hand, in the absence of a field, we show that only a quite strong dilution of nematic sites is capable of destabilizing the Landau point.

Keywords: biaxial nematic transition; field behavior; diluted nematic systems

---

1. Introduction

In a number of recent investigations [1–6], we have performed some calculations for simple statistical lattice models to characterize the biaxial structures in liquid crystalline systems. These statistical models are based on fully-connected Maier–Saupe pair interactions, with a restricted choice of orientational degrees of freedom. They are amenable to detailed calculations, and have been shown to account for most of the qualitative features of the nematic phase diagrams, including the well-known sequences of biaxial nematic, uniaxial nematic, and isotropic structures, as the temperature is raised. At a special value of temperature and of a parameter gauging the degree of biaxiality, model calculations point out the existence of a highly symmetric Landau multicritical point, with a direct transition from a biaxial nematic to an isotropic phase.

In this work, we revisit the same simplified lattice models with a view to perform detailed calculations to investigate the effects produced by an applied field and by the dilution of nematogenic elements. Besides assuming a generalized form of the Maier–Saupe interactions between nematogenic units on a lattice site, we still make a discrete choice of the orientations of the microscopic nematic directors. This special choice, which is reminiscent of an old work of Zwanzig to treat the Onsager model of rigid cylinders, has also been shown to lead to the well-known first-order transition between simple uniaxial nematic and isotropic phases.

The effects of a magnetic field on the phase diagram of nematogenic molecules exhibiting biaxial phases have been the subject of a number of investigations over the years. For micellar lyotropic systems such as potassium laurate/1-decanol/water mixtures, in which shape anisotropy of the micelles is presumably strongly dependent on the temperature and on the concentration of the various components, the existence of a biaxial nematic phase is known since the pioneering work of Yu and Saupe [7]. Besides the biaxial phase, these systems also exhibit rod-like and disk-like uniaxial nematic phases, and, in general, upon heating, these uniaxial phases exhibit a discontinuous transition to an isotropic phase. Nevertheless, the three nematic phases and the isotropic phase become equal at the Landau multicritical point. In the neighborhood of the Landau point, experiments performed for three different values of concentration [8] show that application of a magnetic field turns the transition between the rod-like uniaxial and the isotropic phases into a crossover, while preserving the discontinuous nature of the transition between the disk-like uniaxial and the isotropic phases.

For mineral lyotropic systems, the combination of an intrinsic magnetic moment and a negative anisotropy of the magnetic susceptibility leads to a very strong induced magnetic birefringence [9] and to phase separation into a uniaxial nematic and a biaxial nematic due to entropic effects associated with polydispersity [10].

In thermotropic systems, prompted by experimental results [11] for bent-core, intrinsically biaxial molecules with an essentially fixed shape, there have been studies on the combined effect of biaxiality and a magnetic field on the temperature of the first-order uniaxial to a paranematic phase transition, on the stability of the various phases, and on the presence of multicritical points [11–16]. In particular, within the phenomenological Landau–de Gennes theory, Mukherjee and Rahman [13] have identified the destabilization of the Landau point and its splitting into one critical and two tricritical points.

In the present work, we go beyond the analysis of a Landau expansion. As we start from a microscopic statistical model, we are able to perform calculations to analyze the global phase diagram, which includes considerations on the stability of the Landau multicritical point and also illustrations of the behavior of the uniaxial and biaxial order parameters as temperature and degree of biaxiality are changed.

On the other hand, our motivation for investigating the effects of dilution is purely theoretical, as we would like to check whether the phase diagram of our simple model survives in the presence of an ingredient that is inescapable in the real world. Our results show that this is indeed the case, making clear that representing a fluid by a lattice model does not introduce any artifacts in our obtained phase diagrams.

In Section 2, we define the statistical model in terms of a quite general and elegant two-tensor formalism, proposed [17] by Sonnet, Virga, and Durand (SVD), and which includes all physically reasonable pair interactions between nematogenic elements. With a discrete choice of the microscopic directors, this model Hamiltonian is restricted to a six-state model, which we dub SVD6. Furthermore, we choose a relation between two interaction parameters, the "geometric-mean condition", which preserves the main features of the phase diagrams, and which has been widely used in the area. We obtain a number of results for the phase diagram of this SVD6 model, in terms of temperature T and a parameter Δ, which comes from the geometric-mean condition and provides a form of gauging the degree of biaxiality. In particular, we locate the Landau multicritical point. In Section 3, we discuss the inclusion of an external field in the SVD6 Hamiltonian with the geometric-mean condition. The effects of a magnetic field come from an intricate competition between the magnetic free energy, the entropy and the interactions between nematogens, which can be fully accounted for by the elementary statistical model. In general, the presence of an external field breaks the symmetry necessary to stabilize the Landau point. Except for the very special circumstances of a fully isotropic diamagnetic susceptibility, the main result is the splitting of the transition lines and the disappearance of the Landau multicritical point. We then turn to the study of the effects of dilution. In Section 4, we use the SVD6 model, with the recourse to the geometric-mean condition, to show that the presence of dilution, as far as it is not a strong dilution, does not introduce any qualitative changes with respect to the initial phase diagrams. However, in the zero-field limit and for quite large dilutions, the Landau point turns into a first-order transition, which is similar to the behavior of the critical line in a strongly diluted Ising ferromagnet.

2. The SVD Model

The Hamiltonian representing the interactions between the nematogens in the mean-field SVD model [17] can be written as

$$\mathcal{H}_{\text{SVD}} = -\frac{A}{2V} \sum_{\mu,\nu \in \{x,y,z\}} \left[\left(\sum_{i=1}^{V} q_i^{\mu\nu} \right)^2 + 2\gamma \left(\sum_{i=1}^{V} q_i^{\mu\nu} \right) \left(\sum_{i=1}^{V} b_i^{\mu\nu} \right) + \lambda \left(\sum_{i=1}^{V} b_i^{\mu\nu} \right)^2 \right], \tag{1}$$

in which the parameter $A > 0$ gauges the interaction between nematogens, the dimensionless parameters γ and λ are responsible for a dependence of that interaction on the relative orientation of the nematogens, and V is the number of nematogens, while $q_i^{\mu\nu}$ and $b_i^{\mu\nu}$, with $\mu, \nu \in \{1, 2, 3\}$, are the components of two traceless tensors associated with the ith nematogen. Explicitly, for a generic nematogen with principal axes along the unit vectors $\hat{n}_1$, $\hat{n}_2$ and $\hat{n}_3$, we have

$$\mathbf{q} = \frac{3}{2}\hat{n}_1 \otimes \hat{n}_1 - \frac{1}{2}\mathbf{I} \quad \text{and} \quad \mathbf{b} = \frac{3}{2}\left(\hat{n}_2 \otimes \hat{n}_2 - \hat{n}_3 \otimes \hat{n}_3 \right), \tag{2}$$

$\mathbf{I}$ being the 3×3 identity matrix.

For our simplified model, we restrict the directors to point along the Cartesian axes [18,19], so that each nematogen takes only six distinct states, defining a new Hamiltonian

$$\mathcal{H}_{\text{SVD6}} = -\frac{A}{2V} \sum_{\alpha \in \{x,y,z\}} \left[\left(\sum_{i=1}^{V} q_i^{\alpha\alpha} \right)^2 + 2\gamma \left(\sum_{i=1}^{V} q_i^{\alpha\alpha} \right) \left(\sum_{i=1}^{V} b_i^{\alpha\alpha} \right) + \lambda \left(\sum_{i=1}^{V} b_i^{\alpha\alpha} \right)^2 \right]. \tag{3}$$

Labeling the states of a generic nematogen as $\zeta^{(k)}$, $k \in \{1, 2, 3, 4, 5, 6\}$, the values of the diagonal components of the tensors $\mathbf{q}$ and $\mathbf{b}$, as well as of the z component of the first principal axis are given in Table 1. Further restricting ourselves to the case $\lambda = \gamma^2$, the widely used geometric-mean condition [20,21], Equation (3) can be written as

$$\mathcal{H}_{\text{SVD6}} = -\frac{A}{2V} \sum_{\alpha \in \{x,y,z\}} \left[\sum_{i=1}^{V} (q_i^{\alpha\alpha} + \gamma b_i^{\alpha\alpha}) \right]^2. \tag{4}$$

As further discussed in Section 3, the parameter γ appearing in the last expression can be interpreted as a simplified measure of the intrinsic biaxiality of the nematogens. The effect of γ on the interaction energies under different relative orientations of a pair of nematogens replaces the entropic effects due to the different excluded volumes associated with these orientations for a pair of hard biaxial polyhedra, which, in the case of strong shape anisotropy, can also lead to the spontaneous appearance of stable biaxial phases [22].

Table 1. Values of the diagonal components of the tensors q and b and of the z component of the principal axes of a generic nematogen in the six possible states $\zeta^{(k)}$ of the restricted model.

k	q^{xx}	b^{xx}	q^{yy}	b^{yy}	q^{zz}	b^{zz}	$(n_1^z)^2$	$(n_2^z)^2$	$(n_3^z)^2$
1	$-\frac{1}{2}$	$\frac{3}{2}$	$-\frac{1}{2}$	$-\frac{3}{2}$	1	0	1	0	0
2	$-\frac{1}{2}$	$-\frac{3}{2}$	$-\frac{1}{2}$	$\frac{3}{2}$	1	0	1	0	0
3	$-\frac{1}{2}$	$\frac{3}{2}$	1	0	$-\frac{1}{2}$	$-\frac{3}{2}$	0	0	1
4	$-\frac{1}{2}$	$-\frac{3}{2}$	1	0	$-\frac{1}{2}$	$\frac{3}{2}$	0	1	0
5	1	0	$-\frac{1}{2}$	$\frac{3}{2}$	$-\frac{1}{2}$	$-\frac{3}{2}$	0	0	1
6	1	0	$-\frac{1}{2}$	$-\frac{3}{2}$	$-\frac{1}{2}$	$\frac{3}{2}$	0	1	0

3. Effects of a Magnetic Field

In order to consider the effects of a magnetic field, we add to the Hamiltonian of our simplified model the interactions of the nematogens with a uniform and constant field $\vec{H} = H\hat{z}$ applied along the z axis in the laboratory frame [23]. We then write

$$\mathcal{H} = \mathcal{H}_{\text{SVD6}} - \frac{1}{2} \sum_{i=1}^{V} \sum_{j=1}^{3} \chi_j \left(\vec{H} \cdot \hat{n}_{j,i} \right)^2, \tag{5}$$

in which χ_j gauges the coupling with the magnetic field when the jth principal axis $\hat{n}_j$ of an object lies along the field direction, and the index i runs over all objects. In view of the fact that, even under the geometric-mean condition, there are in general different interaction energies associated with distinct relative orientations of the principal axes of a pair of nematogens, no symmetry should be expected between field effects associated with choices of the sets of parameters $\{\chi_1, \chi_2, \chi_3\}$ such as $\{\chi, 0, 0\}$, $\{0, \chi, 0\}$ and $\{0, 0, \chi\}$.

The partition function Z of the model can be evaluated by introducing an auxiliary diagonal tensor $\mathbf{Q} = \text{diag}\{Q_{xx}, Q_{yy}, Q_{zz}\}$ via three Gaussian identities which allow us to eliminate the quadratic terms arising from Equation (4), yielding

$$Z = \sum_{\{\zeta\}} e^{-\beta\mathcal{H}} = \left(\frac{\beta AV}{2\pi} \right)^{3/2} \sum_{\{\zeta\}} \int [dQ_{\alpha\alpha}] \exp \left\{ -\frac{\beta AV}{2} \sum_{\alpha} Q_{\alpha\alpha}^2 \right.$$
$$\left. + \sum_{\alpha,i} \beta A Q_{\alpha\alpha} \left(q_i^{\alpha\alpha} + \gamma b_i^{\alpha\alpha} \right) + \frac{\beta H^2}{2} \sum_{i,j} \chi_j \left(n_{j,i}^z \right)^2 \right\}, \tag{6}$$

in which $[dQ_{\alpha\alpha}] = dQ_{xx}dQ_{yy}dQ_{zz}$ and $\beta = T^{-1}$ is the inverse temperature (we use a system of units for which Boltzmann's constant $k_B = 1$). Evaluating the sums over the nematogen index i and the states of each nematogen, we obtain

$$Z = \left(\frac{\beta AV}{2\pi} \right)^{3/2} \int [dQ_{\alpha\alpha}] \exp \left\{ -\beta V \psi \left(\mathbf{Q} \right) \right\}, \tag{7}$$

with

$$\psi \left(\mathbf{Q} \right) = \frac{A}{2} \sum_{\alpha} Q_{\alpha\alpha}^2 - T \ln 2 - T \ln \phi \left(\mathbf{Q} \right) \tag{8}$$

and

$$\phi \left(\mathbf{Q} \right) = e^{\frac{3}{2}\beta A Q_{xx} + \frac{1}{4}\beta(\chi_2 + \chi_3)H^2} \cosh \left[\frac{3\beta A\gamma}{2} \left(Q_{yy} - Q_{zz} \right) - \frac{\beta \left(\chi_2 - \chi_3 \right) H^2}{4} \right]$$
$$+ e^{\frac{3}{2}\beta A Q_{yy} + \frac{1}{4}\beta(\chi_2 + \chi_3)H^2} \cosh \left[\frac{3\beta A\gamma}{2} \left(Q_{zz} - Q_{xx} \right) + \frac{\beta \left(\chi_2 - \chi_3 \right) H^2}{4} \right]$$
$$+ e^{\frac{3}{2}\beta A Q_{zz} + \frac{1}{2}\beta\chi_1 H^2} \cosh \left[\frac{3\beta A\gamma}{2} \left(Q_{xx} - Q_{yy} \right) \right]. \tag{9}$$

The partition function can be evaluated by means of the steepest descent method. The equilibrium states correspond to those values $\langle \mathbf{Q} \rangle$ of the tensor $\mathbf{Q}$ that minimize the functional $\psi \left(\mathbf{Q} \right)$. As these values are such that $\mathbf{Q}$ is traceless, we can write

$$\langle \mathbf{Q} \rangle = \begin{pmatrix} -\frac{1}{2} \left(S + \eta \right) & 0 & 0 \\ 0 & -\frac{1}{2} \left(S - \eta \right) & 0 \\ 0 & 0 & S \end{pmatrix}, \tag{10}$$

in which S and η play the roles of order parameters associated with the uniaxial and biaxial nematic phases, respectively. Noticing that at the mean-field level $\langle \mathbf{Q} \rangle$ can be identified with $\langle q + \gamma b \rangle$, we also have

$$\langle \mathbf{Q} \rangle = \frac{3}{2}\left[\langle \hat{n}_1 \otimes \hat{n}_1 \rangle + \gamma \left(\langle \hat{n}_2 \otimes \hat{n}_2 \rangle - \langle \hat{n}_3 \otimes \hat{n}_3 \rangle \right) \right] - \frac{1}{2}\mathbf{I}. \tag{11}$$

Substituting Equation (10) in Equation (8) we obtain

$$\psi\left(S, \eta\right) = \frac{3A}{4}S^2 + \frac{A}{4}\eta^2 - T\ln 2 - T\ln\phi\left(S, \eta\right), \tag{12}$$

with

$$\begin{aligned}
\phi\left(S, \eta\right) = {} & e^{-\frac{3}{4}\beta A(S+\eta)+\frac{1}{4}\beta(\chi_2+\chi_3)H^2} \cosh\left[\frac{\beta A\Delta}{4}\left(\eta - 3S\right) - \frac{\beta\left(\chi_2 - \chi_3\right)H^2}{4}\right] \\
& + e^{-\frac{3}{4}\beta A(S-\eta)+\frac{1}{4}\beta(\chi_2+\chi_3)H^2} \cosh\left[\frac{\beta A\Delta}{4}\left(\eta + 3S\right) + \frac{\beta\left(\chi_2 - \chi_3\right)H^2}{4}\right] \\
& + e^{\frac{3}{2}\beta AS + \frac{1}{2}\beta\chi_1 H^2} \cosh\left(\frac{\beta A\Delta}{2}\eta\right).
\end{aligned} \tag{13}$$

Notice that in the above equations we introduced a biaxiality parameter $\Delta = 3\gamma$ so that we can compare our phase diagrams with the one obtained at zero field by Boccara, Medjani and De Seze [24]. If, for the sake of comparison, we assume that the traceless part of the inertia tensor and the $q + \gamma b$ tensor of an object are proportional to each other, we may interpret that, as Δ is increased, the shape of the nematogen changes from calamitic ("rod-like") uniaxial for $\Delta = 0$, to strongly biaxial for $\Delta = 1$, and finally to discotic ("disk-like") uniaxial for $\Delta = 3$ (see the appendix in [4]). In this paper, we always work in the interval $0 \leq \Delta \leq 3$.

Minimizing Equation (12) we arrive at the state equations

$$\begin{aligned}
\frac{\partial\psi}{\partial S} &= \frac{3A}{2}S - T\frac{\partial\ln\phi}{\partial S} = 0 \\[6pt]
\frac{\partial\psi}{\partial\eta} &= \frac{A}{2}\eta - T\frac{\partial\ln\phi}{\partial\eta} = 0
\end{aligned} \tag{14}$$

If there are multiple solutions to Equation (14), the equilibrium state corresponds to the solution with the smallest value of $\psi\left(S, \eta\right)$. At high temperatures and at zero field, there is an isotropic, disordered solution $S = \eta = 0$. At lower temperatures, depending on the choice of Δ and χH^2, there may appear uniaxial solutions, for which $\eta = 0 \neq S$, and biaxial solutions, for which $S \neq 0$ and $\eta \neq 0$. At zero field, due to the six-fold symmetry of the problem, there are degenerate solutions for the ordered phases, so that for every uniaxial solution $(S, \eta) = (S_0, 0)$ there are equivalent solutions $(S, \eta) = \left(-\frac{1}{2}S_0, \pm\frac{3}{2}S_0\right)$. Of course, these do not correspond to biaxial solutions, which are not three-fold but six-fold degenerate. In fact, for each pair of biaxial solutions $(S_0, \pm\eta_0)$, with $\eta \neq 0$, there are equivalent solutions $\left(-\frac{1}{2}S_0 \mp \frac{1}{2}\eta_0, \frac{3}{2}S_0 \mp \frac{1}{2}\eta_0\right)$ and $\left(-\frac{1}{2}S_0 \pm \frac{1}{2}\eta_0, -\frac{3}{2}S_0 \mp \frac{1}{2}\eta_0\right)$. In the biaxial phase, comparison of Equations (10) and (11) in the $T \to 0$ limit indicates that the zero-field values of η range from $\eta = -\max\{\Delta, (\Delta+3)/2\}$ to $\eta = \max\{\Delta, (\Delta+3)/2\}$. When plotting the values of S and η in Figures 1–3, we select for each choice of Δ and T the solution that provides the clearest distinction between the phases.

Phase boundaries and multicritical points can be determined by the behavior of $\psi\left(S, \eta\right)$ and its derivatives at the solutions of Equation (14).

3.1. Behavior at Zero Field

We first offer a short review of the phase diagram at zero field ($H = 0$), shown in Figure 1 and already discussed in [4,5,24].

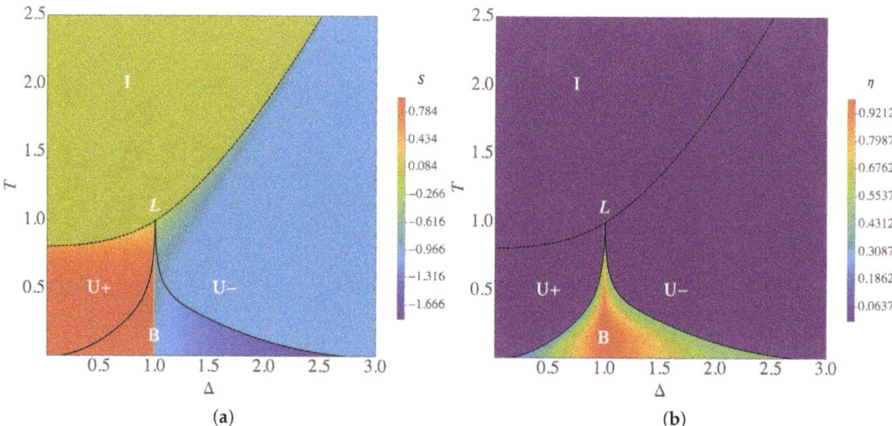

Figure 1. Phase diagram at zero field, in the $T \times \Delta$ plane. The color code shows the behavior of (**a**) the uniaxial order parameter S, which is zero in the high-temperature isotropic phase (I), and (**b**) the biaxial order parameter η, which is nonzero only in the low-temperature biaxial (B) phase. Notice that at low and intermediate temperatures the value of S associated with the smallest equilibrium value of $|\eta|$ changes sign along the line $\Delta = 1$ on which the Landau point $(\Delta_L, T_L) = (1,1)$ is located. Solid (dashed) lines indicate continuous (first-order) transitions. The calamitic (U+) and discotic (U-) uniaxial phases are also indicated.

When $\Delta = 0$, nematogens are intrinsically (calamitic) uniaxial, so that no biaxial phase is expected. Therefore, there is only a first-order transition between the uniaxial and the isotropic phases, which occurs at the temperature for which

$$\left.\frac{\partial \psi}{\partial S}\right|_{(S_0,0)} = 0 \quad \text{and} \quad \psi\left(S_0,0\right) = \psi\left(0,0\right). \tag{15}$$

As the biaxiality parameter Δ is increased, a biaxial solution appears at a sufficiently small temperature. For fixed Δ, the temperature corresponding to the continuous transitions between the biaxial and the uniaxial phases is obtained from

$$\left.\frac{\partial \psi}{\partial S}\right|_{(S_1,0)} = \left.\frac{\partial^2 \psi}{\partial \eta^2}\right|_{(S_1,0)} = 0. \tag{16}$$

The lines of first-order uniaxial-isotropic and continuous biaxial-uniaxial transitions meet at the Landau multicritical point (Δ_L, T_L), the location of which is determined by the conditions

$$\left.\frac{\partial^2 \psi}{\partial S^2}\right|_{(0,0)} = \left.\frac{\partial^3 \psi}{\partial S^3}\right|_{(0,0)} = 0, \tag{17}$$

derived by noticing that at that point S_0 and S_1 both tend smoothly to zero. For $H = 0$, the above equations reduce to

$$\beta A \left(\Delta^2 + 3\right) - 4 = \beta^2 A^2 \left(\Delta^2 - 1\right) = 0, \tag{18}$$

so that, measuring T in units of the energy scale A, the solution to Equation (17) is

$$(\Delta_L, T_L) = (1,1). \tag{19}$$

Further increasing the value of Δ, the temperature of the biaxial–uniaxial transition decreases, becoming zero again at $\Delta = 3$, when the nematogen is once more a (discotic) uniaxial object. At the intermediate-temperature uniaxial phases, the uniaxial order parameter is positive ($S_0 > 0$) if $\Delta < \Delta_L$ and negative ($S_0 < 0$) if $\Delta > \Delta_L$.

We note that this phase diagram qualitatively agrees with the one obtained from computer simulations of a hard biaxial ellipsoid fluid, as a function of the shape anisotropy of the particles [25]. It also reproduces various aspects of phase diagrams of a variety of lyotropic systems [7,26–29], regarding their nematic phases.

3.2. Phase Diagrams in a Field

The effect of a magnetic field is the result of an intricate competition between the magnetic free energy, the entropy and the mutual interactions between nematogens, as dictated by the geometric-mean condition.

The first noticeable effect of a nonzero field is the disappearance of the Landau point, except for the fully symmetric choice $\chi_1 = \chi_2 = \chi_3$, which makes all three principal axes of a nematogen equally likely to lie parallel to the field, having the sole effect of globally shifting all energy levels. This can be checked by fixing H and one of χ_j, say χ_2, and solving for Δ, βA, χ_1 and χ_3 the conditions in Equation (17) supplemented by

$$\left.\frac{\partial \psi}{\partial S}\right|_{(0,0)} = 0 \quad \text{and} \quad \left.\frac{\partial^2 \psi}{\partial \eta^2}\right|_{(0,0)} = 0, \tag{20}$$

the latter conditions being necessary for a continuous transition from a biaxial phase to a stable isotropic phase at a candidate Landau point. The only solution ever obtained is $\Delta = 1$, $\beta A = 1$ and $\chi_1 = \chi_2 = \chi_3$, irrespective of the value of the field.

Nonetheless, for the special choice $\chi_2 < \chi_1 = \chi_3$, which makes the $\hat{n}_2$ principal axis of a nematogen less likely to lie perpendicular to the field, while making the $\hat{n}_1$ and $\hat{n}_3$ principal axes equally likely to lie parallel to the field, the isotropic phase is stable at a point fulfilling the conditions in Equation (17), although the second condition in Equation (20) is not satisfied.

As a typical example of the phase diagram in the presence of a field, let us consider the case $\chi_1 \equiv \chi \neq 0$ and $\chi_2 = \chi_3 = 0$, which is representative of regime $|\chi_1| > |\chi_2|, |\chi_3|$. With this choice, the isotropic phase does not minimize the free energy at any finite temperature, being replaced by a paranematic phase in which $\eta = 0$ but S takes a small but nonzero value. This can be rationalized as follows. Assuming $\chi > 0$, it can be checked that the magnetic energy of a nematogen is minimized when the principal axes $\hat{n}_1$ is parallel to the field direction z. From Equations (10) and (11), taking into account that $\langle \hat{n}_j \otimes \hat{n}_j \rangle^{zz} = \left\langle \left(n_j^z\right)^2 \right\rangle$ and noticing that in a nonzero field the symmetry between $\left\langle \left(n_1^z\right)^2 \right\rangle, \left\langle \left(n_2^z\right)^2 \right\rangle$ and $\left\langle \left(n_3^z\right)^2 \right\rangle$ is explicitly broken, making $\left\langle \left(n_3^z\right)^2 \right\rangle = \left\langle \left(n_2^z\right)^2 \right\rangle < \frac{1}{3}$ and $\left\langle \left(n_1^z\right)^2 \right\rangle > \frac{1}{3}$ at high temperatures, it is clear that $S = \langle Q_{zz} \rangle > 0$. Likewise, assuming $\chi < 0$, so that the magnetic energy of a nematogen is minimized when the principal axes $\hat{n}_1$ is perpendicular to the field direction z, at high temperatures we have $\left\langle \left(n_3^z\right)^2 \right\rangle = \left\langle \left(n_2^z\right)^2 \right\rangle > \frac{1}{3}$ and $\left\langle \left(n_1^z\right)^2 \right\rangle < \frac{1}{3}$, so that $S = \langle Q_{zz} \rangle < 0$.

The general aspect of the phase diagram for both $\chi > 0$ and $\chi < 0$ is shown in Figure 2. In both cases, the Landau point disappears to give rise to a pair of multicritical points: a simple critical point signaling the end of a first-order transition line between the uniaxial and the paranematic phases (which are actually a single phase, in much the same way as the liquid and gaseous phases in simple fluids) and a tricritical point separating regions of first-order and continuous transitions between a biaxial phase and the paranematic phase. The dependence of the uniaxial and biaxial order parameters on Δ and T is illustrated in Figure 3, for the case $\chi > 0$. Notice the extension of the biaxial phase towards the region occupied, at zero field, by the discotic uniaxial phase.

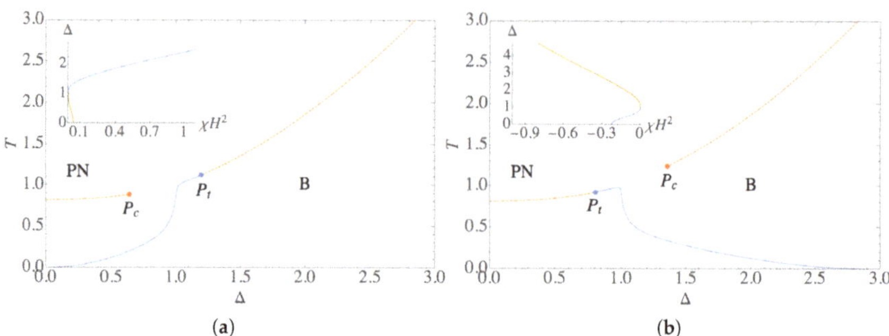

Figure 2. Main plots: Phase diagrams in the presence of a field of intensity $H = 0.1$, with the choice $\chi_1 \equiv \chi \neq 0$ and $\chi_2 = \chi_3 = 0$, both for $\chi = 1$ (**a**) and $\chi = -1$ (**b**). The Landau point is replaced by a simple critical point P_c and a tricritical point P_t. The regions corresponding to the paranematic (PN) and the biaxial (B) phases are also indicated. Insets: Dependence of the values of Δ at P_c (orange curve) and P_t (blue curve) on the product χH^2. Apart from small shifts produced by the field, a combined parametric $T \times \Delta$ plot obtained from the values of T and Δ at P_c and P_t as χH^2 is varied would closely follow the first-order lines in Figure 1.

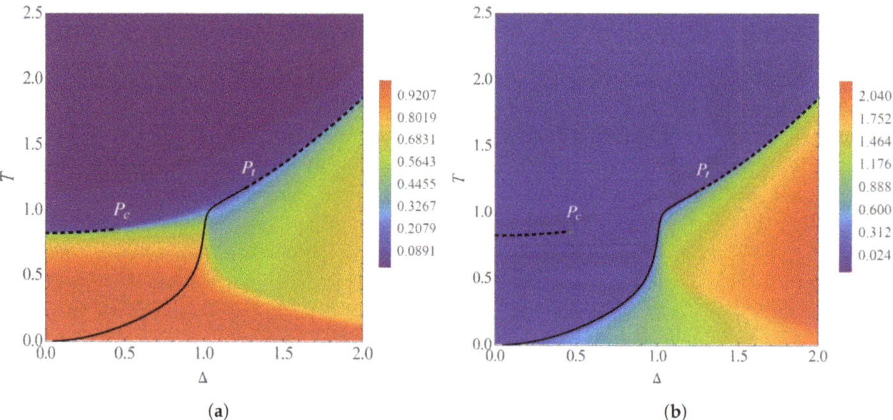

Figure 3. Phase diagram at a field $H = 0.15$, with the choice $\{\chi_1, \chi_2, \chi_3\} = \{1, 0, 0\}$, in the $T \times \Delta$ plane. The color code shows the behavior of (**a**) the uniaxial order parameter S, which is always nonzero in the paranematic phase, which extends up to high temperatures, and (**b**) the biaxial order parameter η, which is nonzero only in the biaxial phase, which now extends to intermediate temperatures when $\Delta > 1$. Notice that both S and η present a steep but nonsingular behavior in the region corresponding, at zero field, to the continuous transition between the biaxial and the discotic uniaxial phases.

The conditions for locating a first-order transition between the uniaxial and the paranematic phases are

$$\left.\frac{\partial \psi}{\partial S}\right|_{(S_0, 0)} = \left.\frac{\partial \psi}{\partial S}\right|_{(S_0', 0)} = 0 \quad \text{and} \quad \psi(S_0, 0) = \psi(S_0', 0), \tag{21}$$

and at the associated critical point the solutions S_0 and S_0' become degenerate, so that

$$\left.\frac{\partial^2 \psi}{\partial S^2}\right|_{(S_0, 0)} = \left.\frac{\partial \psi}{\partial S}\right|_{(S_0, 0)} = 0. \tag{22}$$

A first-order transition between the biaxial and the paranematic phases occurs for

$$\left.\frac{\partial\psi}{\partial\eta}\right|_{(S_2,\eta_2)} = \left.\frac{\partial\psi}{\partial S}\right|_{(S_2,\eta_2)} = \left.\frac{\partial\psi}{\partial S}\right|_{(S_0',0)} = 0 \quad \text{and} \quad \psi\left(S_2,\eta_2\right) = \psi\left(S_0',0\right), \tag{23}$$

while a continuous transition between those same phases satisfies

$$\left.\frac{\partial\psi}{\partial S}\right|_{(S_0',0)} = \left.\frac{\partial^2\psi}{\partial\eta^2}\right|_{(S_0',0)} = 0. \tag{24}$$

Finally, the location of the tricritical point can be determined from the conditions

$$\left.\frac{\partial\psi}{\partial S}\right|_{(S_0',0)} = \left.\frac{d^2\psi}{d\eta^2}\right|_{(S_0',0)} = \left.\frac{d^4\psi}{d\eta^4}\right|_{(S_0',0)} = 0. \tag{25}$$

Notice that these last conditions involve total derivatives of the free-energy functional with respect to η, which means that, while calculating those derivatives, S_0' is implicitly treated as a function of η.

The multicritical points originating from the destabilization of the Landau point become increasingly separated as the magnetic field becomes more intense, as the insets in Figure 2 show. In the limit of an infinite field, only a continuous biaxial-paranematic phase transition remains.

The existence of a tricritical point on the line of biaxial to paranematic phase transitions is compatible with the predictions of [16], which assumes a more restrictive pair interaction independent of a biaxiality parameter, as well as with those of [6], in which nematogens are intrinsically uniaxial. In the latter case, the located tricritical point is observed for $\chi < 0$ and at the value of the magnetic field necessary for our tricritical point to reach the uniaxial limit $\Delta = 0$; see the inset in Figure 2b. This same qualitative behavior for the appearance of a tricritical point in the uniaxial-paranematic transition was predicted for hard rods or hard plates with $\chi < 0$ [30].

Similarly, the critical field associated with a critical end point of the first-order transition line between the uniaxial and the paranematic phases, which has been observed for a fixed biaxiality parameter by various authors [6,11,12,15,31], corresponds to the value of the field, which makes our simple critical point reach the uniaxial limit $\Delta = 0$ for $\chi > 0$; see the inset in Figure 2a. Again, the same qualitative behavior for the appearance of a critical end point in the uniaxial-paranematic transition was predicted for hard rods or hard plates with $\chi > 0$ [32].

Assuming that the biaxiality parameter of a system shows little variation with temperature or the strength of the magnetic field, which would be reasonable for thermotropic liquid crystals, our results suggest that, starting from a biaxial phase and heating the system under a constant field up to the transition to the paranematic phase, the nature of that transition should change from first-order to continuous (or vice-versa, depending on the sign of χ) at a certain value of the field, signaling the tricritical point. Estimating the value of that field would require an estimate of the biaxiality parameter Δ for the system, under the assumption that the geometric mean condition is at least approximately applicable. We expect that lyotropic systems, for which presumably there is a significant dependence of the biaxiality parameter on both the temperature and the composition of the mixture [26,27,29], are better candidates for the observation of a tricritical point.

4. Introducing Dilution

Next we turn our attention to a "nematic lattice gas" model consisting of V sites, which can be either empty or occupied by a single object among $N \leq V$ nematogens. This possibility can be represented at each site i by an occupation variable $t_i \in \{0,1\}$, and if the nematogens interact via the the extension of the Hamiltonian in Equation (4), the canonical partition function of the model at zero field can be written as

$$Z = \sum_{\{t\}}' \sum_{\{\zeta\}} \exp\left\{ \frac{\beta A}{2V} \sum_{\alpha \in \{x,y,z\}} \left[\sum_{i=1}^{V} t_i \left(q_i^{\alpha\alpha} + \gamma b_i^{\alpha\alpha} \right) \right]^2 + \frac{\beta}{2} \sum_{i=1}^{V} \sum_{j=1}^{3} t_i \chi_j \left(\vec{H} \cdot \hat{n}_{j,i} \right)^2 \right\}, \tag{26}$$

in which the prime in the summation over the set $\{t\}$ of occupation variables indicates that it should be restricted to those configurations for which

$$\sum_{i=1}^{V} t_i = N. \tag{27}$$

This restriction can be relaxed by introducing a chemical potential μ (not to be confused with the tensor index in Equation (1)) to enforce that the last equation be satisfied on average. We then have to calculate the grand-partition function

$$\Xi = \sum_{\{t\}} \sum_{\{\zeta\}} \exp\left\{ \frac{\beta A}{2V} \sum_{\alpha \in \{x,y,z\}} \left[\sum_{i=1}^{V} t_i \left(q_i^{\alpha\alpha} + \gamma b_i^{\alpha\alpha} \right) \right]^2 + \frac{\beta}{2} \sum_{i=1}^{V} \sum_{j=1}^{3} t_i \chi_j \left(\vec{H} \cdot \hat{n}_{j,i} \right)^2 + \beta\mu \sum_{i=1}^{V} t_i \right\}. \tag{28}$$

Following the same strategy as in the previous section and introducing an auxiliary diagonal tensor $\mathbf{Q} = \mathrm{diag}\{Q_{xx}, Q_{yy}, Q_{zz}\}$ via three Gaussian identities, which allow us to eliminate the quadratic terms in Ξ, we obtain

$$\Xi = \left(\frac{\beta A V}{2\pi} \right)^{3/2} \sum_{\{t\}} \sum_{\{\zeta\}} \int [dQ_{\alpha\alpha}] \exp\left\{ -\frac{\beta A V}{2} \sum_\alpha Q_{\alpha\alpha}^2 + \right.$$
$$\left. + \sum_{\alpha,i} \beta A Q_{\alpha\alpha} t_i \left(q_i^{\alpha\alpha} + \gamma b_i^{\alpha\alpha} \right) + \frac{\beta}{2} \sum_{i,j} t_i \chi_j \left(\vec{H} \cdot \hat{n}_{j,i} \right)^2 + \beta\mu \sum_{i=1}^{V} t_i \right\}, \tag{29}$$

and performing the sums over the nematogen index i, the occupation variables and the states of each nematogen, we can write

$$\Xi = \left(\frac{\beta A V}{2\pi} \right)^{3/2} \int [dQ_{\alpha\alpha}] \exp\left\{ -\beta V \Psi(\mathbf{Q}) \right\}, \tag{30}$$

with a grand functional

$$\Psi(\mathbf{Q}) = \frac{A}{2} \sum_\alpha Q_{\alpha\alpha}^2 - T \ln \Phi(\mathbf{Q}), \tag{31}$$

in which

$$\Psi(\mathbf{Q}) = 6 + 2e^{\frac{3}{2}\beta A Q_{xx} + \frac{1}{4}\beta(\chi_2+\chi_3)H^2 + \beta\mu} \cosh\left[\frac{3\beta A \gamma}{2}\left(Q_{yy} - Q_{zz} \right) - \frac{\beta(\chi_2 - \chi_3)H^2}{4} \right]$$
$$+ 2e^{\frac{3}{2}\beta A Q_{yy} + \frac{1}{4}\beta(\chi_2+\chi_3)H^2 + \beta\mu} \cosh\left[\frac{3\beta A \gamma}{2}\left(Q_{zz} - Q_{xx} \right) + \frac{\beta(\chi_2 - \chi_3)H^2}{4} \right]$$
$$+ 2e^{\frac{3}{2}\beta A Q_{zz} + \frac{1}{2}\beta\chi_1 H^2 + \beta\mu} \cosh\left[\frac{3\beta A \gamma}{2}\left(Q_{xx} - Q_{yy} \right) \right]. \tag{32}$$

Adopting again the parametrization in Equation (10) we can write the grand functional in terms of S and η, obtaining

$$\Psi(S, \eta) = \frac{3A}{4} S^2 + \frac{A}{4} \eta^2 - T \ln \Phi(S, \eta), \tag{33}$$

with

$$\Phi\left(S,\eta\right) = 2e^{-\frac{3}{4}\beta A(S+\eta)+\frac{1}{4}\beta(\chi_2+\chi_3)H^2+\beta\mu}\cosh\left[\frac{\beta A\Delta}{4}\left(\eta-3S\right)-\frac{\beta\left(\chi_2-\chi_3\right)H^2}{4}\right]$$

$$+\,2e^{-\frac{3}{4}\beta A(S-\eta)+\frac{1}{4}\beta(\chi_2+\chi_3)H^2+\beta\mu}\cosh\left[\frac{\beta A\Delta}{4}\left(\eta+3S\right)+\frac{\beta\left(\chi_2-\chi_3\right)H^2}{4}\right]$$

$$+\,2e^{\frac{3}{2}\beta AS+\frac{1}{2}\beta\chi_1 H^2+\beta\mu}\cosh\left(\frac{\beta A\Delta}{2}\eta\right)+6. \tag{34}$$

In the limit $\beta\mu \gg 1$, as expected for a fully occupied lattice, Equation (33) reduces to

$$\Psi\left(S,\eta\right) = \psi\left(S,\eta\right) - \mu, \tag{35}$$

in which $\psi\left(S,\eta\right)$ is given by Equation (12).

The equilibrium values of S and η are obtained by minimizing $\Psi\left(S,\eta\right)$ with respect to both variables at a fixed chemical potential, yielding the state equations

$$\frac{\partial\Psi}{\partial S} = \frac{3A}{2}S - T\frac{\partial\ln\Psi}{\partial S} = 0$$

$$\frac{\partial\Psi}{\partial\eta} = \frac{A}{2}\eta - T\frac{\partial\ln\Psi}{\partial\eta} = 0 \tag{36}$$

and once again there may exist isotropic, uniaxial and biaxial solutions.

In the presence of a nonzero field, it is straightforward to show that for small dilution we recover the same qualitative results of Section 3 regarding the destabilization of the Landau point, which is replaced by a simple critical point and a tricritical point.

On the other hand, in the limit of zero field ($H = 0$), to which we limit ourselves for the rest of this section, a Landau point occurs when

$$\left.\frac{\partial^2\Psi}{\partial S^2}\right|_{(0,0)} = \left.\frac{\partial^3\Psi}{\partial S^3}\right|_{(0,0)} = 0. \tag{37}$$

The above condition on the third derivative yields $\Delta_L = 1$, while the remaining condition leads to

$$\left(\beta A - 1\right)e^{\beta\mu} - 1 = 0, \tag{38}$$

the solution of which defines a line of Landau points corresponding to a varying chemical potential. For $\beta\mu \gg 1$, we recover the result $\beta_L A = 1$ obtained at zero field in Equation (19). As μ is reduced, the temperature corresponding to the solution of Equation (38) is also reduced, as it can be checked by calculating the implicit derivative of β with respect to μ from Equation (38).

Besides having $\Delta = 1$ and fulfilling Equation (38), a true Landau point has to represent a stable isotropic phase. In the present case, this requires that this phase be a minimum rather than a saddle point of the grand functional. It turns out that, at the candidate Landau point, the second derivative of $\Psi\left(S,\eta\right)$ with respect to η is always zero at the isotropic phase $(S,\eta) = (0,0)$, but that the corresponding fourth derivative,

$$\left.\frac{\partial^4\Psi}{\partial\eta^4}\right|_{(0,0)} = \frac{27\beta^3 A^4\left(1-e^{-\beta\mu}\right)}{8\left(1+e^{-\beta\mu}\right)^2}, \tag{39}$$

changes from positive for $\mu > 0$ to negative for $\mu < 0$, indicating that the Landau point corresponds to a stable phase only if the chemical potential is nonnegative.

In order to obtain phase diagrams involving the concentration c of occupied sites rather than the chemical potential μ, at every equilibrium point we need to solve for μ the equation

$$c = \left\langle \frac{1}{V} \sum_{i=1}^{V} t_i \right\rangle = \frac{1}{\beta V} \frac{\partial}{\partial \mu} \ln \Xi, \tag{40}$$

which is equivalent to solving

$$c = -\left. \frac{\partial \Psi}{\partial \mu} \right|_{(S,\eta)} = \frac{\Phi(S,\eta) - 6}{\Phi(S,\eta)}, \tag{41}$$

with S and η assuming their equilibrium values and $\Phi(S,\eta)$ given by Equation (34).

In particular, at the Landau point the relation between μ and the concentration c is given by

$$c = \frac{e^{\beta\mu}}{1 + e^{\beta\mu}}, \tag{42}$$

indicating that the Landau point is stable if $c > \frac{1}{2}$ and unstable if $c < \frac{1}{2}$. Therefore, for small dilution the Landau point is always stable. In this limit, the phase diagram is qualitatively the same as in the fully packed limit, as illustrated in Figure 4.

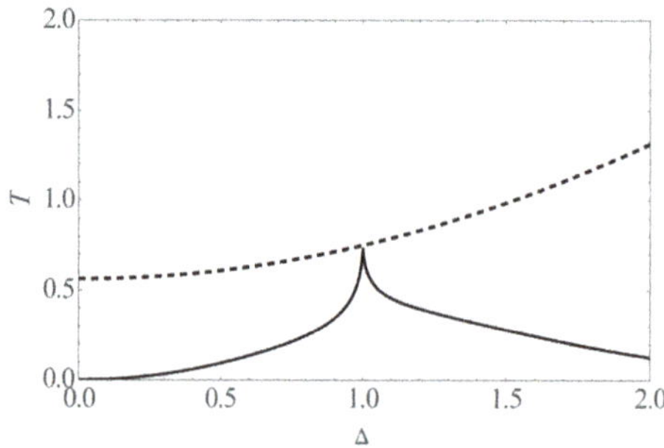

Figure 4. Zero-field phase diagram for the diluted model, when the concentration of nematogens corresponds to $c = 0.75$. Notice the lower temperature at the Landau point and a slightly more asymmetrical uniaxial–biaxial transition line around the line $\Delta = 1$.

5. Conclusions

We use a recently proposed two-tensor formalism, which is based on the physically acceptable pair interactions between nematogenic units, and assume a discrete set of orientations of the microscopic nematic directors, to write a six-state lattice Hamiltonian. The fully-connected version of this system is amenable to detailed statistical mechanics calculations. In the widely used geometric-mean approximation for the model parameters, we regain the well-known phase diagram of nematic biaxial systems, with sequences of biaxial-uniaxial-disordered phase transitions as temperature increases, and the appearance of a well-defined Landau multicritical point. We locate the transition lines and the position of the Landau multicritical point in terms of temperature and a parameter Δ that gauges the degree of biaxiality.

With the addition of an external magnetic field, we show that, except in a quite unusual diamagnetic isotropic case, the topology of this phase diagram is entirely changed. Our calculations indicate that the Landau point can no longer exist. In the phase diagrams, in terms of temperature and the parameter Δ, we show that there appear a pair of multicritical points, and a simple critical point at the end of the first-order border between a uniaxial and a paranematic phase. Also, we show the onset of a tricritical point separating borders of first-order and continuous transitions between a biaxial and a paranematic phase. We expect that lyotropic systems, for which presumably there is a significant dependence of the biaxiality parameter on both the temperature and the composition of the mixture [26,27,29], are good candidates for the observation of both the simple critical point and the tricritical point. We point out that our predictions for $\chi > 0$ are fully compatible with available experimental results for potassium laurate/1-decanol/water mixtures [8], which show, based on measurements performed at three distinct concentrations in the neighborhood of the Landau point, that the application of a magnetic field turns the transition between the rod-like uniaxial and the isotropic phases into a crossover, while preserving the discontinuous nature of the transition between the disk-like uniaxial and the isotropic phases. Locating the simple critical point and the tricritical point in this system would require some extremely fine tuning, but in principle it should be feasible.

On the other hand, in the presence of moderate dilution the zero-field phase diagram is essentially unchanged. For stronger dilution, however, the Landau point also disappears, which is very similar to a phenomenon associated with a dilute ferromagnet.

Finally, we stress that, from a formal point of view, our results are equally applicable to the situation in which a liquid crystal is subject to an electric rather than a magnetic field [33].

Author Contributions: Conceptualization, D.D.R., A.P.V. and S.R.S.; methodology, D.D.R., A.P.V. and S.R.S.; software, D.D.R., A.P.V. and S.R.S.; validation, D.D.R., A.P.V. and S.R.S.; formal analysis, D.D.R., A.P.V. and S.R.S.; investigation, D.D.R., A.P.V. and S.R.S.; resources, A.P.V. and S.R.S.; data curation, D.D.R., A.P.V. and S.R.S.; writing—original draft preparation, A.P.V. and S.R.S.; writing—review and editing, A.P.V. and S.R.S.; visualization, D.D.R., A.P.V. and S.R.S.; supervision, A.P.V. and S.R.S.; project administration, A.P.V. and S.R.S.; funding acquisition, A.P.V. and S.R.S. All authors have read and agreed to the published version of the manuscript.

Funding: This research was funded by INCT/FCx, NAP/FCx, FAPESP (grant 2017/09566-8), and CNPq.

Conflicts of Interest: The authors declare no conflict of interest.

References

1. do Carmo, E.; Liarte, D.B.; Salinas, S.R. Statistical models of mixtures with a biaxial nematic phase. *Phys. Rev. E* **2010**, *81*, 062701. [CrossRef] [PubMed]

2. do Carmo, E.; Vieira, A.P.; Salinas, S.R. Phase diagram of a model for a binary mixture of nematic molecules on a Bethe lattice. *Phys. Rev. E* **2011**, *83*, 011701. [CrossRef] [PubMed]

3. Liarte, D.B.; Salinas, S.R. Enhancement of Nematic Order and Global Phase Diagram of a Lattice Model for Coupled Nematic Systems. *Braz. J. Phys.* **2012**, *42*, 261. [CrossRef]

4. Nascimento, E.S.; Henriques, E.F.; Vieira, A.P.; Salinas, S.R. Maier-Saupe model for a mixture of uniaxial and biaxial molecules. *Phys. Rev. E* **2015**, *92*, 062503. [CrossRef] [PubMed]

5. Nascimento, E.S.; Vieira, A.P.; Salinas, S.R. Lattice Statistical Models for the Nematic Transitions in Liquid-Crystalline Systems. *Braz. J. Phys.* **2016**, *46*, 664. [CrossRef]

6. Petri, A.; Salinas, S.R. Field-induced uniaxial and biaxial nematic phases in the Maier–Saupe–Zwanzig (MSZ) lattice model. *Liquid Cryst.* **2018**, *45*, 980–992. [CrossRef]

7. Yu, L.J.; Saupe, A. Observation of a Biaxial Nematic Phase in Potassium Laurate-1-Decanol-Water Mixtures. *Phys. Rev. Lett.* **1980**, *45*, 1000–1003. [CrossRef]

8. Melnik, G.; Photinos, P.; Saupe, A. Landau point on a nematic-isotropic transition line. *Phys. Rev. A* **1989**, *39*, 1597–1600. [CrossRef]

9. Lemaire, B.J.; Davidson, P.; Ferré, J.; Jamet, J.P.; Panine, P.; Dozov, I.; Jolivet, J.P. Outstanding Magnetic Properties of Nematic Suspensions of Goethite (α-FeOOH) Nanorods. *Phys. Rev. Lett.* **2002**, *88*, 125507. [CrossRef]

10. van den Pol, E.; Lupascu, A.; Diaconeasa, M.A.; Petukhov, A.V.; Byelov, D.V.; Vroege, G.J. Onsager Revisited: Magnetic Field Induced Nematic–Nematic Phase Separation in Dispersions of Goethite Nanorods. *J. Phys. Chem. Lett.* **2010**, *1*, 2174–2178. [CrossRef]

11. Ostapenko, T.; Wiant, D.B.; Sprunt, S.N.; Jákli, A.; Gleeson, J.T. Magnetic-Field Induced Isotropic to Nematic Liquid Crystal Phase Transition. *Phys. Rev. Lett.* **2008**, *101*, 247801. [CrossRef] [PubMed]

12. To, T.B.T.; Sluckin, T.J.; Luckhurst, G.R. Biaxiality-induced magnetic field effects in bent-core nematics: Molecular-field and Landau theory. *Phys. Rev. E* **2013**, *88*, 062506. [CrossRef] [PubMed]

13. Mukherjee, P.K.; Rahman, M. Isotropic to biaxial nematic phase transition in an external magnetic field. *Chem. Phys.* **2013**, *423*, 178. [CrossRef]

14. Aliev, M.; Ugolkova, E.; Kuzminyh, N. Effect of an external magnetic field on the phase behavior of the thermotropic melt of V-shaped molecules. *J. Mol. Liquids* **2019**, *292*, 111395. [CrossRef]

15. Matsuyama, A.; Arikawa, S.; Wada, M.; Fukutomi, N. Uniaxial and biaxial nematic phases of banana-shaped molecules and the effects of an external field. *Liquid Cryst.* **2019**, *46*, 1672. [CrossRef]

16. Mukherjee, P.K.; De, A.K.; Mandal, A. Mean-field theory of isotropic-uniaxial nematic-biaxial nematic phase transitions in an external field. *Phys. Scr.* **2019**, *94*, 025702. [CrossRef]

17. Sonnet, A.M.; Virga, E.G.; Durand, G.E. Dielectric shape dispersion and biaxial transitions in nematic liquid crystals. *Phys. Rev. E* **2003**, *67*, 061701. [CrossRef]

18. de Oliveira, M.J.; Figueiredo Neto, A.M. Reentrant isotropic-nematic transition in lyotropic liquid crystals. *Phys. Rev. A* **1986**, *34*, 3481. [CrossRef]

19. Sauerwein, R.A.; de Oliveira, M.J. Lattice model for biaxial and uniaxial nematic liquid crystals. *J. Chem. Phys.* **2016**, *144*, 194904. [CrossRef]

20. Luckhurst, G.R.; Zannoni, C.; Nordio, P.L.; Segre, U. A molecular field theory for uniaxial nematic liquid crystals formed by non-cylindrically symmetric molecules. *Mol. Phys.* **1975**, *30*, 1345. [CrossRef]

21. Luckhurst, G.R.; Naemura, S.; Sluckin, T.J.; Thomas, K.S.; Turzi, S.S. Molecular-field-theory approach to the Landau theory of liquid crystals: Uniaxial and biaxial nematics. *Phys. Rev. E* **2012**, *85*, 031705. [CrossRef] [PubMed]

22. Dussi, S.; Tasios, N.; Drwenski, T.; van Roij, R.; Dijkstra, M. Hard Competition: Stabilizing the Elusive Biaxial Nematic Phase in Suspensions of Colloidal Particles with Extreme Lengths. *Phys. Rev. Lett.* **2018**, *120*, 177801. [CrossRef] [PubMed]

23. Photinos, D.J. Alignment of biaxial nematics. In *Biaxial Nematic Liquid Crystals: Theory, Simulation, and Experiment*; Luckhurst, G.R., Sluckin, T.J., Eds.; Wiley: West Sussex, UK, 2015; p. 205.

24. Boccara, N.; Mejdani, R.; De Seze, L. Solvable model exhibiting a first-order phase transition. *J. Phys. France* **1977**, *38*, 149. [CrossRef]

25. Camp, P.J.; Allen, M.P. Phase diagram of the hard biaxial ellipsoid fluid. *J. Chem. Phys.* **1997**, *106*, 6681–6688. [CrossRef]

26. Figueiredo Neto, A.M.; Galerne, Y.; Levelut, A.M.; Liebert, L. Pseudo-lamellar ordering in uniaxial and biaxial lyotropic nematics: A synchrotron X-ray diffraction experiment. *J. Phys. Lett.* **1985**, *46*, 499. [CrossRef]

27. Galerne, Y.; Figueiredo Neto, A.M.; Liébert, L. Microscopical structure of the uniaxial and biaxial lyotropic nematics. *J. Chem. Phys.* **1987**, *87*, 1851–1856. [CrossRef]

28. Oliveira, E.A.; Liebert, L.; Neto, A.M.F. A new soap/detergent/water lyotropic liquid crystal with a biaxial nematic phase. *Liquid Cryst.* **1989**, *5*, 1669–1675. [CrossRef]

29. Akpinar, E.; Reis, D.; Figueiredo Neto, A.M. Effect of alkyl chain length of alcohols on nematic uniaxial-to-biaxial phase transitions in a potassium laurate/alcohol/K_2SO_4/water lyotropic mixture. *Liquid Cryst.* **2012**, *39*, 881–888. [CrossRef]

30. Varga, S.; Kronome, G.; Szalai, I. External field induced tricritical phenomenon in the isotropicnematic phase transition of hard non-spherical particle systems. *Mol. Phys.* **2000**, *98*, 911–915. [CrossRef]

31. Vause, C.A. Connection between the isotropic-nematic Landau point and the paranematic-nematic critical point. *Phys. Lett. A* **1986**, *114*, 485–490. [CrossRef]

32. Varga, S.; Jackson, G.; Szalai, I. External field induced paranematic–nematic phase transitions in rod-like systems. *Mol. Phys.* **1998**, *93*, 377–387.

33. Lelidis, I.; Durand, G. Electric-field-induced isotropic-nematic phase transition. *Phys. Rev. E* **1993**, *48*, 3822–3824. [CrossRef] [PubMed]

Article

Ordering of Rods near Surfaces: Concentration Effects

Dora Izzo

Instituto de Física, Universidade Federal do Rio de Janeiro, C.P. 68528, Rio de Janeiro 21941-972, Brazil;
izzo@if.ufrj.br

Received: 24 February 2019; Accepted: 16 May 2019; Published: 21 May 2019

Abstract: We study the orientation of rods in the neighborhood of a surface. A semi-infinite region
in two different situations is considered: (i) the rods are located close to a flat wall and (ii) the rods
occupy the space that surrounds a sphere. In a recent paper we investigated a similar problem:
the interior of a sphere, with a fixed concentration of rods. Here, we allow for varying concentration,
the rods are driven from a reservoir to the neighborhood of the surface by means of a tunable chemical
potential. In the planar case, the particle dimensions are irrelevant. In the curved case, we consider
cylinders with dimensions comparable to the radius of curvature of the sphere; as they come close to
the surface, they have to accommodate to fill the available space, leading to a rich orientational profile.
These systems are studied by a mapping onto a three-state Potts model with annealed disorder on
a semi-infinite lattice; two order parameters describe the system: the occupancy and the orientation.
The Hamiltonian is solved using a mean-field approach producing recurrence relations that are
iterated numerically and we obtain various interesting results: the system undergoes a first order
transition just as in the bulk case; the profiles do not have a smooth decay but may present a step and
we search for the factors that determine their shape. The prediction of such steps may be relevant in
the field of self-assembly of colloids and nanotechnology.

Keywords: rods; curved surface; Potts

1. Introduction

In the past fifty years, a great deal of attention has been devoted to understanding how liquid
crystals are ordered in small cells. Applications in technology have pushed forward the interest in such
systems: optical devices with specific properties depend on the anchoring conditions of the mesogens
to the surfaces [1]; coating spherical colloidal particles with thin layers of liquid crystals in specific
ways provides flexibility for tuning directional interactions between colloids [2]. In that scenario,
particles are of the order of a few nanometeres, much smaller than their confining volumes, with sizes
in the micrometer range: a description in terms of continuum theories [3] is completely satisfactory.

More recently, larger particles, commonly named macroscopic liquid crystals, have also
captured attention. Examples of such structures are liquid crystals formed by viruses [4,5],
filamentous biopolymers [6] and carbon nanotubes [7,8], with dimensions of the order of hundreds of
nanometers. Not only is confinement a relevant issue, but also self-assembly of colloids mediated by
large particles is a new problem to be investigated. From a theoretical point of view, continuum theories
break down and one looks to understand how these large particles accommodate finding their optimal
conformation when mutual packing and alignment to boundaries compete at the same length scale.
Some theoretical descriptions have addressed such problems using Monte Carlo simulations and
molecular dynamics [9–12].

Our system consists of a colloidal liquid crystal represented by a set of rod-like particles. They are
driven from a reservoir by means of a chemical potential and we are primarily interested in describing

how these rods self-organize around a curved surface. The concentration is not fixed and vacancies come into play, leading to *annealed* disorder. Excluded volume interactions are considered and assist organization because the particle dimensions are sizeable with respect to the surface curvature radius, typically on the order of 1/10 (as in the case of filamentous bacteriophage fd-viruses in microchambers); Figure 1 illustrates this situation.

Figure 1. Representation of the distribution of rods around a sphere; here, we show a cross section of the three-dimensional system. For large particle concentrations, the rods are forced to align around the sphere; this alignment decreases towards the bulk.

For a given value of a chemical potential (large enough to guarantee minimal occupancy just next to the sphere), the orientation profile can be understood as follows: (i) just outside the sphere there is enough room to accommodate a given number of particles, anchored homeotropically; (ii) beyond this region, in the upper rows, more room is available, which amounts to a decrease in the excluded volume interactions and therefore a reduction in ordering; (iii) for rows further away from the sphere, the "squeezing" effect vanishes and an isotropic configuration sets in.

We model this system in terms of a lattice surrounding a sphere. Particles (rods) occupy the lattice sites, but vacancies may occur. We follow Oliveira and Figueiredo Neto [13] and associate the orientation of each rod to a microscopic state of the three-state Potts model and solve it on a lattice. Nevertheless, the lattice description of the curved three-dimensional system is unfeasible (discussed in the next session) and we are led to solve the two-dimensional version of the problem.

The system described above is the main subject of this paper. Nevertheless, we will also address a related problem: the effect of *quenched* disorder. In this case, it is more realistic to consider the system studied in a previous work [14], in which large rods are located inside a sphere and all lattice sites are occupied. We predict the behavior of this system in terms of a qualitative discussion.

This paper is organized as follows: in Section 2, we define the two-dimensional version of the problem and describe the mean-field approach for the three-state Potts model with annealed disorder; in Section 3, the results are presented, first for a simpler problem, the planar surface (PS) case, and then for the curved surface (CS) case; these results are discussed in Section 4, where we comment on the effect of introducing quenched disorder in a similar system; conclusions are left for Section 5.

2. The Problem

2.1. The Two-Dimensional Version

We model the colloidal liquid crystal as a system of rods in a lattice. A single rod is located on a lattice site but not all sites must be occupied. Because it is not possible to embed a curved object into a cubic lattice, we reduced our problem to two-dimensions redefining the lattice structure and the rods' geometry as follows.

The three-dimensional version of the planar surface case (PS) consists of a cubic lattice bounded by a flat surface. The analogous two-dimensional problem is just a square lattice bounded by a line.

In order to study the two-dimensional analogue of the curved surface case (CS), the sphere is replaced by a disk. We choose a generic cross section of the three-dimensional space that surrounds a sphere with a planar lattice network that represents the symmetry of the problem as illustrated in Figure 2.

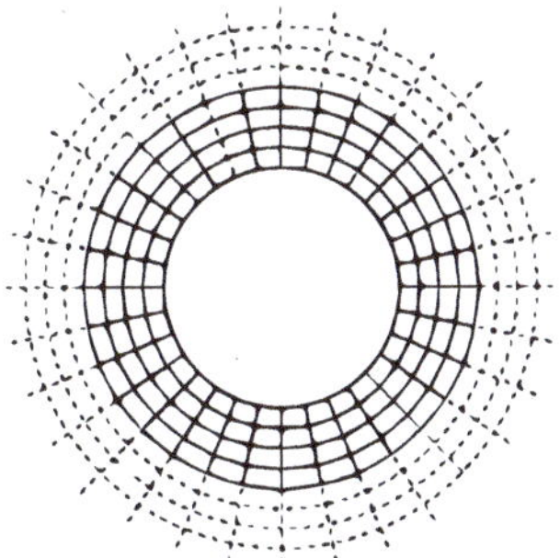

Figure 2. The network of lattice sites around a disk.

Accordingly, the rods must be replaced by rectangles with linear dimensions b and L ($b < L$). Every geometric center is fixed on a lattice site and rotates on the lattice plane. Because only excluded volume interactions are considered, the alignment of a given rod may be constrained only by its nearest neighbors' orientation. Next, we discuss the geometric constraints and show that, in the CS case, the geometry is responsible for an extra channel of disorder towards the bulk.

Consider a planar lattice such as the one in Figure 2; let R be the radius of the disk and N the number of lattice sites by row. Accommodating homeotropic anchored rods around the circle requires: $R > N\rho(0)b$. Towards the bulk, the increase in distance between nearest neighbor rods amounts to a decrease in excluded volume interactions (mutual alignment) and rotation is gradually allowed; at $\bar{r} = N\rho(\bar{r})L$, excluded volume interactions finally die out, which characterizes the "bulk". Therefore, the region of interest is $R < r < \bar{r}$.

2.2. The Three-State Potts Model with Annealed Disorder

We consider a plane (either PS or CS) and use the mapping proposed by Oliveira and Figueiredo Neto [13]. Because we aim to describe occupancy and ordering, we need two order parameters.

The concentration profile is determined by a set of order parameters $\{\rho_\ell\}$ associated with the fraction of occupied sites in the system. ρ_ℓ is an average over a microscopic parameter η_ℓ on the ℓth row: $\rho_\ell \equiv \langle \eta_\ell \rangle_0$.

The directional order in each row is described by another set of order parameters defined as follows. The orientation of each rod refers to the surface director; in the PS case, the surface is a straight line; in the CS case, it is a circle. We use the three-state Potts model to describe the system: one of the three states corresponds to alignment perpendicular to the surface (homeotropic), whereas the other two are degenerate and refer to alignment in the other orthogonal directions parallel to the surface (planar). The state of ordering is thus described by the set $\{q_\ell\}$, obtained by an average over the microscopic parameters η_ℓ and s_ℓ on the ℓth row: $q_\ell = \langle \eta_\ell . s_\ell \rangle_0$, where the variable s_ℓ describes the microscopic directional state on row ℓ. In fact, q_l specifies the corresponding averaged moment of quadrupole with respect to a homeotropic alignment in the rod frame of reference Λ_ℓ [15]:

$$\Lambda_\ell = \begin{bmatrix} -q_\ell/2 & 0 & 0 \\ 0 & -q_\ell/2 & 0 \\ 0 & 0 & q_\ell \end{bmatrix}. \tag{1}$$

In a mean-field description of the annealed three-state Potts model, the expression for the energy per column of a given plane, either in the PS or in the CS cases, is given by [16]:

$$F = -\mu \sum_{\ell=0}^{\infty} \rho_l - \sum_{\ell=0}^{\infty} \epsilon_\ell q_\ell q_{\ell+1} - \sum_{\ell=0}^{\infty} \epsilon_\ell (q_\ell)^2 + \frac{k_B T}{3} \sum_{\ell=0}^{\infty} [\ln(\rho_\ell + 2q_\ell) + 2\ln(\rho_\ell - q_\ell) - 3\ln 3(1 - \rho_\ell)] . \quad (2)$$

The index ℓ labels the rows: $\ell = 1$ refers to the first row, next to the surface; ρ_0 and q_0 fix the anchoring conditions as specified below. Each site has two nearest neighbors on the same row and one nearest neighbor on the row just above it. μ is a common chemical potential that drives particles from the reservoir and ϵ_ℓ is the interaction energy between a rod on row ℓ and its nearest neighbors.

Following the general procedure of the mean-field approach, we minimize expression (2) with respect to ρ_ℓ and q_ℓ, which yeld the recurrence relations:

$$\begin{cases} -\mu + \frac{k_B T}{3}[\ln(\rho_\ell + 2q_\ell) + 2\ln(\rho_\ell - q_\ell) - 3\ln 3(1 - \rho_\ell)] = 0, \\[2mm] -\epsilon_\ell q_{\ell+1} - \epsilon_{\ell-1} q_{\ell-1} - 2\epsilon_\ell q_\ell + \frac{2}{3} k_B T \ln \frac{\rho_\ell + 2q_\ell}{\rho_\ell - q_\ell} = 0. \end{cases} \quad (3)$$

This set is solved numerically under constraints in the bulk and on the surface as follows. In the bulk, ρ_∞ and q_∞ are obtained from Equation (3), by fixing $\rho_\ell = \rho_{\ell+1} = \rho_{\ell-1} = \rho_\infty$ and $q_\ell = q_{\ell+1} = q_{\ell-1} = q_\infty$, leading to:

$$\begin{cases} \rho_\infty = \frac{(1-\rho_\infty)^3 \exp 3\mu/k_B T}{(\rho_\infty - q_\infty)^2} - 2q_\infty, \\[2mm] q_\infty = \frac{k_B T}{6\epsilon_\infty} \ln \frac{\rho_\infty + 2q_\infty}{\rho_\infty - q_\infty}. \end{cases} \quad (4)$$

On the surface, we must specify the anchoring conditions: ρ_0 and q_0. Equation (3) set the limits for stability:

$$\begin{cases} 0 < \rho_\ell < 1 \qquad \text{and} \\[2mm] -1/2 < q_\ell < \rho_\ell < 1. \end{cases} \quad (5)$$

The first of the above equations is trivial; the second shows that, for a given occupancy, the upper and lower bounds of q_0 correspond to homeotropic and planar anchoring, respectively, while $q_0 = 0$ is associated with a disordered state on the surface.

3. Numerical Procedure and Results

We iterate the equations presented above; three hundred layers are enough to obtain a full picture of the profile. The region far from the surface, but still under the effect of the chemical potential, shal be termed "bulk"; the particle reservoir, free from the chemical potential, is not shown in the profiles below. In what follows, we use reduced non-dimensional parameters: the temperature $\theta \equiv k_B T/\epsilon$ and the chemical potential $a \equiv \mu/\epsilon$.

Here, similarly to what was reported previously [14], the specific values of the homeotropic anchoring conditions ρ_0 and q_0 affect only the immediate vicinity of the surfaces (not shown here); the profiles are determined primarily by the bulk configuration. Moreover we notice that, for $\mu = 0$, the high temperature values of ρ and q are 0.75 and 0.0, respectively.

We start by presenting the results for the PS case and then turn to the CS case.

3.1. The Planar Surface (PS) Case

The existence of a first order transition is well known in analogous three-dimensional systems. Figure 3 shows the phase diagram describing the orientational ordering transition in the case of non-confinement: regions *I* and *II* correspond to the ordered (nematic) and isotropic phases, respectively.

In the presence of a surface, the first order phase transition persists; we find that ordering transition. for $\mu = 0$, the discontinuous transitions set in for $1.61 < \theta < 1.72$, and occur for ρ and q

simultaneously. This is not the case for $\mu \neq 0$, where ρ and q have transitions in different windows of temperature. Figure 4 shows their coexistence profiles for $\mu = 0$ in the beginning ($\theta = 1.62$) and in the end ($\theta = 1.69$) of the first order transition window; they show a persistence step in the disordered solution reminiscent from the homeotropic boundary condition on the surface; the steps are wide for $\theta = 1.62$, which is closer to the ordered end of the transition, while, for $\theta = 1.69$, the steps almost disappear. We emphasize that θ is associated with the bulk degree of order, which means that the step width is determined by the bulk configuration.

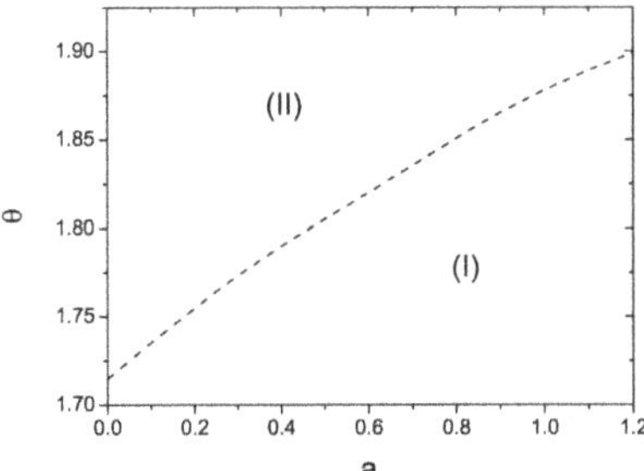

Figure 3. Phase diagram for the bulk ordering transition, described by the non-dimensional chemical potential and temperature. The nematic and isotropic phases correspond to regions *I* and *II*, respectively. The dashed line represents the discontinuous phase transition.

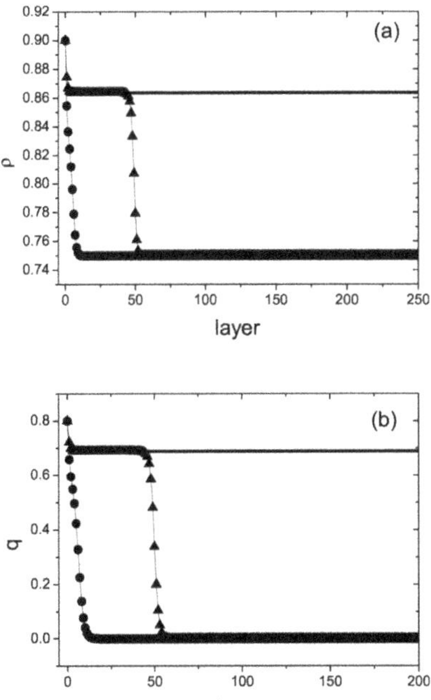

Figure 4. PS case profiles for $a = 0$: (**a**) occupancy and (**b**) orientational order parameter. Ordered solutions for $\theta = 1.62$ (continuous lines) and disordered solutions (circles and triangles). Circles: $\theta = 1.69$, triangles: $\theta = 1.62$, corresponding to the high and low temperature ends of the first order window.

Next, we study the phase coexistence for fixed temperature and varying chemical potential. Similarly to what was reported above, the mean-field solution predicts a window of coexistence, rather than a single point. Figure 5 illustrates this aspect of the transitions: we fix $\theta = 1.69$ and vary a. In Figure 5a, we notice that ρ changes discontinuously; this jump occurs somewhere between $a \simeq 0.20$ and $a \simeq 0.82$. In Figure 5b, a similar jump is observed in q, within exactly the same a range.

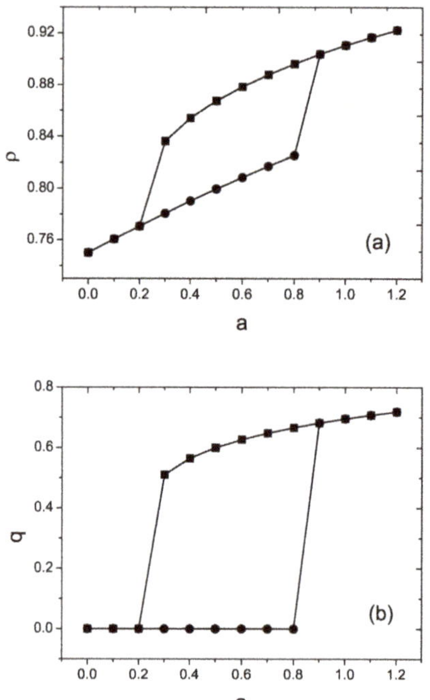

Figure 5. PS case. First order transition as a function of $a \equiv \mu/\epsilon$ for fixed $\theta = 1.69$. (**a**) occupancy and (**b**) orientational order parameter. Circles: disordered solution, squares: ordered solution. Lines are just guides to the eye.

An important issue to be investigated is the influence of μ on the steps shape. Figure 6 illustrates this behavior: the steps go further into the bulk with increasing μ. We also verified that, rather than increasing slowly, at $a = 0.07$, the steps disrupt into full plateaus (both for ρ and q), a manifestation of the first order transition. Concerning the steps' height, the ρ profile is more sensitive to the increase in μ than the q profile.

These results show that the width of the nematic film increases with a, lending strong support for complete orientational wetting transition in the limit $a \to a_c$, where $a_c = 0.07$. Therefore, we studied the increase of the adsorption Γ for a system of N layers, defined as:

$$\Gamma = \frac{1}{N} \sum_{\ell=1}^{N} [\rho_\ell - \rho_\infty] \tag{6}$$

as a function of $(a - a_c)/a_c$, shown in Figure 7, confirming the complete orientational wetting scenario.

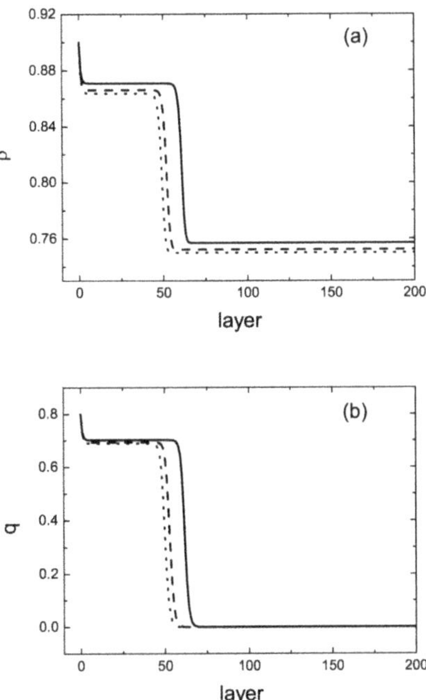

Figure 6. PS case. Effect of increasing chemical potential on profiles step for $\theta = 1.62$. (a) occupancy and (b) orientational order parameter. Dotted line: $a = 0.00$, dashed line: $a = 0.02$ and continuous line $a = 0.06$.

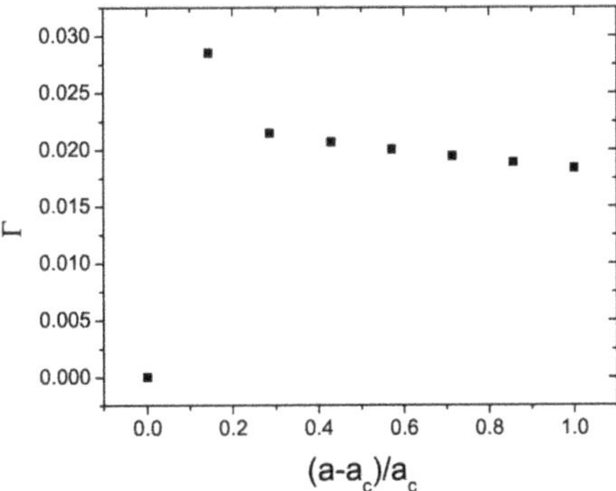

Figure 7. PS case. Adsorption Γ as a function of the difference between the reduced chemical potentials below the transition and at the transition ($\theta = 1.62$).

We also checked the effect of imposing planar anchoring. Figure 8 shows profiles that illustrate the situation: in this case, the boundary condition on the wall is strongly planar ($q(0) = -0.49, \rho(0) = 0.9$), $a = 0$ and we considered two temperatures in the vicinity of the first order transition, $\theta = 1.60$ and $\theta = 1.70$. For the lower temperature (below the transition), the first layers show planar ordering and eventually reach an condition; for the higher temperature (above the transition), the first layers show also planar ordering that decay into an isotropic state towards the bulk. It seems that no long range planar ordering can be reached throughout the sample: this is because the mean-field approach for Potts models always favours one direction.

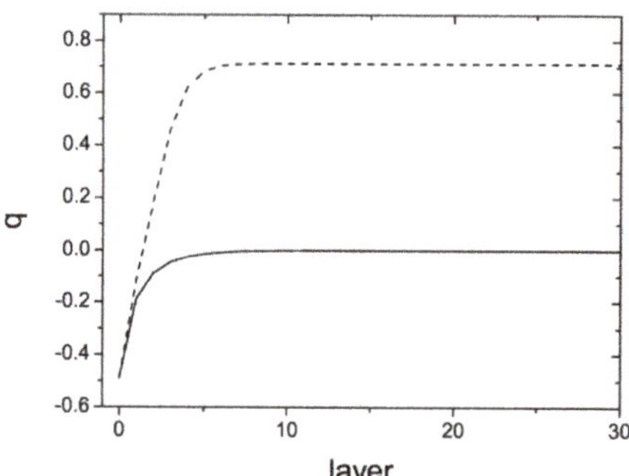

Figure 8. PS case. Profile for the orientational order in the case of strongly planar anchoring, $a = 0.00$. Continuous line: $\theta = 1.60$, just below the transition; dashed line: $\theta = 1.60$, just above the transition.

Next, we turn to the more interesting situation: the curved surface case.

3.2. The Curved Surface (CS) Case

Here, we show the results for decreasing interactions towards the bulk: we fix the interactions values on the surface ($\theta_s \equiv \epsilon_1 / k_B T$) and in the bulk ($\theta_b \equiv \epsilon_\infty / k_B T$); we may also impose the interaction decrease rate.

Here, similarly to the last case, first order transitions are obtained within a window of parameters: both for the bulk temperature at fixed chemical potential and for the chemical potential at fixed bulk temperature (not shown here).

In the vicinity of the surface, for sufficiently strong interparticle interactions (low θ_s), an ordered configuration sets in. In order to describe how the system evolves to a less ordered, or even disordered state in the bulk, we vary the interaction decrease rate (θ increase rate). Figure 9 shows profiles for $\epsilon_\ell \sim \ell^{-1}$ and $\epsilon_\ell \sim \ell^{-0.7}$, for $\theta_s = 1.69$, $\theta_b = 1.80$ and $a = 1.0$. The occupancy (Figure 9a) decreases to a finite value in the bulk, whereas the orientational order parameter (Figure 9b) decreases to zero (isotropic state). Both order parameter profiles reach the corresponding bulk values presenting a step associated with an ordered region close to the surface; the width of the steps increases as the interaction decay rate becomes smoother.

The dependence of ϵ_ℓ with ℓ describes the alignment strength between a rod in row ℓ and its neighbors. Let R be the circle (surface) radius, a the distance between adjacent rows, d_ℓ and d_0 the distance between neighboring sites on row ℓ and on the surface, respectively. Then, $(R + \ell.a)/N \simeq d_\ell$. If $\epsilon \sim \ell^\gamma$, where γ is a generic exponent, then $\epsilon(d_\ell) \sim (d_\ell / d_0 - 1)^{-\gamma}$. This means that the alignment strength ϵ depends on a non-dimensional distance between rods $\bar{d}$ like $(\bar{d} - 1)^{-\gamma}$.

Another aspect regarding steps is that they occur just in situations where the bulk degree of disorder is not too high. As the bulk temperature increases, the step width decreases, up to a situation where the bulk disorder washes out the step, that is, no order is observed throughout the system. Figure 10 illustrates the dependence of the step width on the bulk temperature; the region next to the surface is ordered ($\theta_s = 1.69$), $a = 1.0$ and the interaction decays as $\epsilon_\ell \sim \ell^{-1}$.

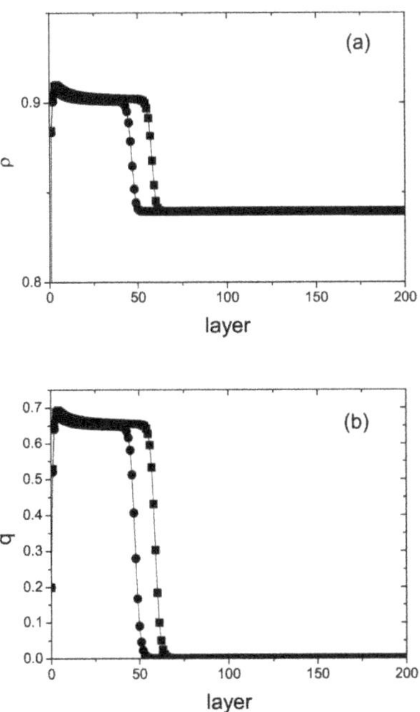

Figure 9. CS case. Profiles for: (**a**) occupancy and (**b**) orientational order parameter. Circles: $\epsilon_\ell \sim \ell^{-1}$; squares: $\epsilon_\ell \sim \ell^{-0.7}$. $\theta_s = 1.69$, $\theta_b = 1.80$ and $a = 1.0$. Lines are just guides to the eye.

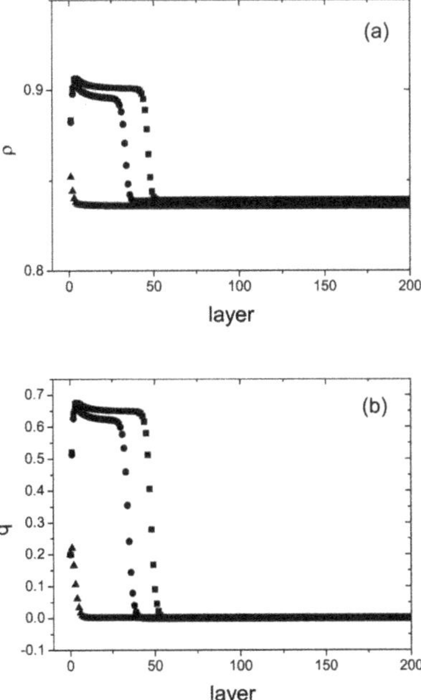

Figure 10. CS case. Profiles for: (**a**) occupancy and (**b**) orientational order parameter. $\epsilon_\ell \sim \ell^{-1}$; $\theta_s = 1.69$; $a = 1.0$; squares: $\theta_b = 1.80$, circles: $\theta_b = 1.82$, and triangles: $\theta_b = 1.90$.

We also verified the opposite case: a disordered bulk and an ordered surface, such that the bulk temperature is fixed at a high value ($\theta_b = 1.82$) and we keep decreasing the surface temperature, starting from $\theta_s = 1.65$. Appreciable changes in the step height and width are reached only at much lower values of θ_s, of the order of 1.25. With these results (not shown here), we conclude that the steps shape is much more sensitive to the bulk than to the surface temperature.

Concerning the dependence of the step shape on μ, we observed that the CS case responds similarly to what was reported for the PS case (Figure 6).

4. Discussion

4.1. Quenched Disorder

In order to compare the effects of disorder in related systems, we start this section by recalling a problem considered in our previous work [14]. There we had rods confined *inside* a disk-shaped cavity and all lattice sites were occupied, that is, we had a fixed concentration of particles. In order to allow for varying concentration, we introduce here permanent vacant sites: that is, a system with site *quenched* disorder, as opposed to *annealed* disorder studied so far. To our knowledge, there is no complete solution for the Potts model with site disorder in the literature [17], therefore we analyse this problem with an approximate argument.

We recall the expression for the free energy per column:

$$F = \sum_{\ell=0}^{\infty} \mu_\ell q_1 - \sum_{\ell=0}^{\infty} \epsilon_\ell q_\ell q_{\ell+1} - \sum_{\ell=0}^{\infty} \epsilon_\ell (q_\ell)^2 + \frac{k_B T}{3} \sum_{\ell=0}^{\infty} [2(1-q_\ell)\ln(1-q_\ell) + (1+2q_\ell)\ln(1+2q_\ell) - 3\ln 3] , \quad (7)$$

where μ_ℓ is an external potential which vanishes everywhere, except for $\ell = 1$ (next to the surface).

The corresponding recurrence relations for the order parameter are:

$$q_\ell = \frac{\mu_\ell}{2\epsilon_\ell} - \frac{1}{2}(q_{\ell+1} + q_{\ell-1}) + \frac{k_B T}{3\epsilon_\ell} \ln \frac{1+2q_\ell}{1-q_\ell}. \quad (8)$$

One way to introduce quenched dilution is to claim that the number or nearest neighbor sites is reduced by a factor h, where $0 < h < 1$. In that order, we replace all ϵ_ℓ's by $h.\epsilon_\ell$'s in Equations (7) and (8), yielding the modified recurrence relations:

$$q_\ell = \frac{\mu_\ell}{2h\epsilon_\ell} - \frac{1}{2}(q_{\ell+1} + q_{\ell-1}) + \frac{k_B T}{3h\epsilon_\ell} \ln \frac{1+2q_\ell}{1-q_\ell}. \quad (9)$$

These last relations show that, in this case of quenched disorder, the introduction of vacancies by means of h brings no qualitative novelty: first order transitions are expected as a function of the concentration (or h); moreover, the results obtained formerly [14] will be modified by rescaled T's and μ_ℓ's.

4.2. Annealed Disorder

The present study, which takes into account annealed disorder, produces a much richer dependence on the concentration, to be discussed next.

A first order transition is observed both in the PS and CS cases, as expected. The profiles depend on the chemical potential (or concentration) and on the strength of the interparticle interactions; nevertheless, the shape of the steps is determined primarily on the degree of ordering (temperature) in the bulk. All results are robust with respect to the strength of the homeotropic anchoring, they do not affect the profiles, except for the immediate neighborhood of the surfaces. Such behavior is not observed in simulations of similar systems [10–12], where profiles are strongly dependent on the anchoring condition. The reason for this discrepancy is that the mean-field approach intrinsically favors a preferential direction.

In the PS case, for vannishing μ, we obtained profiles associated with different temperature values, all of them in the first order transition temperature range. The disordered solution in the low temperature end of the coexiste window (unstable solution) has a step, reflecting the presence of the nearby surface; the disordered solution in the high temperature end of coexistence (stable solution) presents just a tiny step. Temperature is associated with the degree of order in the bulk; the step, which arises because of the anchoring at the nearby surface, can be destroyed for sufficiently high bulk disorder. We also characterize the isotropic nematic first order transition in terms as a complete orientational wetting scenario. The results for planar anchoring are presented; nevertheless, they are not very reliable due to the details of the mean-field approach.

In the CS case, the effect of decreasing interactions towards the bulk produces very interesting effects. For an ordered surface and a disordered bulk, the profile may present a step. A first observation regarding the steps is that their widths depend on the slope of the interaction profile, being wider for smoother profiles. We also showed that, similarly to what happens in the PS case, steps are destroyed when interactions in the bulk are weak enough. On the other hand, for a fixed degree of disorder in the bulk, enhancing the interactions' strength on the surfac, produces just a feeble effect on the magnitude of the order parameters and on the step width. These two observations lead us to conclude that the step height depends weakly on the surface temperature and their width is mainly determined by the bulk degree of disorder.

It is possible to compare our results for the bulk densities at the nematic isotropic with those of de Las Heras end Velasco [10]; they obtain a global packing fraction of the order or 0.29, to be compared to ours $\rho - \rho_0$ ($\rho_0 = 0.75$), which is approximately 0.15 for $a = 1.0$.

We also studied the effect of varying the chemical potential on the step shape. Increasing μ amounts to enhancing the overall concentration and produces wider steps. This behavior has also been observed in previous numerical studies on spherocylinders near flat [9] and curved walls [10].

Our motivation was to describe the distribution of colloids near surfaces, a three-dimensional problem. Solving the two-dimensional problem amounts to disregarding interactions between rods in neighboring planes. In a mean-field description, it consists of considering a smaller number of first neighbors. Having obtained the orientation profile on a given plane, we claim that this simplification reduces the three-dimensional ρ_ℓ's and q_ℓ's by a common factor. This affects the corresponding step widths and heights; nevertheless, the qualitative features of the profiles are the same. This is certainly true for flat surfaces and a good approximation for curved surfaces of large radii.

5. Conclusions

Motivated by recent experiments on colloidal particles with dimensions comparable to the confining volumes, we extend a model proposed in a recent work to describe the orientation of rods on planes bounded by straight or curved lines, considering the concentration as an extra parameter: it corresponds to the fraction of occupied sites. The particles are driven from a reservoir to the system by means of a chemical potential, so we allow for annealed disorder. They occupy sites on a semi-infinite lattice above a line or around a disk. We propose a lattice model to describe such systems and map it onto a three-state Potts model with vacancies, which is solved within a mean-field approach. In the curved case, the effect of the curvature is reproduced by layer dependent interparticle interactions. For both types of surfaces, we obtained the occupancy and the orientational profiles, which were studied under various conditions.

The main outcome of our study is to predict the non-smooth behavior of the profiles: the occurrence of steps both in the occupancy and in the orientational order parameters profiles. They may occur because the surfaces impose specific anchoring; nevertheless, their stability is primarily determined by the degree of order in the bulk. Steps like these have been also predicted in the literature, mainly in simulations, albeit with no systematic analysis. Furthermore, we believe that it may be of interest for applications in nanotechnology.

A shortcoming of our study is the weak influence of the anchoring on the overall profile. Moreover, we considered a fixed chemical potential throughout the sample; a more realistic procedure would be to model it as a decreasing function from the surface towards the bulk; this procedure would imply a more complicated interpretation of the results. At this time, an extension of this problem is underway: biaxial particles are being considered.

Funding: This research received no external funding.

Acknowledgments: We thank M. J. de Oliveira for very useful discussions and comments.

Conflicts of Interest: The authors declare no conflicts of interest.

References

1. Chen, R.H. *Liquid Crystals Displays: Fundamental Physics and Technology*; Wiley Series in Display Technology: New York, NY, USA, 2011; pp. 171–192.
2. Arsenault, A.; Fournier-Bidoz, S.; Hatton, B.; Miguez, H.; Tetrault, N.; Verkis, E.; Wong, S.; Yang, S.; Kitaev, V.; Ozin, G. Towards the synthetic all-optical computer:science fiction or reality? *J. Mater. Chem.* **2004**, *14*, 781–794. [CrossRef]
3. Kleman, M.; Lavrentovich, O.D. Topological point defects in nematic liquid crystals. *Philos. Mag.* **2006**, *86*, 4117–4137. [CrossRef]
4. Tombolato, F.; Ferrarini, A.; Grelet, E. Chiral nematic phase of suspensions of rodlike viruses: Left-handed phase helicity from a right-handed molecular helix. *Phys. Rev. Lett.* **2006**, *96*, 258302. [CrossRef] [PubMed]
5. Adams, M.; Dogic, Z.; Keller, S.L.; Fraden, S. Entropically driven microphase transitions in mixtures of colloidal rods and spheres. *Nature* **1998**, *393*, 349–352. [CrossRef]
6. Janmey, P.A.; Slochower, D.R.; Wang, Y.-H.; Wen, Q.; Cibērs, A. Polyelectrolyte properties of filamentous biopolymers and their consequences in biological fluids. *Soft Matter* **2014**, *10*, 1439–1449. [CrossRef] [PubMed]
7. Song, W.; Kinloch, I.A.; Windle, A. Nematic liquid crystallinity of multiwall carbon nanotubes. *Science* **2003**, *302*, 1363. [CrossRef] [PubMed]
8. Zamora-Ledezema, C.; Blanc, C.; Maugey, M.; Zakri, C.; Poulin, P.; Anglaret, E. Anisotropic films of single-wall carbon nanotubes from aligned lyotropic nematic suspensions. *Nano Lett.* **2008**, *8*, 4103–4107. [CrossRef] [PubMed]
9. Dijkstra, M.; van Roij, R.; Evans, R. Wetting and capillary nematization of a hard-rod fluid: A simulation study. *Phys. Rev. E* **2001**, *63*, 051703. [CrossRef] [PubMed]
10. De las Heras, D.; Velasco, E. Domain walls in two-dimensional nematics confined in a small circular cavity. *Soft Matter* **2014**, *10*, 1758–1766. [CrossRef] [PubMed]
11. Smallenburg, F.; Lowen, H. Close packing of rods on spherical surfaces. *J. Chem. Phys.* **2016**, *144*, 164903. [CrossRef] [PubMed]
12. Allahyarov, E.; Voigt, A.; Lowen, H. Smectic monolayer confined on a sphere: Topology at the particle scale. *Soft Matter* **2017**, *13*, 8120–8135. [CrossRef] [PubMed]
13. De Oliveira, M.J.; Figueiredo Neto, A.M. Reentrant isotropic-nematic transition in lyotropic liquid crystals. *Phys. Rev. A* **1986**, *34*, 3481–3482. [CrossRef]
14. Izzo, D.; de Oliveira, M.J. Ordering of rods near planar and curved surfaces. *AIP Adv.* **2018**, *8*, 015216. [CrossRef]
15. Luckhurst, G.R. Biaxial nematics: Order parameters and distribution functions. In *Biaxial Nematic Liquid Crystals—Theory, Simulation and Experiment*; Luckhurst, G.R., Sluckin, T.J., Eds.; John Wiley: West Sussex, UK, 2015; pp. 25–53.
16. Wu, F.Y. The Potts model. *Rev. Mod. Phys.* **1982**, *54*, 235–268. [CrossRef]
17. Binder, K.; Young, A.P. Spin glasses: Experimental facts, theoretical concepts, and open questions. *Rev. Mod. Phys.* **1982**, *58*, 801–976. [CrossRef]

MDPI
St. Alban-Anlage 66
4052 Basel
Switzerland
Tel. +41 61 683 77 34
Fax +41 61 302 89 18
www.mdpi.com

Crystals Editorial Office
E-mail: crystals@mdpi.com
www.mdpi.com/journal/crystals